S0-EKK-214

FLOODING AND PLANT GROWTH

PHYSIOLOGICAL ECOLOGY

A Series of Monographs, Texts, and Treatises

A complete list of titles in this series appears at the end of this volume.

FLOODING AND PLANT GROWTH

Edited by

T. T. KOZLOWSKI

Department of Forestry
University of Wisconsin
Madison, Wisconsin

1984

ACADEMIC PRESS, INC.

(Harcourt Brace Jovanovich, Publishers)

Orlando San Diego San Francisco New York London
Toronto Montreal Sydney Tokyo São Paulo

ACADEMIC PRESS, INC.
Orlando, Florida 32887

United Kingdom Edition published by
ACADEMIC PRESS, INC. (LONDON) LTD.
24/28 Oval Road, London NW1 7DX

Library of Congress Cataloging in Publication Data

Main entry under title:

Flooding and plant growth.

(Physiological ecology)
Includes index.
1. Plants, Effect of floods on. 2. Floodplain flora. 3. Floods. I. Kozlowski, T. T. (Theodore Thomas), Date . II. Series.
QK870.F58 1984 581.5'263 83-15811
ISBN 0-12-424120-4 (alk. paper)

PRINTED IN THE UNITED STATES OF AMERICA

84 85 86 87 9 8 7 6 5 4 3 2 1

To P. J. Kramer

Contents

4. Responses of Woody Plants to Flooding

T. T. KOZLOWSKI

5. Effect of Flooding on Water, Carbohydrate, and Mineral Relations

T. T. KOZLOWSKI AND S. G. PALLARDY

6. Effects of Flooding on Hormone Relations

D. M. REID AND K. J. BRADFORD

7. Effects of Flooding on Plant Disease

L. H. STOLZY AND R. E. SOJKA

Contributors

Numbers in parentheses indicate the pages on which the authors' contributions begin.

K. J. Bradford (195), Department of Vegetable Crops, University of California, Davis, California 95616

Malcolm C. Drew (47), Agricultural Research Council Letcombe Laboratory, Wantage, Oxfordshire OX12 9JT, England

Donal D. Hook (265), Department of Forestry, Clemson University, Clemson, South Carolina 29632

Michael B. Jackson (47), Agricultural Research Council Letcombe Laboratory, Wantage, Oxfordshire OX12 9JT, England

T. T. Kozlowski (1, 129, 165), Department of Forestry, University of Wisconsin, Madison, Wisconsin 53706

S. G. Pallardy (165), School of Forestry, University of Missouri, Columbia, Missouri 65211

F. N. Ponnamperuma (9), International Rice Research Institute, Los Baños, Laguna, Phillipines

D. M. Reid (195), Plant Physiology Research Group, Biology Department, University of Calgary, Calgary, Alberta T2N 1N4, Canada

R. E. Sojka (221), Coastal Plains Soil and Water Conservation Research Center, United States Department of Agriculture, Florence, South Carolina 29502

L. H. Stolzy (221), Department of Soil and Environmental Sciences, University of California, Riverside, California 92521

S. J. Wainwright (295), Department of Botany and Microbiology, University College, Swansea, Swansea SA2 8PP, Wales

Preface

Growth and distribution of plants are controlled chiefly by too little or too much water. Although many comprehensive volumes have been published on effects of drought on plants, the responses of plants to flooding have received relatively little attention. Yet temporary or continuous flooding with fresh or salt water is very common throughout the world. Soil inundation is variously traceable to overflowing of rivers, storms, overirrigation, inadequate drainage, and impoundment of water by dams. Flooding leads to rapid depletion of soil oxygen and changes in physiological processes of plants that markedly influence their growth and survival. Flooding affects our well-being not only by extensive destruction and impairment of goods and services but also by restricting yields of foods and fibers. With these important considerations in mind, this work was planned to bring together in one volume the current state of knowledge and opinion on the effects of flooding of soil with fresh or salt water on metabolism and growth of herbaceous and woody plants.

The contributors to this volume were chosen for their demonstrated competence and research productivity. Hence, the book is both authoritative and comprehensive. The opening chapter discusses the causes and extent of flooding and introduces the reader to the general nature of plant responses to flooding. Subsequent chapters address in depth the various effects of flooding on soil; metabolism and growth of herbaceous plants; growth and composition of communities of woody plants; physiological processes of plants, with particular emphasis on water, carbohydrate, mineral, and hormone relations; and plant diseases. The final two chapters describe various physiological and morphological adaptations of plants to flooding with fresh and salt water.

The book was planned as text or reference material for upper-level undergraduate students, graduate students, investigators, and growers. Its interdisciplinary scope will make it useful to academics, growers of plants, and those interested in plants for aesthetic reasons. The subject matter will be of particular interest to agronomists, biochemists, plant ecologists, engineers, foresters, horticulturists, plant anatomists, meteorologists, geneticists, plant breeders, plant physiologists, and landscape architects.

I am indebted to each contributor for scholarly work, patience, and attention to detail during the production phases.

T. T. Kozlowski

CHAPTER 1

Extent, Causes, and Impacts of Flooding

T. T. KOZLOWSKI
Department of Forestry
University of Wisconsin
Madison, Wisconsin

I. INTRODUCTION

Temporary or continuous flooding is common throughout the world, with about 72% of the earth's surface covered by submerged soils or sediments (Ponnamperuma, 1972). Much flooding is the result of overflowing of river banks, and no continent, except Antarctica, is free of such flooding. Records of floods along the Nile go back thousands of years. Perhaps the worst floods in history have been caused by China's Yellow River, which has been recorded as overflowing its banks more than 1500 times. The most disastrous Yellow River flood occurred in 1887 when more than 900,000 people were drowned (Briggs, 1973). India is traversed by numerous river systems that overflow their banks, primarily during the southwest monsoon season in July, August, and September. Although floods in October are less frequent than during monsoons, they cause much damage because of the preceding soil-moisture conditions. In northern India damage by flooding is counteracted by increased soil fertility as a result of deposition of silt. In contrast, in areas with steep slopes, damage is caused by erosion and cutting of banks, resulting in reduced crop yields for many years.

Along the Amazon River in Brazil ~60,000 km^2 are intermittently flooded (Falesi, 1972). In addition, extensive deforestation in Amazonia has caused record high flood crests (Gentry and Lopez-Parodi, 1980). All the major rivers in Europe overflow their banks periodically. Flooding also occurs on other continents, including Australia (Beadle, 1962), and New Zealand (Holloway, 1962).

ISBN 0-12-424120-4

In the United States records are available of more than 10,000 floods. Periodic overflowing of the Mississippi River in the south may inundate more than 800,000 ha of forested land (Broadfoot and Williston, 1973). In the state of Mississippi alone ~1.6 million hectares of land are flooded during the spring and early summer months (Kennedy, 1970). Briggs (1973) has provided a good historical account of flooding of major rivers in the United States, including the Mississippi, Missouri, Ohio, Connecticut, Red, Sacramento, Willamette, Tennessee, Wabash, and Susquehanna, among others.

Much catastrophic flooding is associated with storms. Floods on islands in the western Pacific are most frequently the result of typhoons. During severe storms the North Sea periodically goes over dikes in Holland and floods low-lying areas. In eastern Mexico, the Gulf Coast of Texas, and some parts of the United States, flooding is often associated with hurricanes. Temporary waterlogging of the soil may also be the result of overirrigation, seepage from irrigation channels, movement of water by underground aquifers, and impoundment by flood-control dams.

II. FLOODPLAINS

The term floodplain refers to land adjacent to streams that has been flooded naturally in historic time. Many floodplain plants grow in areas that are covered with water much of the year; others grow where they are temporarily flooded.

The terms river bottom, bottomland, hardwood bottom, and floodplain forest often are used synonymously. However, it is more appropriate to consider swamps as permanently flooded except during droughts, and floodplain forests as periodically flooded, usually in the late winter or spring. Swamps in bottomlands are called river or alluvial swamps. They occur in the lowest parts of the bottoms, either adjacent to a river or between a floodplain forest and uplands. Swamps at some distance from bottomland are classed as inland or nonalluvial swamps (Meanley, 1972).

Swamp and bottomland forests are very extensive. In the southeastern United States the Great Dismal Swamp of Virginia and North Carolina covers nearly 200,000 ha; the Okefenokee Swamp in the southeastern corner of Georgia comprises some 120,000 ha (Meanley, 1972). The Everglades of southern Florida make up one of the largest freshwater marshes on the North American continent, covering more than 1 million ha of seasonally flooded land, with scattered elevated spots supporting trees (Loveless, 1959).

III. RESERVOIRS

The occurrence of catastrophic floods has led to physical control of floodwaters by walling them off with levees and preventing their overflowing on entire floodplains. However, levees have often been unsatisfactory. Therefore, extensive systems of dams have been created behind which waters can accumulate until river channels can carry the water with safety (Maas, 1951). Storage of water in reservoirs is effective in reducing flood peaks in the rush of streams just below a dam, but the effect decreases rapidly with distance downstream. Also, the percentage of reduction in peak flow possible by a given amount of reservoir storage is greater for rapidly rising and falling floods than when the flow is spread out over a longer period of time. A disadvantage of reservoirs for flood control is that they accumulate the sediment carried by streams (Leopold and Maddock, 1954).

A characteristic of reservoirs is that they necessarily flood certain lands to protect others. Such impoundments have the advantage of creating reservation sites, habitats for wildfowl, etc. As many studies have shown, however, impoundment by reservoirs often creates adverse sites for upland plants (Hall and Smith, 1955; Loucks and Keen, 1973).

Dam collapse is merely one cause of floods, but the results are both catastrophic and dramatic because collapse usually occurs suddenly and without warning. Briggs (1973) provided much information on specific disasters associated with collapse of dams in the United States. Other important examples include the 1959 collapse of a dam across the Tera River in northwestern Spain, which completely destroyed the village of Rivaldelego and killed 123 people, and collapse of the Malpasset dam in 1959 on the Reyran River in France. Water from the Valant dam in northern Italy killed ~3000 people. Although the dam itself did not break, large portions of a mountain collapsed and millions of tons of rock and mud came down in a landslide, forcing the reservoir to overflow.

IV. CONSEQUENCES OF FLOODING

Flooding results in extensive destruction or impairment of goods, services, health, and crops. Despite huge expenditures for protective flood-control works such as levees and dams, flood losses have continued at a very high level over the years. Progressive increases in flood losses reflect changes in price levels, improvement in assessment of losses, increase by humans in occupancy of floodplains, and changes in the frequency and magnitude of flooding (Holmes, 1961). Estimates of annual damage from floods have ranged from one-fourth to more than one billion dollars (Langbein and Hoyt, 1959).

Much flood loss is the result of the attractiveness of floodplains, where humans elect to congregate and build cities in areas that rivers sometimes inundate. Man's affinity for floodplains reflects the high productivity of alluvial soils, the feasibility of establishing highways on floodplains, and the availability of rivers as sources of water and means of waste disposal. Aquatic areas along streams and floodplains are also aesthetically attractive (Maddock, 1976).

When a soil is flooded, gas exchange between the soil and air is drastically reduced (Armstrong, 1979). Shortly after a soil is inundated, microorganisms consume practically all of the oxygen in the water and soil. However, poor soil aeration is not limited to flooded soils, but is a problem with many unflooded, fine-textured soils (Stolzy *et al.*, 1975). Such soils contain little air, and gas exchange between them and the atmosphere is very slow.

The poor soil aeration associated with flooding induces a number of changes in the soil and in plants that usually adversely influence growth. A wide variety of toxic compounds accumulate in waterlogged soils (Ponnamperuma, 1972). Shortly after they are flooded, plants exhibit sequential changes in metabolism and physiological processes. Reduced water absorption and closure of stomata leading to a lowered rate of photosynthesis are among the earliest plant responses to flooding (Kozlowski, 1982). Subsequent changes include decreased permeability of roots (Kramer and Jackson, 1954); reduced mineral uptake (Greenwood, 1969; Epstein, 1972); alterations in growth-hormone balances (Reid, 1977); leaf epinasty, chlorosis, and abscission (Reid, 1977); and arrested vegetative and reproductive growth (Gill, 1970; Rowe and Beardsell, 1973). Morphological changes include hypertrophy of lenticels as well as formation of aerenchyma tissue, adventitious roots, and knee roots and pneumatophores (Sena Gomes and Kozlowski, 1980; Tang and Kozlowski, 1982; Newsome *et al.*, 1982; Kozlowski, 1982). When flooding is severe and prolonged, plants often are killed (Whitlow and Harris, 1979; Kozlowski, 1982). Because flood tolerance varies widely among plant species and cultivars, flooding may be expected to greatly alter species composition (Buma and Day, 1975).

Waterlogging of soil is not restricted to areas of heavy rainfall, but also occurs periodically in arid regions that are irrigated. Under certain conditions the flooding of soil by irrigation adversely affects plants by decreasing soil aeration and causing erosion and salt problems. Donman and Houston (1967) estimated that one-half to one-third of the world's acreage of irrigated land has drainage problems. Sometimes soil oxygen deficiency created by irrigation stunts plant growth, and because of slow recovery after adequate aeration, it decreases crop yield or increases the length of time required for high yield (Letey *et al.*, 1967). Often a relatively short period of irrigation adversely influences physiological processes in plants. The effects of 3 hr of low soil O_2 were still evident 24 hr after the plants were subjected to adequate O_2 (Stolzy *et al.*, 1964). One day of O_2 deficiency reduced yield of peas, with reduction greatest when the deficiency occurred just before or during bloom (Erickson and van Duren, 1960).

The use of irrigation systems to increase crop yield has been in vogue for centuries. For example, irrigation from mountain streams by use of diversion canals was practiced in Japan even before 600 B.C. Irrigation was introduced into Egypt more than 5000 years ago. Embankments were built along the Nile River with cross banks to hold floodwaters and channels were dug to inundate vast areas during the high-river stage. By the thirteenth century ~3 million acres were irrigated each year. In China construction of the 700-mile-long Imperial Canal in A.D. 700 provided irrigation for vast areas of crop-producing land. Gulhati and Smith (1967) provided a good review of the history of irrigation for agricultural production throughout the world.

Irrigation may cause soil-aeration problems for deep-rooted perennial crops if the soil is kept very wet. In a chlorotic California citrus orchard, water was generally applied when the surface soil dried. Unfortunately, the duration of irrigation wet the subsoil, which did not need water; hence the subsoil was always wet. Decreasing the duration of irrigation to maintain only the upper part of the soil profile in a wet condition resulted in disappearance of the chlorosis (Letey *et al.*, 1967). Leakage from irrigation ditches also causes soil waterlogging. For example, in western Canada, where more than 280,000 ha of land are irrigated, ~24,000 ha are permanently waterlogged because of water seepage from irrigation channels (Reid, 1977).

Irrigation sometimes leads to excessive soil erosion, especially on steep slopes. Erosion is critical on irrigated lands with slopes in excess of 1% (Tovey *et al.*, 1962). Erosion may be the result of uncontrolled concentration of runoff water, excessive flow of water in furrows, or normal flow of water when the soil is friable. With furrow irrigation most erosion occurs at the upper end of the run, where flow is greatest. Meck and Smith (1967) discussed measures for reducing erosion on irrigated land.

The remainder of this volume addresses in more detail the effects of flooding on soils, the physiological responses of plants, and the growth and yield of food- and fiber-producing plants. Mechanisms of flooding injury and adaptations of plants to soil inundation are also discussed.

REFERENCES

Armstrong, W. (1979). Aeration in higher plants. *Adv. Bot. Res.* **7,** 226–332.

Beadle, N. C. W. (1962). Soil phosphate and the delineation of plant communities in eastern Australia. *Ecology* **43,** 281–288.

Briggs, P. (1973). "Rampage: The Story of Disastrous Floods, Broken Dams, and Human Fallibility." McKay, New York.

Broadfoot, W. M., and Williston, H. L. (1973). Flooding effects on southern forests. *J. For.* **71,** 584–587.

Buma, P. G., and Day, J. C. (1975). Reservoir induced plant community changes: a methodological explanation. *J. Environ. Manage.* **3,** 219–250.

Donman, W. W., and Houston, C. E. (1967). Drainage related to irrigation management. *In* "Drainage of Agricultural Lands" (R. W. Hagan, H. R. Haise, and T. W. Edminster, eds.), pp. 974–987. Am. Soc. Agronomy, Madison, Wisconsin.

Epstein, E. (1972). "Mineral Nutrition of Plants. Principles and Perspectives." Wiley, New York.

Erickson, A. E., and van Duren, D. M. (1960). The relation of plant growth and yield to soil oxygen availability. *Trans. Int. Congr. Soil Sci. 7th* **4,** 428–434.

Falesi, I. C. (1972). Estudo atual dos conhecimentos sobre es solos da Amazonia Brasileira. *IPEAN–Ministerio Agric. Bol. Tec.* No. 54.

Gentry, A. H., and Lopez-Parodi, J. (1980). Deforestation and increased flooding of the upper Amazon. *Science (Washington, D.C.)* **210,** 1354–1356.

Gill, C. J. (1970). The flooding tolerance of woody species—a review. *For. Abstr.* **31,** 671–688.

Greenwood, D. J. (1969). Effect of oxygen distribution in the soil on plant growth. *In* "Root Growth" (W. J. Whittington, ed.), pp. 202–223. Butterworth, London.

Gulhati, N. D., and Smith, W. C. (1967). Irrigated agriculture: an historical review. *In* "Irrigation of Agricultural Lands" (R. M. Hagan, H. R. Haise, and T. W. Edminster, eds.), pp. 3–11. Am. Soc. Agronomy, Madison, Wisconsin.

Hall, T. F., and Smith, G. E. (1955). Effects of flooding on woody plants, West Sandy dewatering project, Kentucky Reservoir. *J. For.* **53,** 281–285.

Holloway, J. T. (1962). Forests and water—the New Zealand problem. *N. Z. For. Serv. Info. Ser.* No. 36.

Holmes, R. C. (1961). Composition and size of flood losses. *In* "Papers on Flood Problems" (G. F. White, ed.), pp. 9–20. Univ. of Chicago Press, Chicago, Illinois.

Kennedy, H. E. (1970). Growth of newly planted water tupelo seedlings after flooding and siltation. *For. Sci.* **16,** 250–256.

Kozlowski, T. T. (1982). Water supply and tree growth. II. Flooding. *For. Abstr.* **43,** 145–161.

Kramer, P. J., and Jackson, W. T. (1954). Causes of injury to flooded tobacco plants. *Plant Physiol.* **29,** 241–245.

Langbein, W. B., and Hoyt, W. G. (1959). "Water Facts for the Nation's Future." Ronald, New York.

Leopold, L., and Maddock, T. (1954). "The Flood Control Controversy: Big Dams, Little Dams, and Land Management." Ronald, New York.

Letey, J., Stolzy, L. H., and Kemper, W. D. (1967). Soil aeration. *In* "Irrigation of Agricultural Lands" (R. M. Hagan, H. R. Haise, and T. W. Edminster, eds.), pp. 941–949. Am. Soc. Agronomy, Madison, Wisconsin.

Loucks, W. L., and Keen, R. A. (1973). Submersion tolerance of selected seedling trees. *J. For.* **71,** 496–497.

Loveless, C. M. (1959). A study of the vegetation in the Florida everglades. *Ecology* **40,** 1–9.

Maas, A. (1951). "Muddy Waters, the Army Engineers and the Nation's Rivers." Harvard Univ. Press, Cambridge, Massachusetts.

Maddock, T. (1976). A primer on floodplain dynamics. *J. Soil Water Conserv.* **31,** 44–47.

Meanley, B. (1972). "Swamps, River Bottoms, and Canebreaks." Barre Publ., Barre, Massachusetts.

Meck, S. J., and Smith, D. D. (1967). Water erosion under irrigation. *In* "Irrigation of Agricultural Lands" (R. M. Hagan, H. R. Haise, and T. W. Edminster, eds.), pp. 950–963. Am. Soc. Agronomy, Madison, Wisconsin.

Newsome, R. D., Kozlowski, T. T., and Tang, Z. C. (1982). Responses of *Ulmus americana* seedlings to flooding of soil. *Can. J. Bot.*, **60,** 1688–1695.

Ponnamperuma, F. N. (1972). The chemistry of submerged soils. *Adv. Agron.* **24,** 29–95.

Reid, D. M. (1977). Crop response to water-logging. *In* "Physiological Aspects of Crop Nutrition and Resistance" (U. S. Gupta, ed.), pp. 251–287. Atma Ram, Delhi, India.

Rowe, R. N., and Beardsell, D. V. (1973). Waterlogging of fruit trees. *Hortic. Abstr.* **43,** 533–548.

Sena Gomes, A. R., and Kozlowski, T. T. (1980). Growth responses and adaptations of *Fraxinus pennsylvanica* seedlings to flooding. *Plant Physiol.* **66,** 267–271.

Stolzy, L. H., Taylor, O. C., Dugger, W. M., Jr., and Mersereau, J. P. (1964). Physiological changes in the ozone susceptibility of the tomato plant after short periods of inadequate oxygen diffusion to the roots. *Soil Sci. Soc. Am. Proc.* **28,** 305–308.

Stolzy, L. H., Zentmyer, G. A., and Rowlier, M. H. (1975). Dynamics and measurement of oxygen diffusion and concentration in the root zone and other microsites. *In* "Biology and Control of Soil-Borne Pathogens" (G. A. Bruehl, ed.), pp. 50–54. Am. Phytopathol. Soc., St. Paul, Minnesota.

Tang, Z. C., and Kozlowski, T. T. (1982). Some physiological and growth responses of *Betula papyrifera* seedlings to flooding. *Physiol. Plant,* **55,** 415–420.

Tovey, R., Meyers, V. I., and Martin, J. W. (1962). Furrow erosion on steep irrigated lands. *Idaho Agric. Exp. Stn. Bull.* No. 53.

Whitlow, T. H., and Harris, R. W. (1979). Flood tolerance in plants: a state of the art review. *U.S. Army Corps Eng. Tech. Rep.* No. E–79–2. U.S.A.C.E. Waterways Exp. Stn. Environ. Lab., Vicksburg, Mississippi.

CHAPTER 2

Effects of Flooding on Soils

F. N. Ponnamperuma
International Rice Research Institute
Los Baños, Laguna, Philippines

ISBN 0-12-424120-4

I. INTRODUCTION

Flooding or submerging an air-dry soil in water sets in motion a series of physical, chemical, and biological processes that profoundly influence the quality of a soil as a medium for plant growth. The nature, pattern, and extent of the processes depend on the physical and chemical properties of the soil and on duration of submergence. Draining and drying a flooded soil reverse most of those changes.

II. PHYSICAL EFFECTS

A. Restriction of Gas Exchange

1. Gas Exchange in Well-Drained Soils

Ten to sixty percent of the volume of soils supporting dryland plants is gas. The proportion of gas to water and solids is high in free-draining, arable soils with good structure. The composition of gas in well-drained soil is fairly stable, in spite of oxygen consumption, carbon dioxide production, and nitrogen fixation by soil organisms, because of rapid exchange of gases between soil and air.

Gas exchange is the result of mass flow caused by variations in temperature, pressure, wind speed, or rainfall and to diffusion. Gaseous diffusion is by far the more important of the two processes (Baver *et al.*, 1972; Russell, 1973).

Application of Fick's law

$$dq = -DA\left(\frac{dc}{dx}\right)_T dt \tag{1}$$

where D is the diffusion coefficient and dq the number of moles of the substance diffusing in time dt across a cross-sectional area A under a concentration gradient of dc/dx at temperature T, to gaseous diffusion in soils has shown that the gas in porous, well-drained soil can be maintained at near equilibrium with the atmosphere (Baver *et al.*, 1972). This is supported by measurements of O_2 and CO_2 fluxes (Greenwood, 1970).

2. Gas Exchange in Flooded Soils

Increasing the water content of air-dry soil drastically decreases A in Eq. (1). Thus gaseous diffusion virtually ceases when the gaseous pore space drops to about 10% (Wesseling, 1974). This occurs at soil moisture tensions of about 20–30 cm. Flooding or waterlogging a soil virtually eliminates gas-filled pores and limits gas exchange between soil and air to molecular diffusion in soil water. This process is $\sim 10^4$ times slower than in air, because D_{water}/D_{air} for both O_2 and CO_2 is 1.13×10^{-4} (Grable, 1966). Consequently, the O_2 supply of the soil is cut off, and gases formed by soil metabolism accumulate.

3. *Depletion of Molecular Oxygen*

Within a few hours of flooding, microorganisms and roots use up the O_2 present in the water or trapped in the soil and render submerged soil practically devoid of the gas. Evidence for absence of O_2 in flooded soils and sediments has been summarized by Ponnamperuma (1972).

Flooded soil, however, is not uniformly devoid of O_2. The O_2 concentration may be high in the surface film or layer, not more than a few millimeters thick, in contact with the oxygenated surface water. The thickness of the layer represents a balance between O_2 diffusion into the soil and its consumption chemically and biochemically. The oxidized layer has the chemical characteristics of an aerobic soil and acts as a sink for substances diffusing upward from the anaerobic bulk of the soil. The aerobic–anaerobic interface has important implications for the nitrogen economy of flooded soils.

4. *Accumulation of Soil Gases*

Drastic restriction of gas exchange between flooded soil and the atmosphere leads to accumulation in the soil of nitrogen, carbon dioxide, methane, and hydrogen. The gases build up pressure and escape as bubbles. Analyses of such bubbles from flooded rice fields at different times during a season revealed wide variations in composition, as follows: nitrogen, 10 to 95%; methane, 15 to 75%; carbon dioxide, 1 to 20%; hydrogen, 0 to 10% (Harrison and Aiyer, 1913). More recent data indicate that CO_2 concentrations as high as 50% may persist for several weeks in cold, acid soils that are flooded. (Ponnamperuma, 1976).

B. Thermal Effects

1. *Net Solar Radiation Absorbed*

Flooding affects net solar radiation absorbed by soil, heat fluxes in and out of a given volume of soil, heat capacity of the soil, and soil temperature. Net solar radiation absorbed by soil (H) is related to the incident irradiance R, the albedo s of the surface, and long-wave radiation B according to

$$H = R(1 - s) - B \tag{2}$$

Because water darkens soil color and because water surfaces have low reflectancy, flooding reduces the value of s in Eq. (2). Some albedo values adapted from Baver *et al.* (1972) are as follows: dry clay, 0.23; wet clay, 0.16; water, 0.03–0.10.

2. *Heat Capacity and Heat Flux*

The specific heat capacity C of a soil can be related to the volume fractions of soil minerals (V_1), organic matter (V_2), and water (V_3) and their specific heats by

$$C = 0.46V_1 + 0.60V_2 + V_3 \tag{3}$$

and the heat flux density dQ/dt can be determined by

$$dQ/dt = -kdt/dx \tag{4}$$

Water content affects thermal conductivity k; consequently, the thermal diffusivity k/C depends on water content. Usually k/C is lessened by flooding, and both diurnal and seasonal fluctuations of soil temperature are consequently damped.

3. *Soil Temperature*

Interactions of net radiation, absorption, heat fluxes, thermal diffusivity, and temperature of the incoming flood water determine soil temperature. Figure 1 shows the temperature profile of a flooded rice field (Kondo, 1952). The mean

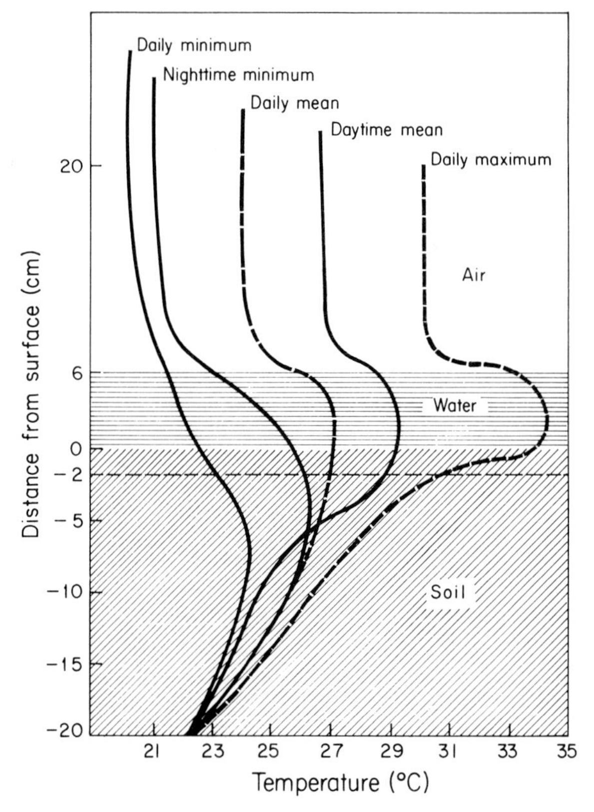

Fig. 1. Temperature profile in a flooded rice field. From Kondo (1952).

temperature of the surface layer was 6°C lower for flooded soils than for well-drained soils (Bonneau, 1982).

The temperature of flooded soils markedly affects the velocity and pattern of chemical and electrochemical changes initiated by flooding, release of nutrients, production of toxins, and plant growth (Ponnamperuma, 1976).

C. Swelling of Colloids

When dry soil is flooded, soil colloids absorb water and the soil swells. The rate of water sorption and volume increase of mineral soils depend on the clay content, type of clay mineral, and nature of the exchangeable cations. Swelling is usually complete in 1–3 days. The higher the clay content, the greater the swelling. The expanding-lattice types of clay (smectite and beidellite) swell more than the fixed-lattice type (kaolinite and halloysite). Sodic soils swell more than calcareous or acid soils.

D. Rheological Changes

1. Consistency

As the moisture content of dry soil is increased, cohesion of water films around soil particles causes them to stick together, rendering the soil plastic. In this plastic condition, soils are easily puddled. At higher water contents (as in flooded soils), cohesion decreases rapidly. The rapid decrease is the result of increase in the thickness of water films between soil particles, as shown by

$$F = \kappa 4 \pi r T / d \qquad (5)$$

where F is cohesion, κ a constant, r the radius of the particles, T the surface tension, and d the distance between particles.

Loss of cohesion decreases soil strength, rendering the use of heavy machinery on flooded soils impractical. When flooded soil is drained and dried, cohesion increases, and the soil shrinks and cracks. Alternate flooding and drying help aggregate formation.

2. Plasticity

Soil colloids absorb water, swell, and lubricate the coarser particles, which make a soil moldable or plastic. The plastic limit is the moisture content at which the soil becomes plastic; the liquid limit signifies a moisture content at which the water films are so thick that the soil mass flows under slight applied pressure (Brown, 1977). Increase in clay content increases the limits, as do increases in smectite, exchangeable sodium, and organic matter.

Flooded soils are well above the liquid limit. Hence flooded soils have low shear strength (resistance to slipping and sliding of soil particles one over the other) and also low compression (change in volume under an applied stress). Subjecting flooded soil to compression or shearing forces aggravates the structure impairment that follows the flooding of dry soil.

E. Destruction of Structure

Good soil structure is vital to dryland crops because it helps aeration and drainage. Flooding a dry soil destroys soil structure by disrupting the aggregates.

The breakdown of aggregates is the result of reduction in cohesion with increase in water content, deflocculation of clay as a result of dilution of the soil solution, pressure of entrapped air, stresses caused by uneven swelling, and destruction of cementing agents. Sodic soils show marked breakdown of aggregates on flooding, whereas soils high in iron and aluminum oxides or organic matter undergo little aggregate destruction (Sanchez, 1976). On drying and reoxidation, structure is partially restored through soil cracking and cementing by hydrated ferric oxides.

F. Water Movement

Water moves out of the root zone of flooded soils by evapotranspiration and by percolation and seepage. Evaporation from a free water surface depends on net radiant energy available at the surface, atmospheric temperature and humidity, and wind speed; plants growing in an expanse of water have a slight influence on the rate of evaporation of that water.

Percolation is the downward movement of water under the influence of gravity. Seepage is the lateral movement of subsurface water. Because water movement in flooded soils is a combination of the two processes, Wickham and Singh (1978) used seepage and percolation collectively.

Factors that favor water movement in flooded soils under a fixed head of water are coarse texture, granulated structure, low bulk density, 1:1–type clays, calcium saturation, organic matter, soil cracking, absence of plow pans, and deep water table (Wickham and Singh, 1978). Flooding decreases water movement in soils of low permeability because of dispersion of soil particles, swelling, aggregate destruction, and clogging of pores by microbial slime. In porous, nonswelling soils, flooding (by providing a greater head of water and by increasing hydraulic conductivity) increases percolation.

A percolation rate of more than 1.5 m/day is necessary to meet the oxygen demands of roots in flooded soil (Grable, 1966). Percolation at more than 10 mm/day improves rice yields (Greenland, 1981). This is attributed to removal of toxins. Percolation at 2–3 cm/day does not oxygenate more than 1 cm of the surface layer of flooded soil (Houng, 1981).

III. BIOTIC ZONATION

The presence of a layer of standing water and differentiation of the soil profile into aerobic and anaerobic zones have profound consequences on the ecology of flooded soils.

A. Standing Water

The surface water layer becomes the habitat of heterotrophic flora and fauna, as well as algae and macrophytes. Consequently, it is a zone of intense physiological activity.

The bacteria in the floodwater are aerobic and the fauna largely zooplankton (Watanabe and Furusaka, 1980). The algae include phytoplankton, filamentous algae, and higher algae growing as anchored species (Roger and Watanabe, 1984). Physiologically, they are either nitrogen fixers or non–nitrogen fixers. The macrophytes are of three kinds: surface plants both rooted and free floating, submerged rooted plants growing mostly below the surface, and plants growing in shallow water or wet soil. The biomass production in the standing water of fallow tropical rice fields may be as high as 12 tons of fresh matter per hectare per 100 days (Roger and Watanabe, 1984).

The main physiological activities in the standing water are photosynthesis, respiration, ammonification, nitrification, and nitrogen fixation. Because of photosynthesis by algae and macrophytes during the daytime, the water becomes supersaturated with O_2 and unsaturated with CO_2. Depletion of CO_2 may increase pH to 10.0 (De Datta, 1981) during the daytime. At night the O_2 concentration decreases because of consumption by respiration. Cyanobacteria (blue-green algae) may fix as much as 0.5 kg N/ha/day (Stewart *et al.*, 1979). Ammonia formed by mineralization of organic matter or diffusing upward from the anaerobic zone is nitrified.

B. Aerobic–Anaerobic Interfaces

Flooded soil has three types of aerobic–anaerobic interfaces: the surface oxidized layer in contact with oxygenated water and the anaerobic soil matrix, the rhizosphere of marsh plants, and the oxidized subsoil.

The surface oxidized layer receives plant debris and mineral matter from the supernatant water and substances diffusing upward from the anaerobic zone. This leads to enrichment of 1 cm of the surface soil with organic matter, nitrogen, phosphorus, iron, manganese, and other substances. Aerobic bacteria, algae, and photosynthetic bacteria are present. In addition, methane-oxidizing bacteria oxidize methane moving into the surface layer from the anaerobic layer (Harrison and Aiyer, 1915; De Bout *et al.*, 1978).

Because of O_2 diffusion from the roots of marsh plants, the rhizosphere is oxidized (Armstrong, 1978). Thus the rhizosphere of rice has a positive effect on aerobic bacteria (including heterotrophic nitrogen fixers) and a negative effect on *Clostridium,* sulfate reducers, and denitrifiers (Watanabe and Furusaka, 1980). Yoshida (1978) cited evidence for nitrogen fixation in the rhizosphere of other aquatic plants and microbial sulfide oxidation in rice. Diffusion of organic substrates and nitrogen from rice roots helps nitrogen fixation in flooded rice fields (Ponnamperuma, 1972; Yoshida, 1978; Buresh *et al.*, 1980).

Formation of manganic oxide deposits below the plow sole of flooded rice fields has been attributed to the presence of manganese-oxidizing bacteria (Watanabe and Furusaka, 1980).

C. The Anaerobic Zone

Within a few hours of flooding an air-dry soil, the bulk of the soil is rendered practically devoid of O_2. Consequently, aerobes are replaced by facultative anaerobes, which in turn are superseded by strict anaerobes (Yoshida, 1978). Fungi and actinomycetes are suppressed and bacteria predominate. Strict anaerobes in a paddy field constituted only 10% of the count of facultative anaerobes (Watanabe and Furusaka, 1980). Soil metabolism that causes denitrification and reduction of manganese, iron, and sulfate as well as nitrogen fixation and methane formation is the work of anaerobic bacteria in the O_2-free zone of flooded soils (Yoshida, 1978; Watanabe, 1984).

IV. ELECTROCHEMICAL CHANGES

Flooding air-dry soils causes direct and indirect electrochemical changes. One direct and almost instantaneous change is dilution of the soil solution. This increases pH, decreases electrical conductance, and alters the diffuse double layer of colloidal particles. But these changes are insignificant compared with the drastic changes in redox potential, pH, electrical conductance, ionic strength, ion exchange, sorption, and desorption caused by soil reduction.

Flooding cuts off a soil's O_2 supply. After the aerobic organisms have used up the O_2 present in the soil, facultative and obligate anaerobes proliferate, using oxidized soil components and dissimilation products of organic matter as electron acceptors in their respiration and reducing a soil in thermodynamic sequence (Table I). The reduced state of a flooded soil is shown by its low redox potential (0.2 to −0.4 V compared with 0.8–0.3 V for aerobic soils), by the absence of NO_3^-, and by the presence of Fe^{2+}, Mn^{2+}, NH_4^+, and S^{2-}.

TABLE I

Thermodynamic Sequence of Soil Reduction[a]

System	E_0^{7b}	pE_0^{7c}
$O_2 + 4\,H^+ + 4\,e^- = 2\,H_2O$	0.814	13.80
$2\,NO_3^- + 12\,H^+ + 10\,e^- = N_2 + 6\,H_2O$	0.741	12.66
$MnO_2 + 4\,H^+ + 2\,e^- = Mn^{2+} + 2\,H_2O$	0.401	6.80
$CH_3COCOOH + 2\,H^+ + 2\,e^- = CH_3CHOHCOOH$	−0.158	−2.67
$Fe(OH)_3 + 3\,H^+ + e^- = Fe^{2+} + 3\,H_2O$	−0.185	−3.13
$SO_4^{2-} + 10\,H^+ + 8\,e^- = H_2S + 4\,H_2O$	−0.214	−3.63
$CO_2 + 8\,H^+ + 8\,e^- = CH_4 + 2\,H_2O$	−0.244	−4.14
$N_2 + 8\,H^+ + 6\,e^- = 2\,NH_4^+$	−0.278	−4.69
$NADP^+ + H^+ + 2\,e^- = NADPH$	−0.317	−5.29
$NAD^+ + H^+ + 2\,e^- = NADH$	−0.329	−5.58
$2\,H^+ + 2\,e^- = H_2$	−0.413	−7.00
Ferredoxin (Fe^{3+}) + e^- = Ferredoxin (Fe^{2+})	−0.431	−7.31

[a]From Ponnamperuma (1977a).
[b]E_0 corrected to pH 7.0
[c]pE_0 corrected to pH 7.0.

A. Decrease in Redox Potential

1. *Potential Changes*

Redox potential (E_h) falls rapidly after flooding, reaches a minimum within a few days, rises rapidly to a maximum, and then decreases asymptotically with time (Figs. 2 and 3). The presence of organic matter and a temperature of 35°C favor E_h decrease, and values as low as −0.25 V may be reached within 2 weeks after submergence in soils low in nitrate and in manganic and ferric oxides (Ponnamperuma, 1972, 1981).

Gambrell and Patrick (1978) have given the potentials at pH 7.0 for the observations listed in the following tabulation:

Observation	E_h (V)
Disappearance of O_2	0.33
Disappearance of NO_3^-	0.22
Appearance of Mn^{2+}	0.20
Appearance of Fe^{2+}	0.12
Disappearance of SO_4^{2-}	−0.15
Appearance of CH_4	−0.25

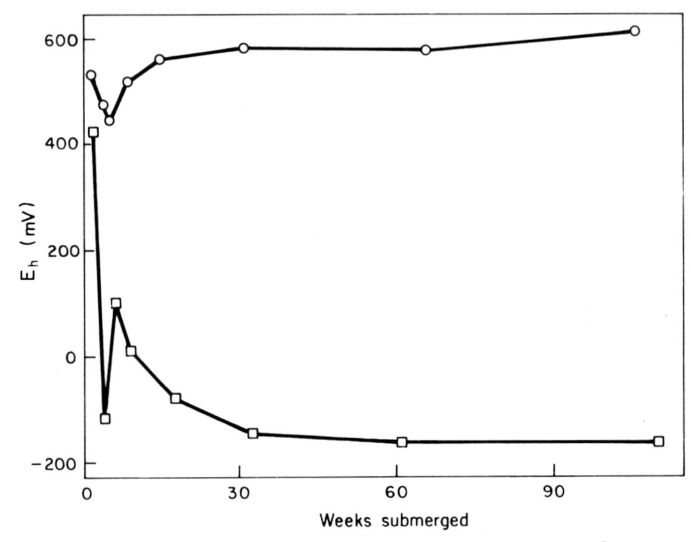

Fig. 2. Changes of E_h with time in a well-drained soil and a submerged soil. ○, Well drained; □, submerged, no drainage. From Ponnamperuma (1955).

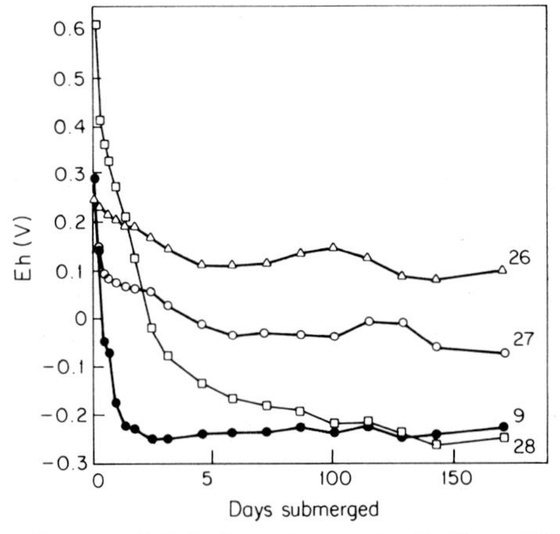

Fig. 3. Changes of E_h in four submerged soils. From IRRI (1963).

Soil no.	pH	Organic matter (%)	Active Fe (%)	Active Mn (%)
9	6.0	3.9	0.65	0.005
26	7.5	1.1	1.13	0.063
27	6.6	2.0	1.25	0.223
28	4.6	2.9	4.10	0.050

The redox potential is high in the floodwater and in the first few millimeters of the surface soil. Then it drops rapidly with soil depth to negative values. Positive values are rapidly resumed below the plow sole, where Mn(IV) compounds are present (De Gee, 1950).

The E_h is a useful guide to the sequence of soil reduction, but it does not indicate the capacity of the soil to resist E_h changes. Buffering agents in the oxidizing range are oxygen and nitrate, whereas in the reducing range ferric compounds dominate (Patrick, 1981).

2. *Theoretical Considerations*

Although redox potentials of flooded soils have been used in ecological studies (Pearsall and Mortimer, 1939; Bass-Becking *et al.*, 1960; Aomine, 1962; Ponnamperuma, 1981), they are of limited theoretical or practical value. But potentials of interstitial waters of flooded soils and sediments are thermodynamically meaningful and have been used successfully to explain the changes in concentration of ecologically important ions in soil solutions (Ponnamperuma, 1972; Yamane, 1978; Lindsay, 1979; Rowell, 1981).

For the equilibrium

$$\text{Ox} + n\,e^- + m\,\text{H}^+ = \text{Red} \tag{6}$$

$$E_h = E_0 + 2.303\frac{RT}{nF}\frac{(\text{Ox})}{(\text{Red})} - 2.303\frac{RTm}{nF}\text{pH}$$

or

$$\text{p}E = \text{p}E_0 - \frac{1}{n}\text{p(Ox)} + \frac{1}{n}\text{p(Red)} - \frac{m}{n}\text{pH} \tag{7}$$

where (Ox) and (Red) are the activities of the oxidized and reduced phases, E_h is the redox potential, pE (electron activity) = $E_h/2.303RTF^{-1}$ = $E_h/0.0591$ at 25°C, R is the gas constant, T the absolute temperature, and F the Faraday constant.

The solutions of flooded soils have E_h values of −60 to 170 mV or pE values of −1 to 3 at pH 7. The E_h or pE values of the solutions of most acid, mineral soils that have undergone reduction conform to

$$E_h = 1.06 - 0.059 \log \text{Fe}^{2+} - 0.177\text{pH} \tag{8}$$

or

$$\text{p}E = 17.87 + \text{pFe}^{2+} - 3\text{pH} \tag{9}$$

B. Changes in pH

1. *The pH Values of Flooded Soils*

When an acid soil is kept flooded its pH increases, whereas the opposite happens in alkaline soils (Fig. 4). Thus the pH values of most flooded mineral

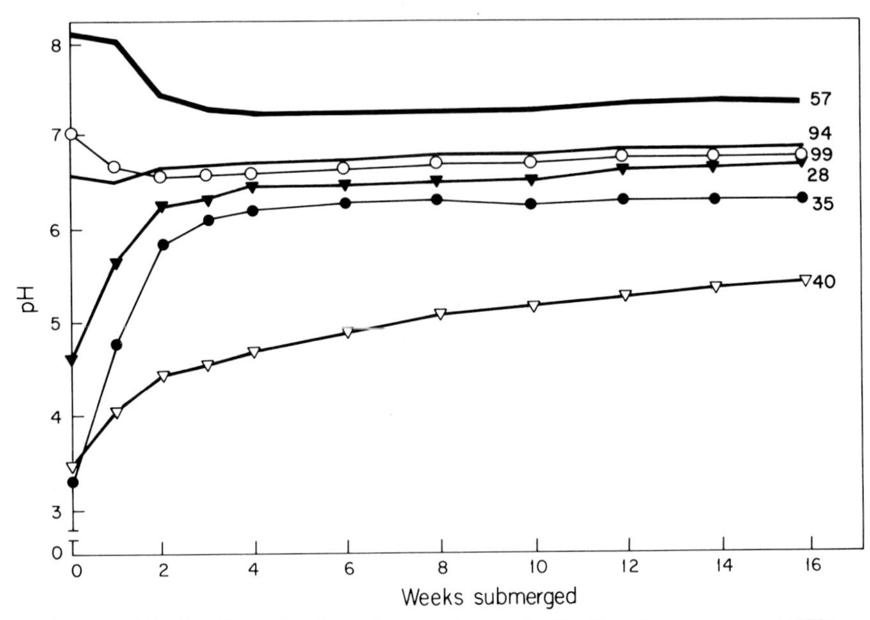

Fig. 4. Kinetics of the pH values of some submerged soils. From Ponnamperuma (1972).

Soil no.	Texture	pH	Organic matter (%)	Active Fe (%)	Active Mn (%)
28	Clay	4.9	2.9	4.70	0.08
35	Clay	3.4	6.6	2.60	0.01
40	Clay	3.8	7.2	0.08	0.00
57	Clay loam	8.7	2.2	0.63	0.07
94	Clay	6.7	2.6	0.96	0.09
99	Clay loam	7.7	4.8	1.55	0.08

soils are between 6.7 and 7.2, and those of their interstitial solutions between 6.5 and 7.0 (Ponnamperuma, 1972). But if a soil's reducible iron content is low, its pH may not rise above 5.0 even after months of submergence, and if the organic matter content of a high-pH soil is low, the pH may not decrease below 8.0.

The increase in pH of acid soils is due mainly to reduction of Fe(III) to Fe(II). It can be described for most acid, mineral soils by Eq. (8) or (9). The decrease in pH of sodic and calcareous soils and the check on the pH rise of acid soils are the results of accumulation of CO_2. Some observed relationships between pH and the partial pressure of CO_2 (P_{CO_2}) are (Ponnamperuma *et al.*, 1969)

$$\text{Sodic soil} \qquad \text{pH} = 6.4 - 1.00 \log P_{CO_2} \tag{10}$$

Calcareous soil	$\mathrm{pH} = 6.1 - 0.66 \log P_{CO_2}$	(11)
Neutral soil	$\mathrm{pH} = 6.1 - 0.64 \log P_{CO_2}$	(12)
Ferruginous acid soil	$\mathrm{pH} = 6.1 - 0.57 \log P_{CO_2}$	(13)

According to Eqs. (10)–(13), at a P_{CO_2} of 0.1 bar (a common value in flooded soils) the pH of the soils just listed is 6.7–7.4.

2. *Effects of pH on Chemical Equilibria in Flooded Soils*

A soil's pH markedly affects the concentration in the soil solution of ecologically important ions through the involvement of hydrogen ions in chemical equilibria (Ponnamperuma, 1972, 1978, 1981).

The effect of pH on hydroxide equilibria can be described by

$$\mathrm{pH} + \tfrac{1}{2} \log M^{2+} - K \qquad (14)$$

where M is Fe, AlOH, Zn, or Cu; and $K = 5.4$ for Fe, 2.2 for AlOH, 3.0 for Zn, and 1.6 for Cu.

According to Eq. (14) a pH increase of 1 unit should decrease the concentration of M 100 times. Thus an increase of pH from 4.0 to 5.0 depresses the concentration of water-soluble aluminum from 7 to 0.07 mg/liter and abolishes aluminum toxicity. On the other hand, a pH decrease from 7.0 to 6.5 in ferruginous soil may increase the concentration of water-soluble iron from 35 to 350 mg/liter and cause iron toxicity.

The effect of pH on solubility of calcium and manganese is given by

$$\mathrm{pH} + \tfrac{1}{2} \log M^{2+} + \tfrac{1}{2} \log P_{CO_2} = K \qquad (15)$$

where M is Ca or Mn, and $K = 4.92$ for Ca and 4.06 for Mn. At a high pH (8.5) and low P_{CO_2} (0.05 bar), the concentrations of calcium and manganese are 0.08 and 0.003 mg/liter, respectively. These are inadequate for plant growth. But if as a result of flooding pH decreases to 7.0, the concentrations reach the adequate values of 80 and 3 mg/liter, respectively.

C. Changes in Specific Conductance and Ionic Strength

1. *Specific Conductance*

The specific conductance κ of an aqueous solution at constant temperature depends on the nature and content of ions present in solution. In flooded soils κ represents a balance between ion production during soil metabolism and ion inactivation or replacement by slower-moving ions.

The specific conductance of the interstitial solutions of flooded soils increases during the first few weeks after flooding, reaches a peak, and then declines to a

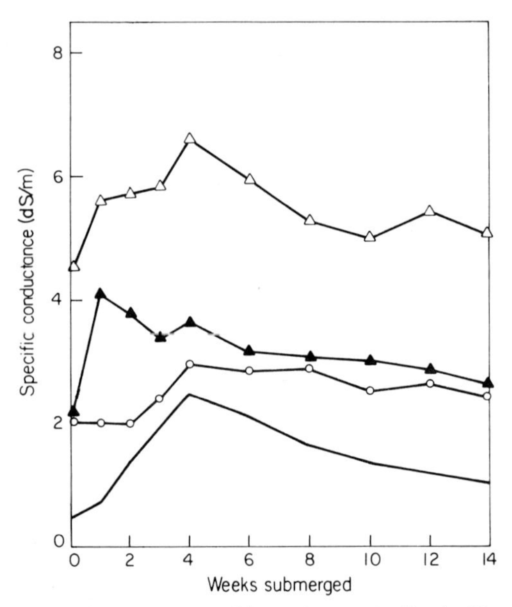

Fig. 5. Kinetics of the specific conductance of four submerged soils. △, Silty clay loam (pH 8.7, O.M. 2.2%); ▲, clay loam (pH 6.7, O.M. 2.6%); ○, clay (pH 7.3, O.M. 1.2%); ———, clay (pH 4.6; O.M. 2.9%) (O.M., organic matter). From Ponnamperuma (1977a).

fairly stable value, but its kinetics vary with the soil (Fig. 5). The peak increase is 1–2 dS/m.

The increase in κ during the first few weeks of submergence is the result of production of NH_4^+, HCO_3^-, $RCOO^-$, Mn^{2+}, and Fe^{2+}, followed by displacement of Na^+, K^+, Ca^{2+}, and Mg^{2+} from the soil colloids by Mn^{2+} and Fe^{2+}. The decrease is the result of removal of HCO_3^-, conversion of $RCOO^-$ and HCO_3^- to CH_4, and precipitation of Mn^{2+} as $MnCO_3$ and Fe^{2+} as $Fe_3(OH)_8$. The close similarity between the kinetics of alkalinity, cations, and κ shows the role of the ions just listed in controlling the changes in κ (Fig. 6).

2. *Ionic Strength*

The ionic strength I of an aqueous solution is related to the concentration c and valence z of the ions present by

$$I = \frac{1}{2} \sum c_i z_i^2 \tag{16}$$

Flooding increases the concentration of NH_4^+ and HCO_3^-. It also converts insoluble Mn(IV) and Fe(III) compounds to Mn(II) and Fe(II) forms, which are much more soluble in water. These processes increase ionic strength.

Because the kind and concentration of ions influence both specific conduc-

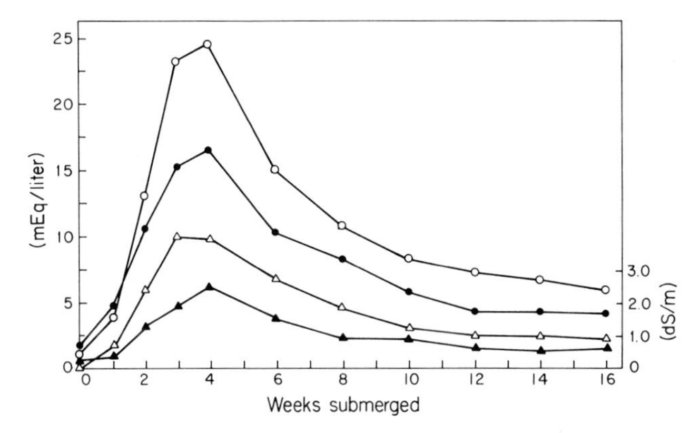

Fig. 6. Kinetics of specific conductance and cation concentrations in a submerged ferrallitic soil. ○, Total alkalinity (mEq/liter); ●, $Ca^{2+} + Mg^{2+} + NH_4^+ + Na^+ + K^+$ (mEq/liter); △, Fe^{2+} + Mn^{2+} (mEq/liter); ▲, specific conductance (dS/m). From Ponnamperuma (1972).

tance and ionic strength, κ and I should be related. For the solutions of reduced soils, I (mol/liter) is numerically equal to 16κ (dS/m) up to ionic strengths of 0.05 (Ponnamperuma *et al.*, 1966; Yamane, 1978).

D. Ion Exchange

Ion exchange is the replacement by ions in the soil solution of cations and anions held by electrostatic attraction on soil colloids. Soil reduction that follows flooding affects ion-exchange reactions by altering the electric charge on soil colloids and by increasing markedly the concentrations in the soil solution of Fe^{2+} and Mn^{2+} in acid soils, Ca^{2+} in alkaline soils, and HCO_3^- in all soils.

1. Cation Exchange

Cations are held on soil colloids by permanent negative charges residing on the particles and by pH-dependent negative charges. The permanent negative charge on a mineral colloid is constant in the pH range 2.5–10 (Talibudeen, 1981). But the pH-dependent negative charges, arising from the ionization of carboxyl and phenolic groups of soil organic matter and the dissociation of surface hydroxyls in silica and silicates, increase as pH increases.

The increase in pH of acid soils on flooding from less than 3 to ~7 should increase ionization of carboxyl groups (whose pK_a values are generally between 3 and 5) and of hydroxyls attached to silicon (whose pK_a values range from 2 to 9.7, depending on the mineral). The decrease in pH of alkaline soils should depress ionization of both groups of colloids. Consequently, flooding should increase cation-exchange capacity of acid soils and decrease that of alkaline soils.

The Fe^{2+}, Mn^{2+}, and NH_4^+ concentrations in the solutions of flooded soils may attain values as high as 10, 2, and 8 m*M*, respectively. These ions displace Na^+, K^+, Ca^{2+}, and Mg^{2+} into the soil solution. The close parallelism between the kinetics of water-soluble Fe^{2+} and Mn^{2+} and other cations illustrates this (Fig. 6). The findings of Sims and Patrick (1978) that greater amounts of exchangeable iron, manganese, zinc, and copper are present at low E_h and low pH than at high E_h and pH confirm the role of Fe^{2+} and Mn^{2+} in cation exchange in flooded soils. Low E_h and low pH favor mobilization of Fe^{2+} and Mn^{2+}.

2. *Anion Exchange*

At pH values below their point of zero charge, hydrous oxides of aluminum, iron, and manganese are positively charged. Below pH 6, exposed oxygen and hydroxyl groups in 1:1 and 2:1 lattice clays accept protons and acquire a positive charge. Amino groups of soil organic matter carry positive charges below their pK_b values (Talibudeen, 1981; Mott, 1981).

The positive sites attract anions. Of common anions, NO_3^- and Cl^- are held by simple electrostatic attraction and are said to be *nonspecifically* adsorbed. All other anions, including organic anions, are *specifically* adsorbed and do not readily undergo ion exchange (Mott, 1981).

Flooding affects anion exchange properties of soils by increasing pH of acid soils and decreasing that of alkaline soils, reducing Fe(III) and Mn(IV) hydrous oxides to the Fe(II) and Mn(II) forms, converting NO_3^- to N_2 and SO_4^{2-} to insoluble sulfides, and loading the soil solution with up to 50 m*M* HCO_3^-.

The increase in concentration of water-soluble phosphate and silica and desorption of sulfate caused by flooding soils (Ponnamperuma, 1972) may be the result of decrease in anion-exchange capacity and increase in the bicarbonate concentration.

E. Sorption and Desorption

Sorption is the concentration of gases, liquids, or solutes in a layer of molecular dimensions on surfaces of solids with which they are in contact. Sorption is due to cation or anion exchange, covalent bonding, action of van der Waals forces, or isomorphous substitution. When the mechanism is known, the process is called adsorption. Sorbed substances are in equilibrium with the soil solution and may undergo desorption. Sorbing agents are clay; hydrous oxides of aluminum, iron, manganese, and silicon; or organic matter (Ellis and Knezek, 1972; Stevenson and Ardakani, 1972).

Sorbates of ecological interest include the cationic forms of copper, zinc, and metallic pollutants; the anionic forms of boron, molybdenum, phosphorus, and sulfur; and organic substances, including pesticides. The Freundlich and Langmuir isotherms are used to describe sorption (Ellis and Knezek, 1972). According to both equations, sorption increases with concentration of the substance sorbed.

Flooding dilutes the soil solution. It should therefore favor desorption. But the reduction of Fe(III) and Mn(IV) hydrous oxides drastically alters their surface properties and produces large amounts of water-soluble Fe^{2+} and Mn^{2+}. Thus ions sorbed on them may be desorbed, and exchangeable cations may also be displaced into the solution. The increase in pH of acid soils and the decrease in pH of alkaline soils affect the surface properties of clays, hydrous oxides, and organic matter. Changes in pH may increase sorption or desorption, depending on the point of zero charge of the sorbent.

Sorption and desorption play an important role in controlling the availability of plant nutrients and pollution by toxic metals and organic pesticides (Burchill *et al.*, 1981).

V. CHEMICAL TRANSFORMATIONS

The main chemical changes brought about by flooding or waterlogging an air-dry soil are the disappearance of O_2, accumulation of CO_2, anaerobic decomposition of organic matter, transformations of nitrogen, and reduction of Fe(III), Mn(IV), and SO_4^{2-} (Ponnamperuma, 1972; Patrick and Reddy, 1978).

A. Disappearance of Oxygen[1]

B. Accumulation of Carbon Dioxide

Up to 3 tons of CO_2 per hectare may accumulate in the plowed layer of a flooded soil. Because CO_2 is soluble in water and chemically reactive, it forms carbonic acid and bicarbonates and reacts with divalent cations to form insoluble carbonates. The partial pressure of CO_2 (P_{CO_2}) is a good measure of CO_2 accumulation.

The P_{CO_2} in a soil increases on flooding, reaches a peak of 0.2–0.5 bar 1–3 weeks later, and declines to a fairly stable value of 0.05–0.2 bar (Fig. 7). Acid soils high in organic matter but low in iron and manganese show a rapid increase in P_{CO_2} to ~0.5 bar within 1–2 weeks after flooding, followed by a slow decline

[1]See Section II,A.

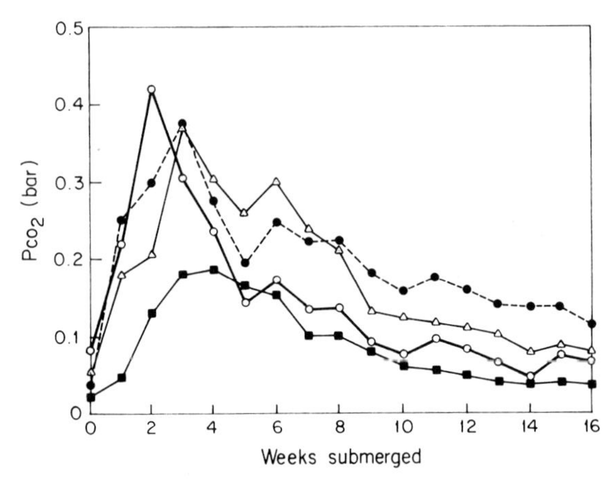

Fig. 7. Kinetics of P_{CO_2} in four submerged soils. From IRRI (1965).

Soil	Texture	pH	Organic matter (%)	Active Fe (%)	Active Mn (%)
■	Clay	5.7	1.3	0.79	0.36
○	Clay	5.9	3.3	1.67	0.33
△	Clay	5.4	1.5	1.67	0.20
●	Clay	6.3	2.4	1.27	0.20

to ~0.3 bar. Neutral soils low in organic matter give P_{CO_2} values that are less than 0.2 bar (Ponnamperuma, 1977a). An increase in P_{CO_2} depresses pH and increases the concentration of water-soluble iron, manganese, and calcium.

C. Anaerobic Decomposition of Organic Matter

Decomposition of organic matter in flooded soils differs from that in a well-drained soil in speed, pathways, and end products.

1. Speed of Decomposition

Organic matter decomposes more slowly in flooded soil than in well-drained soil, as shown by accumulation of organic matter in marshes and peat in water-logged depressions.

Decomposition of organic matter in well-drained soil is accomplished by actinomycetes, fungi, and a wide range of bacteria, assisted by soil fauna. Because of the high energy release associated with aerobic respiration of soil organisms, decomposition of substrate and synthesis of cell substance proceed rapidly. Hence organic matter disappears rapidly in warm, moist soils. In flooded soils the decomposition is almost entirely the work of anaerobic bacteria. Because they are less diverse than aerobic microorganisms, and because they function at the much lower energy level of fermentation, both decomposition and assimilation are slow.

Gambrell and Patrick (1978) reported that anaerobic conditions reduced the rate of decomposition of native organic matter in soil to half that under aerobic conditions, and that of wheat straw to 13%. A long-term experiment at the International Rice Research Institute (IRRI) showed that growing two crops of wetland rice per year for 6 years on a soil previously used for dryland crops increased the organic matter content from 2.59 to 3.22% in the dry-fallowed plots and to 3.50% in continuously flooded plots.

High C:N ratio of substrate, soil acidity, and low temperature retard organic matter decomposition. Although natural organic substances decompose slowly in anaerobic soils, reducing conditions accelerate the degradation of insecticides, herbicides, and fungicides (Sethunathan and Siddaramappa, 1978).

2. *Decomposition Pathways*

Decomposition of organic matter in soil is a process in which microorganisms use the substrate to derive energy and carbon for cell synthesis. It is accomplished by respiration.

Aerobic and anaerobic respiration follow a common pathway to the formation of pyruvic acid, the key intermediate in metabolism of carbohydrates by bacteria (Doelle, 1975). In aerobic respiration pyruvic acid is oxidized to CO_2 through the tricarboxylic acid (TCA) cycle, with oxygen as terminal electron acceptor. In the absence of O_2, facultative and obligate anaerobes use NO_3^-, Mn(IV), Fe(III), SO_4^{2-}, dissimilation products of carbohydrates and proteins, CO_2, N_2, and even H^+ ions as electron acceptors in their respiration. If the electron acceptors are inorganic, the process is termed anaerobic respiration; if they are organic substances, fermentation (Doelle, 1975). Fermentation produces an array of reaction products, most of which have been identified in anaerobic soils (Ponnamperuma, 1972; Yoshida, 1978; Tsutsuki, 1984; Watanabe, 1984).

Aerobic and anaerobic breakdown of proteins follow common pathways to the formation of amino acids. In aerobic oxidation, aerobic and facultative aerobic bacteria oxidize amino acids, using them as their sole source of carbon, nitrogen, and energy (Doelle, 1975), liberating ammonia, and producing CO_2 or carbon moieties that can be oxidized through the TCA cycle. In fermentation, amino acids are converted to ammonia, organic acids (including pyruvic acid), CO_2, methane, amines, and hydrogen sulfide.

The main pathways of pesticide degradation in flooded soils are reductive dechlorination, dehydrochlorination, hydrolysis, and reduction (Sethunathan and Siddaramappa, 1978).

3. End Products of Anaerobic Decomposition of Organic Matter

The stable end products in the decomposition of organic matter in well-drained soils are CO_2 and humic materials. The CO_2 escapes into the atmosphere, whereas the humic materials remain in the soil, bound to clay and hydrous oxides of aluminum and iron. The nitrogen that is released as ammonia is converted to nitrate, and sulfur compounds are oxidized to sulfate.

The major end products of anaerobic decomposition of organic matter in soils are CO_2, methane, and humic materials. Ammonium is the stable form of nitrogen, and sulfide that of sulfur. But the anaerobic metabolism of bacteria produces an array of substances, many of them transitory, not found in well-aerated soils. Some of them, gleaned from various sources (IRRI, 1966; Ponnamperuma, 1972; Yoshida, 1978; Neue and Scharpenseel, 1984; Tsutsuki, 1984; Watanabe, 1984), are listed here:

1. Gases: ammonia, carbon dioxide, carbon monoxide, hydrogen, hydrogen sulfide, methylamine, nitrogen, nitric oxide, nitrogen dioxide, nitrous oxide, phosphine, and sulfur dioxide
2. Hydrocarbons: butane, butadiene, ethane, ethylene, methane, propane, propadiene, and propylene
3. Alcohols: *n*-butanol, 2-butanol, 2,3-butanediol, ethanol, methanol, *n*-propanol, and 2-propanol
4. Carbonyls: acetaldehyde, acetone, butyraldehyde, formaldehyde, methyl ethyl ketone, propionaldehyde, and *n*-valeraldehyde
5. Volatile fatty acids: acetic, butyric, formic, isovaleric, propionic, and valeric
6. Nonvolatile acids: capric, fumaric, lactic, malonic, and oxaloacetic
7. Phenolic acids: *p*-coumaric, ferulic, *p*-hydroxybenzoic, sinapic, and vanillic
8. Volatile sulfur compounds: carbon disulfide, carbonyl sulfide, ethyl mercaptan, methyl mercaptan, and methyl disulfide

Humus of soils that are alternately flooded and drained, as in rice cultivation, differs from that of dryland soils in the following respects (Kuwatsuka *et al.*, 1978; Tsutsuki and Kuwatsuka, 1978; Tsutsuki and Kumada, 1980): the degree of humification is less; the content of hydrogen and nitrogen is higher and the degree of unsaturation lower; the content of carboxyl and phenolic groups is lower, whereas that of alcoholic and methoxyl groups is higher; and the molecular weight is higher.

D. Transformations of Nitrogen

The main nitrogen transformations in flooded soils are mineralization, nitrification–denitrification, immobilization, volatilization, and biological fixation.

1. Mineralization

Conversion of organic nitrogen to inorganic forms in flooded soils stops at the ammonium stage because of the lack of oxygen to carry the process through to nitrate. Hence ammonium accumulates in flooded soils. Ammonium is derived from anaerobic deamination of amino acids, degradation of purines, and hydrolysis of urea. Less than 1% comes from nitrate reduction. Although aerobic deamination may be more rapid than the anaerobic process, inorganic nitrogen is released in larger quantities and faster in anaerobic soils than in aerobic soils because less immobilization of nitrogen occurs in anaerobic soils (Broadbent and Reyes, 1971).

Soils vary widely in their capacity to produce ammonium (Fig. 8). Soils rich in organic matter rapidly release ammonium and attain concentrations exceeding 300 mg/kg in the soil within 2 weeks after submergence (at 25–35°C). Soils low in organic matter may release as little as 40 mg/kg even after several weeks of submergence. The soil solution may contain from 5 mg/liter of ammonium for clay soils to more than 100 mg/liter for sandy soils high in organic matter.

Water management and temperature markedly influence the rate and amount of ammonium released when soil is kept flooded. Dry soil that is flooded releases

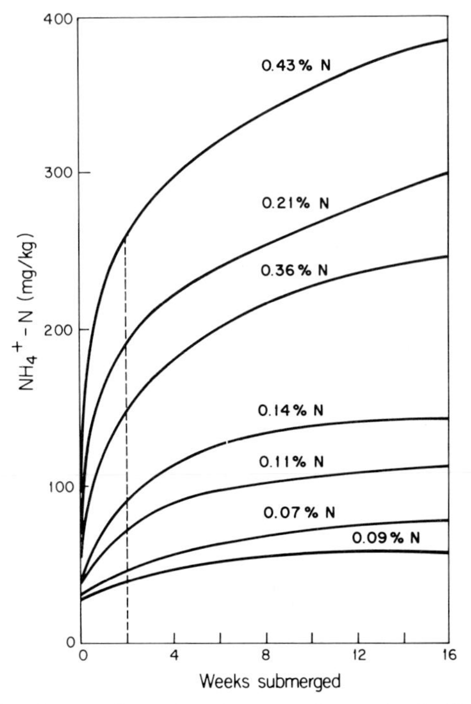

Fig. 8. Kinetics of ammonium release in seven submerged soils. From IRRI (1965).

more ammonium than does a wet soil (Dei and Yamasaki, 1979). Intermittent drying and flooding increase ammonium production (Reddy and Patrick, 1975). The Q_{10} for mineralization of soil nitrogen was 2.0 in the range 5–35°C (Stanford *et al.*, 1973). Dei and Yamasaki (1979) found that the amount of ammonium released in flooded soils can be described by

$$Y = k[(T - 15)D]^n \quad (17)$$

where Y is the amount of nitrogen mineralized, T the daily soil temperature, D the number of days, $T - 15$ the effective temperature, k a coefficient, and n a constant related to the mineralization pattern.

2. *Nitrification–Denitrification*

Ammonium fertilizers broadcast on the surface, ammonium formed by mineralization of nitrogen in the surface oxidized layer, and ammonium diffusing to that zone from the reduced soil are nitrified. The nitrate formed moves by mass flow and diffusion into the reduced zone, where it is highly unstable (Ponnamperuma, 1972). Here it is denitrified (converted to nitrous oxide or nitrogen) and lost from the system.

Air-dry soils may contain as much as 350 mg/kg of nitrate nitrogen. Within a few days of submergence most of this nitrogen is lost by denitrification. High

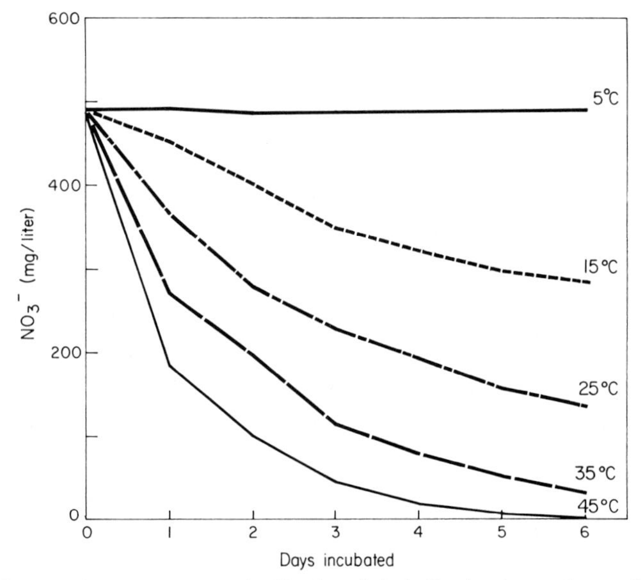

Fig. 9. Influence of temperature on the kinetics of denitrification in a submerged soil. From Ponnamperuma (1977a).

organic-matter content favors denitrification; low temperature retards it (Fig. 9). Alternate wetting and drying favor the formation and subsequent destruction of nitrate (Reddy and Patrick, 1975).

Nitrite, which is an intermediate product of both aerobic nitrification and anaerobic denitrification, is unstable in both aerobic and anaerobic media. Nitrite may be an important intermediate in denitrification in soils that are alternately flooded and dried (Ponnamperuma, 1972).

3. *Immobilization of Nitrogen*

Nitrogen can be fixed in soils chemically and biologically and rendered temporarily unavailable to plants. The high concentration of ammonium ions present in flooded soils should favor both processes.

a. Chemical Fixation. Ammonium may be fixed in the clay lattice, be specifically adsorbed on soil colloids, or participate in humus synthesis. The following factors favor ammonium fixation: high content of 2:1–type clays, high concentration of ammonium, alternate wetting and drying, and high soil pH.

Grewal and Kanwar (1973) reported that the ammonium fixing capacity of 23 wet soils treated with 100 mg N/kg was 8.0–30.5%. Pasricha (1976) suggested that ammonium was held at specific adsorption sites in two reduced clays. Ammonium and amino acids condense with phenolic substances in soils to form humus (Stevenson, 1979).

b. Biological Immobilization. The amount of inorganic nitrogen immobilized biologically in soil represents a balance between mineralization of organic nitrogen compounds and synthesis of microbial tissue. The balance between the two processes is determined by water regime, duration of submergence, C:N ratio of the organic matter, and temperature.

Because microbial activity is less intense in anaerobic soils than in aerobic soils (Section V,C), the nitrogen requirement of the microflora is lower, and less nitrogen is immobilized in flooded soils. As active proliferation of anaerobic bacteria induced by flooding slows down, the demand for nitrogen decreases. Although the nitrogen requirement for decomposition of organic matter is much less anaerobically than aerobically, materials with widely variable C:N ratios, such as straw, can temporarily immobilize nitrogen, whereas leguminous green manures release the greater part of the nitrogen in 2–3 weeks (Ponnamperuma, 1972). Nitrogen immobilization rates are independent of temperature in the range 10–40°C (Kai and Wada, 1979).

4. *Volatilization*

The role of the standing water in loss of nitrogen as ammonia from flooded soils has received increasing attention (Mikkelsen *et al.*, 1978; De Datta, 1981;

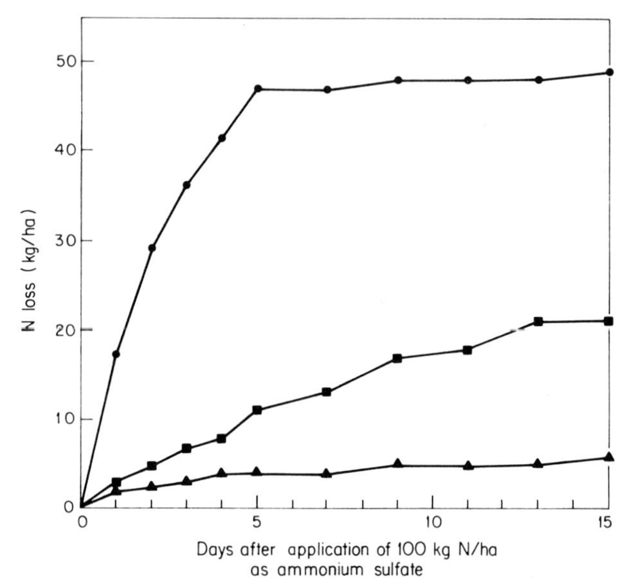

Fig. 10. Influence of soil properties on loss of ammonia from three flooded soils. ●, Sodic soil (pH 8.6); ■, Maahas clay (pH 6.6); ▲, Luisiana clay (pH 4.7). From IRRI (1977).

Savant and De Datta, 1983). Field studies suggest that up to 20% of nitrogen in urea or ammonium sulfate broadcast on flooded rice fields may be lost in a season as ammonia gas. High concentration of ammonium in the flood water, high soil pH, high photosynthetic activity by algae and submerged plants, and high temperature favor ammonia volatilization.

Loss of ammonia is not limited to fertilized fields, for ammonium is continuously formed in the floodwater by organic matter decomposition and is transferred by diffusion to the floodwater from the anaerobic soil. Figure 10 shows the effect of pH on nitrogen loss.

5. *Biological Nitrogen Fixation*

Flooding increases a soil's capacity to fix nitrogen biologically. Biological nitrogen fixation involves reduction of N_2 to ammonium. Photosynthesis and respiration provide the powerful reductants and ATP needed for the process (Ponnamperuma, 1972). Flooded soil is an ideal medium for nitrogen fixation because blue-green algae living free or in association with *Azolla* thrive in the floodwater and on the soil surface, while aerobic nitrogen fixers function at the aerobic–anaerobic interfaces and anaerobic bacteria flourish in the reduced layer sustained by intermediate products of the anaerobic decomposition of organic matter.

A pH above 7, the presence of available phosphate and iron, and adequate amounts of CO_2 in the floodwater favor growth of blue-green algae (Roger and Reynaud, 1979). Blue-green algae (Cyanobacteria) fix ~0.5 kg N/ha/day (Stewart *et al.*, 1979).

The heterotrophic nitrogen fixers, both aerobic and anaerobic, contribute to nitrogen fixation in flooded soils (Watanabe, 1978). Their activity is favored by anaerobic conditions, presence of decomposable organic matter, and rice plants. They may account for as much as 60% of the nitrogen-fixing activity of wetland rice soils (Matsuguchi, 1979).

Azolla, a water fern, with a blue-green alga (*Anabaena azollae*) living in symbiosis, can fix up to 400 kg N/ha/year under field conditions (Becking, 1979). Abundant phosphorus, calcium, iron, and molybdenum favor nitrogen fixation by *Azolla.*

Apparently because of autotrophic and heterotrophic nitrogen fixation, the nitrogen content of a continuously flooded soil at IRRI increased from 0.153 to 0.189% in 6 years.

E. Reduction of Manganese, Iron, and Sulfate

When the oxygen and nitrate present in a flooded soil are consumed, anaerobic bacteria use oxidized soil components such as hydrous oxides of Mn(IV) and Fe(III) and SO_4^{2-} as electron acceptors in their respiration, converting them to their reduced counterparts Mn(II), Fe(II), and S^{2-}.

1. Manganese

Within 1 to 3 weeks after flooding almost all the active (EDTA-dithionite-extractable) manganese present in a soil is reduced to Mn(II) compounds that are more soluble in water than are those of Mn(IV). The reduction is both chemical and biochemical and precedes that of iron (Yoshida, 1978).

Acid soils high in manganese and organic matter build up water-soluble Mn^{2+} concentrations as high as 90 mg/liter, at 25–35°C, within 1–2 weeks after flooding (Fig. 11). Alkaline soils and soils low in manganese rarely contain more than 10 mg/liter of water-soluble manganese at any stage of submergence.

The decrease in concentration of water-soluble manganese is the result of its precipitation as manganous carbonate. The activity of Mn^{2+} after the peak is given by

$$\text{pH} + \tfrac{1}{2} \log \text{Mn}^{2+} + \tfrac{1}{2} \log P_{CO_2} = 4.4$$

2. Iron

The most conspicuous chemical change that takes place when soil is flooded is reduction of Fe(III) to Fe(II). The brown color of iron-rich soils changes to

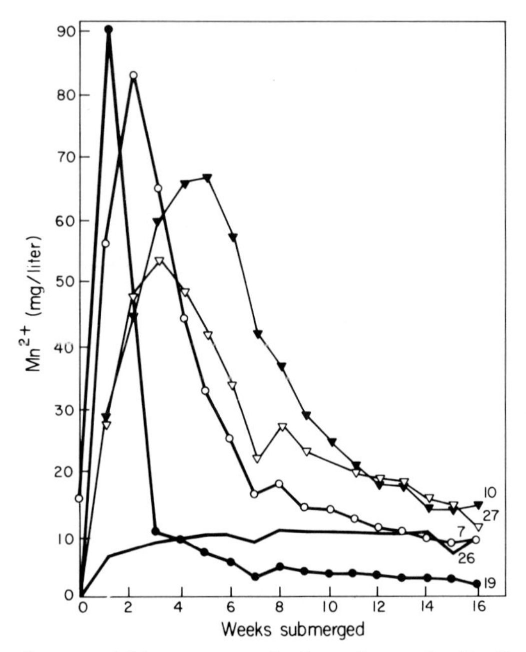

Fig. 11. Kinetics of water-soluble manganese in five submerged soils. From Ponnamperuma (1977b).

Soil no.	pH	Organic matter (%)	Active Mn (%)
7	5.9	3.3	0.33
19	5.5	4.2	0.13
10	5.4	1.5	0.20
27	6.6	2.0	0.31
26	7.6	1.5	0.06

shades of gray, green, or blue, and large amounts of iron are brought into solution. The reduction of iron is a consequence of anaerobic metabolism and appears to involve chemical reduction by bacterial metabolites, although direct reduction coupled with respiration may also be involved (Yoshida, 1978). Five to fifty percent of active Fe(III) present in a soil may be reduced within a few weeks of submergence, depending on temperature, organic-matter content, nitrate concentration, and crystallinity of the oxides. The reduced iron acts as a sink for oxygen diffusing into the soil and is a source of Fe^{2+} ions. The solid phase of reduced iron includes green hydrated magnetite ($Fe_3O_4 \cdot nH_2O$), black hydrotroilite ($FeS \cdot nH_2O$), white siderite ($FeCO_3$), and blue vivianite ($Fe_3PO_4 \cdot 8H_2O$).

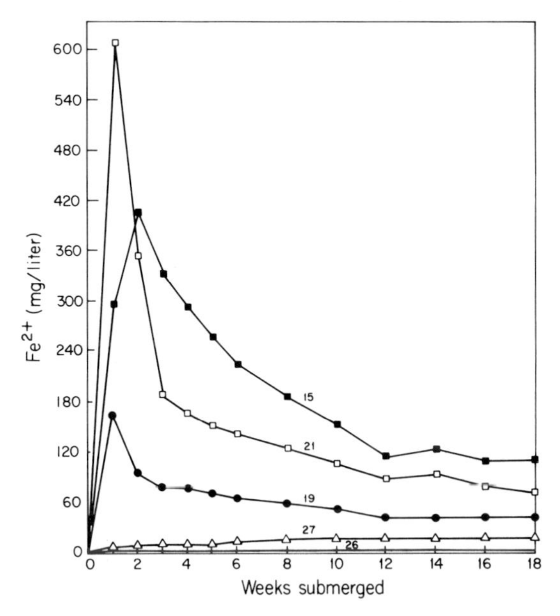

Fig. 12. Kinetics of water-soluble iron in five submerged soils. From Ponnamperuma (1977a).

Soil no.	Texture	pH	Organic matter (%)	Active Fe (%)	Active Mn (%)
21	Clay loam	4.6	4.1	2.78	0.02
15	Clay loam	5.3	2.5	0.91	0.05
19	Clay loam	5.5	4.2	2.30	0.13
27	Clay	6.6	2.0	1.60	0.31
26	Clay loam	7.6	1.5	0.30	0.06

The kinetics of water-soluble iron depend on soil properties and temperature. Acid soils high in organic matter and iron build up concentrations as high as 600 mg/liter within 1–3 weeks after flooding and show a roughly exponential decrease to levels of 50–100 mg/liter, which persist for several months (Fig. 12). Low temperature retards the peak and broadens the area under it (Ponnamperuma, 1976). Soils high in organic matter but low in iron give high concentrations that persist for several months. In neutral and calcareous soils low in organic matter, the concentration of water-soluble iron may be less than 1 mg/liter.

The activity of water-soluble Fe^{2+} in reduced soils is given by

$$pH + \tfrac{1}{2} \log Fe^{2+} = K \tag{19}$$

and K varies from the theoretical value of 5.4 to 4.8 (Ponnamperuma, 1977a).

3. *Sulfate*

Reduction of sulfate is accomplished by a small group of obligate anaerobic bacteria of the genera *Desulfovibrio* and *Desulfotomaculum*. These bacteria use a variety of fermentation products and H_2 to reduce SO_4^{2-}. Strong acidity, nitrate, and low temperature retard sulfate reduction.

In acid soils the concentration of water-soluble sulfate increases, reaches a peak, and then decreases. The initial increase is the result of desorption of sulfate adsorbed on hydrous oxides following an increase in pH (Section III,E). The subsequent decrease is the result of reduction. Nearly neutral pH, presence of organic matter, and temperatures of 25–45°C favor sulfate reduction (Ponnamperuma, 1972, 1981).

The hydrogen sulfide formed by sulfate reduction dissociates as follows:

$$H_2S = H^+ + HS^- \quad (20)$$

$$HS^- = H^+ + S^{2-} \quad (21)$$

At the pH values of flooded soils, the bulk of the water-soluble hydrogen sulfide is present as H_2S and HS^-, but the activity of S^{2-} is sufficient to precipitate Fe^{2+} and Zn^{2+} as insoluble sulfides (Ayotade, 1977). This keeps the concentration of H_2S in the soil solution below 0.1 mg/liter, the toxic limit for rice.

VI. EFFECTS ON FERTILITY

Soil fertility is the status of a soil with respect to its capacity to provide plants with nutrients in the amount, rate, and balance needed for optimum growth. Fertility depends on (1) the presence of water, oxygen, and adequate nutrients in forms the plant can absorb; (2) soil capacity to deliver oxygen and nutrients by mass flow and diffusion to the root surface; (3) presence of a favorable ionic composition; and (4) absence of substances that interfere with the movement of nutrients in balanced amounts into roots. The physical, chemical, electrochemical, and biological changes brought about by flooding markedly influence soil fertility. The effects depend on soil properties, especially temperature and duration of submergence.

A. Physical Effects

Restriction of oxygen entry into the soil, swelling of soil colloids, and destruction of structure drastically reduce the rate of respiration and nutrient absorption in plants not possessing a mechanism for the internal transfer of oxygen to the roots. For most crop plants, except rice, root growth is reduced when the oxygen diffusion rate (ODR) falls below 0.2 mg/cm/min (Meek and Stolzy, 1978). Puddling and compaction by traffic on a flooded soil aggravate the adverse effects. Lowering of soil temperature by flooding may retard germination, seedling growth, and nutrient absorption even after a flooded soil is drained. In highly permeable soils, flooding increases the loss of nutrients by leaching.

B. Electrochemical Effects

A decrease in redox potential below 0.2 V destroys nitrate; favors accumulation of ammonium and nitrogen fixation; increases availability of phosphorus, silicon, iron, and manganese; decreases availability of sulfur, zinc, and copper; and generates substances that interfere with nutrient uptake.

A decrease in redox potential benefits marsh plants such as rice, because it

TABLE II

Phytotoxicity of Some Products of the Anaerobic Decomposition of Organic Matter Identified in Soils

Product	Toxic concentration	Seedling	Reference
Gases			
Ammonia	17.5 m*M*	Rice	Bonner (1946)
Carbon dioxide	0.2–0.3 bar	Barley	Vlamis and Davis (1944)
	0.05–0.1 bar	Rice	Saito and Takahashi (1954)
Ethylene	1–10 μl/liter	Barley, maize, rice, wheat	Cannell and Lynch (1984)
Hydrogen sulfide	2 μ*M*	Rice	Mitsui (1960)
Alcohols			
Ethanol	40 m*M*	Rice	Takijima (1963)
n-Butanol	10 m*M*	Rice	Takijima (1963)
Carbonyls			
Acetaldehyde	4 m*M*	Rice	Takijima (1963)
Acetone	40 m*M*	Rice	Takijima (1963)
Volatile fatty acids			
Acetic acid	1–5 m*M*	Rice	Rao and Mikkelsen (1977)
n-Butyric acid	1–5 m*M*	Rice	Rao and Mikkelsen (1977)
Formic acid	6 m*M*	Rice	Takijima (1963)
Propionic acid	1–5 m*M*	Rice	Rao and Mikkelsen (1977)
Phenolic acids			
p-Coumaric acid	15 m*M*	Wheat	Cannell and Lynch (1984)
p-Hydroxybenzoic acid	7–70 m*M*	Wheat, maize, sorghum	Cannell and Lynch (1984)
Thiol			
Ethyl mercaptan	2 m*M*	Rice	Takijima (1963)

prevents iron deficiency in neutral and alkaline soils and increases availability of nutrients on most soils.

An increase in pH of acid soils, a decrease in pH of alkaline soils, and pH stabilization at 7 should markedly influence soil fertility through effects on nutrient absorption, concentration of nutrients or toxins, and microbial processes that release or destroy plant nutrients, or that generate toxins.

A pH of 6.5–7.0 in the soil solution favors nutrient absorption, increases availability of phosphorus, and favors mineralization of organic matter, regardless of the soil pH before flooding. Benefits related to initial pH include elimination of aluminum and manganese toxicities and alleviation of toxicities of iron, organic acids, and CO_2 in acid soils; and increasing the availability of iron, manganese, zinc, and calcium in alkaline soils (Ponnamperuma, 1978).

Flooding decreases the specific conductance of saline soils and increases that of nonsaline soils. A specific conductance exceeding 4 dS/m hinders nutrient and water uptake and injures plants (Maas and Hoffman, 1977).

The increase in ionic strength caused by flooding decreases the activity coefficients of ions in the soil solution. As a result, chemical and electrochemical equilibria that regulate the concentration of nutrient ions are affected.

The high concentrations of cations and bicarbonate ions produced by soil

TABLE III

Kinetics of Water-Soluble Boron, Copper, Molybdenum, and Zinc in Three Submerged Soils[a]

Soil and element	Concentration (mg/liter)			
	1 week	2 weeks	4 weeks	6 weeks
Luisiana clay (pH 4.8, O.M.[b] 2.8%)				
B	0.55	0.49	0.46	0.30
Cu	0.11	0.11	0.04	0.02
Mo	0.18	0.06	0.15	0.24
Zn	0.30	0.09	0.08	0.03
Maahas clay (pH 6.6, O.M. 2.8%)				
B	1.80	1.01	1.15	1.18
Cu	0.06	0.05	0.03	0.02
Mo	0.04	0.09	0.17	0.12
Zn	0.18	0.08	0.06	0.03
Keelung silt loam (pH 7.7, O.M. 6.9%)				
B	0.48	0.55	0.68	0.52
Cu	0.05	0.05	0.04	0.03
Mo	0.09	0.10	0.17	0.27
Zn	0.14	0.08	0.04	0.03

[a]From Ponnamperuma (1977b).
[b]O.M., Organic matter.

metabolism displace ions from soil colloids. Thus the concentration of potassium, calcium, magnesium, and phosphorus in the soil solution increases, and that benefits plants. One disadvantage is that these ions may be removed from the root zone by percolating water. Flooding causes impoverishment of sandy soils. The increase in concentration of water-soluble phosphate and silica caused by flooding is due partly to desorption (Ponnamperuma, 1972).

C. Chemical Effects

The beneficial chemical effects of flooding on soil fertility include an influx of dissolved and suspended nutrients, accumulation of nitrogen, increase in solubility of phosphorus and silicon, and an increase in concentration of potassium in the soil solution (Ponnamperuma, 1977a). Among the disadvantages are depletion of O_2, accumulation of CO_2, destruction of nitrate, and generation of high concentrations of water-soluble iron and manganese. In acid soils high in organic matter, volatile fatty acids and excess water-soluble iron may persist for several weeks after submergence. The effects of the products of the anaerobic decomposition of organic matter on plant growth are summarized in Table II. Flooding increases the availability of molybdenum but decreases that of zinc and copper (Table III).

VII. SOIL FORMATION

The water regime is an important soil-forming factor. Thus properties associated with wetness are used in soil classification (Soil Survey Staff, 1975).

A. Aquic Suborders

Transformation of weathered rock material into soil is brought about by the accretion of organic matter, leaching of soluble salts, transformation and translocation of organic and inorganic substances, and synthesis of secondary minerals. The balance among these processes determines the character of the soil that develops. Aquic soils owe some distinguishing character to flooding or waterlogging. They include the suborders Aqualfs, Aquents, Aquepts, Aquods, Aquols, Aquox, and Aquults. These soils are usually found in flat or concave landscapes and have one feature in common—soil reduction due to temporary exclusion of O_2 by water.

During the period of waterlogging or flooding, the soil is reduced (Section IV). When the water table recedes and O_2 reenters the soil, the chemical changes caused by flooding are largely reversed.

Aquic soils have three main horizons: a partially oxidized A horizon in which organic matter accumulates, a zone in which oxidizing and reducing conditions alternate, and a zone of permanent reduction. Because the soil is intermittently saturated with water, oxidation of organic matter is slow and organic matter accumulates in the A horizon. In the second horizon, iron and manganese are alternately reduced and oxidized. During oxidation, iron and manganese oxides are deposited as rusty mottles, streaks, or concretions. The zone of permanent waterlogging is bluish green because ferrous compounds are present.

B. Paddy Soils

Paddy soils are managed in a special way for wet cultivation of rice. Management practices include leveling the land, construction of levees to impound water, puddling (plowing and harrowing the wet soil), maintaining 5–10 cm of standing water during the 4–5 months the crop is on the land, draining and drying the fields at harvest, and reflooding after an interval that varies from a few days to several months. These operations and oxygen secretion by rice roots lead to the development of certain features peculiar to paddy soils (Ponnamperuma, 1972).

During the period of submergence, iron and manganese are reduced and the soil turns dark gray. When the soil is drained at harvest, the entire profile above the water table is reoxidized, giving it a highly mottled appearance, with vertical streaks corresponding to root channels. Reduced iron and manganese move out of the root zone in percolating waters and are deposited below the plow sole. Cations displaced from soil colloids by Fe^{2+} and Mn^{2+} migrate out of the root zone and are lost by "ferrolysis" (Brinkman, 1979).

C. Acid Sulfate Soils

Acid sulfate soils are extremely acid (pH < 3.5) and are derived from sediments high in sulfidic materials and low in bases. The first phase of their formation, the buildup of pyrites, proceeds in tidal swamps. The main steps in pyrite formation are reduction of sulfate to hydrogen sulfide by *Desulfovibrio* and *Desulfotomaculum,* precipitation of iron monosulfide, oxidation of hydrogen sulfide to sulfur chemically and biochemically, and formation of pyrite, thus

$$SO_4^{2-} + 10\ H^+ + 8\ e^- = H_2S + 4\ H_2O \quad (22)$$

$$H_2S = H^+ + HS^- \quad (23)$$

$$HS^- = H^+ + S^{2-} \quad (24)$$

$$Fe^{2+} + S^{2-} = FeS \quad (25)$$

$$H_2S + \tfrac{1}{2}\ O_2 = S + H_2O \quad (26)$$

$$FeS + S = FeS_2 \quad (27)$$

While submerged, the reduced sediment is nearly neutral in reaction and supports salt-tolerant plants. But when drained, O_2 enters the soil and oxidizes the pyrite and sulfur present chemically and biochemically to sulfuric acid (Van Breemen and Pons, 1978), rendering the soil extremely acid and producing jarosite, a yellow mineral characteristic of acid sulfate soils. Submergence reverses most of the changes.

D. Sodic Soils

Sodic soils contain sufficient exchangeable sodium to interfere with plant growth. Their pH is usually above 8.5, and the sodium adsorption ratio of the saturation extract is 15 or more. The alkalinity is the result of the presence of sodium carbonate and bicarbonate. Sodic soils are found in arid areas in depressions that undergo temporary waterlogging or flooding.

Sodium carbonate is of chemical and biological origin. The biological process is the result of reduction of sulfate and loss of hydrogen sulfide by volatilization:

$$Na_2SO_4 + 8\,H^+ + 8\,e^- = Na_2S + 4\,H_2O \tag{28}$$

$$Na_2S + 2\,H_2O + 2\,CO_2 = 2\,NaHCO_3 + H_2S \tag{29}$$

$$2\,NaHCO_3 = Na_2CO_3 + H_2O + CO_2 \tag{30}$$

E. Saline Soils

Saline soils contain sufficient salts in the root zone to interfere with plant growth. Two common causes of salinity are intrusion of surface or underground saline waters and concentration of salt in the surface soil by evaporation of groundwater rich in salts. Thus irrigation in arid areas without drainage leads to the buildup of salinity and alkalinity. The problem is aggravated if the irrigation water itself is saline (Shalhevet and Kamburov, 1976).

F. Peat Soils

Continuous waterlogging or flooding provides conditions conducive to peat formation. Marsh vegetation that thrives uninterrupted because of the availability of water generates organic matter, whose bulk accumulates because of slow decomposition of organic matter under anaerobic conditions (Section V,C). As one generation of plants follows another, layer upon layer of organic residues is deposited. Accompanying this process is a succession of vegetation, climaxing in forest trees (Brady, 1974). Swamp forests cover 32×10^6 ha in the rain-forest belts of the tropics. Those of southeast Asia are 4000–5000 years old (Driessen, 1978).

ACKNOWLEDGMENTS

I thank Dr. I. Watanabe, Dr. T. Woodhead, Mrs. M. T. C. Cayton, and Mrs. R. S. Lantin for their valuable comments.

REFERENCES

Aomine, S. (1962). A review of research on redox potentials of paddy soils in Japan. *Soil Sci.* **94,** 6–13.

Armstrong, W. (1978). Root aeration in the wetland condition. *In* "Plant Life in Anaerobic Environments" (D. D. Hook and R. M. M. Crawford, eds.), pp. 269–297. Ann Arbor Sci. Publ., Ann Arbor, Michigan.

Ayotade, K. A. (1977). Kinetics and reactions of hydrogen sulfide in solution of flooded rice soils. *Plant Soil* **46,** 381–389.

Bass-Becking, L. G., Kaplam, L. R., and Moor, D. (1960). Limits of the natural environment in terms of pH and oxidation–reduction potentials. *J. Geol.* **68,** 243–284.

Baver, L. D., Gardner, W. H., and Gardner, W. R. (1972). "Soil Physics." Wiley, New York.

Becking, J. H. (1979). Environmental requirements of *Azolla* for use in tropical rice production. *In* "Nitrogen and Rice," pp. 345–374. International Rice Research Institute, Los Baños, Philippines.

Bonneau, M. (1982). Soil temperature. *In* "Constituents and Properties of Soils" (M. Bonneau and B. Souchier, eds.), pp. 366–371. Academic Press, New York and London.

Bonner, J. (1946). The role of organic matter, especially manure, in the nutrition of rice. *Bot. Gaz. (Chicago)* **108,** 267–279.

Brady, N. C. (1974). "The Nature and Properties of Soils." Macmillan, New York.

Brinkman, R. (1979). "Ferrolysis a Soil-Forming Process in Hydromorphic Conditions." Centre for Agric. Publ. and Documentation, Wageningen, The Netherlands.

Broadbent, F. E., and Reyes, O. C. (1971). Uptake of soil and fertilizer nitrogen by rice in some Philippine soils. *Soil Sci.* **112,** 200–205.

Brown, K. W. (1977). Shrinkage and swelling of clay, clay strength, and other properties of clay soils and clays. *In* "Minerals in Soil Environments" (J. B. Dixon and S. B. Weed, eds.), pp. 689–708. Soil Sci. Soc. Am., Madison, Wisconsin.

Burchill, S., Hayes, M. H. B., and Greenland, D. J. (1981). Adsorption. *In* "The Chemistry of Soil Processes" (D. J. Greenland and M. H. B. Hayes, eds.), pp. 221–400. Wiley, Chichester.

Buresh, R. J., Casselman, M. E., and Patrick, W. H., Jr. (1980). Nitrogen fixation in flooded soil systems, a review. *Adv. Agron.* **33,** 150–192.

Cannell, R. Q., and Lynch, J. M. (1984). Possible adverse effects of decomposing organic matter on plant growth. *In* "Organic Matter and Rice," in press. International Rice Research Institute, Los Baños, Philippines.

De Bout, J. A. M., Le, K. K., and Bouldin, D. R. (1978). Bacterial oxidation of methane in rice paddy. *In* "Environmental Role of Nitrogen Fixing Blue Green Algae and Asymbiotic Bacteria" (V. Granhall, ed.), pp. 91–99. Ecology Bulletin, Stockholm.

De Datta, S. K. (1981). "Principles and Practices of Rice Production." Wiley, New York.

De Gee, J. C. (1950). Preliminary oxidation potential determination in a "sawah" profile near Bogor (Java). *Trans. Int. Congr. Soil Sci. 4th* **1,** 300–303.

Dei, Y., and Yamasaki, S. (1979). Effect of water and crop management on the nitrogen-supplying capacity of paddy soils. *In* "Nitrogen and Rice," pp. 451–484. International Rice Research Institute, Los Baños, Philippines.

Doelle, H. W. (1975). "Bacterial Metabolism." Academic Press, New York.

Driessen, P. M. (1978). Peat soils. *In* "Soils and Rice," pp. 763–780. International Rice Research Institute, Los Baños, Philippines.

Ellis, B. G., and Knezek, B. D. (1972). Adsorption reaction of micronutrients in soils. *In* ''Micronutrients in Agriculture'' (J. J. Mortvedt, P. M. Giordano, and W. L. Lindsay, eds.), pp. 59–78. Soil Sci. Soc. Am., Madison, Wisconsin.
Gambrell, R. P., and Patrick, W. H., Jr. (1978). Chemical and microbiological properties of anaerobic soils and sediments. *In* ''Plant Life in Anaerobic Environments'' (D. D. Hook and R. M. M. Crawford, eds.), pp. 375–423. Ann Arbor Sci. Publ., Ann Arbor, Michigan.
Grable, A. H. (1966). Soil aeration and plant growth. *Adv. Agron.* **18,** 57–106.
Greenland, D. J. (1981). Recent progress in studies of soil structure, and its relation to properties and management of paddy soils. *In* ''Proceedings of Symposium on Paddy Soils'' (Institute of Soil Science Academia Sinica, eds.), pp. 42–58. Science Press, Beijing, China.
Greenwood, D. J. (1970). Distribution of carbon dioxide in the aqueous phase of aerobic soils. *Soil Sci.* **21,** 314–329.
Grewal, J. S., and Kanwar, J. S. (1973). ''Potassium and Ammonium Fixation in Indian Soils.'' Indian Council of Agric. Res., New Delhi.
Harrison, W. H., and Aiyer, P. A. S. (1913). The gases of swamp rice soils. I. Their composition and relation to the crop. *Mem. Dep. Agric. India Chem. Ser.* **3,** 65–104.
Harrison, W. H., and Aiyer, P. A. S. (1915). The gases of swamp rice soils. II. Their utilization for the aeration of the roots of the crop. *Mem. Dep. Agric. India Chem. Ser.* **4,** 1–7.
Houng, K. H. (1981). A theoretical evaluation of percolating rate on the thickness of oxidizing zone of paddy soils. *Proc. Natl. Sci. Council Repub. China* **5,** 274–278.
International Rice Research Institute (IRRI) (not dated). ''Annual Report for 1963.'' Los Baños, Philippines.
International Rice Research Institute (IRRI) (1965). ''Annual Report for 1964.'' Los Baños, Philippines.
International Rice Research Institute (IRRI) (1966). ''Annual Report for 1965.'' Los Baños, Philippines.
International Rice Research Institute (IRRI) (1977). ''Annual Report for 1976.'' Los Baños, Philippines.
Kai, H., and Wada, K. (1979). Chemical and biological immobilization of nitrogen in paddy soils. *In* ''Nitrogen and Rice,'' pp. 157–174. International Rice Research Institute, Los Baños, Philippines.
Kondo, Y. (1952). Physiological studies on cool-weather resistance of rice varieties. *Nogyo Gijutsu Kenkyusho Hokoku Di Seiri, Iden. Sakumotsu, Ippan (Natl. Inst. Agric. Sci. Bull. Jpn. Ser. D)* **3,** 113–228.
Kuwatsuka, S., Tsutsuki, K., and Kumada, K. (1978). Chemical studies on soil humic acids. I. Elementary composition of humic acids. *Soil Sci. Plant Nutr. (Tokyo)* **23,** 337–347.
Lindsay, W. L. (1979). ''Chemical Equilibria in Soils.'' Wiley, New York.
Maas, E. V., and Hoffman, G. J. (1977). Crop salt tolerance–assessment. *J. Irrig. Drain. Div. ASCE* **103,** No. IR2 (Proc. Pap. No. 12993).
Matsuguchi, T. (1979). Factors affecting heterotrophic nitrogen fixation in submerged rice soils. *In* ''Nitrogen and Rice,'' pp. 207–222. International Rice Research Institute, Los Baños, Philippines.
Meek, B. D., and Stolzy, L. H. (1978). Short-term flooding. *In* ''Plant Life in Anaerobic Environments'' (D. H. Hook and R. M. M. Crawford, eds.), pp. 351–373. Ann Arbor Sci. Publ., Ann Arbor, Michigan.
Mikkelsen, D. S., De Datta, S. K., and Obcemea, W. N. (1978). Ammonia volatilization losses from flooded rice soils. *Soil Sci. Soc. Am. J.* **42,** 725–730.
Mitsui, S. (1960). ''Inorganic Nutrition Fertilization and Soil Amelioration of Lowland Rice.'' Yokendo, Tokyo.
Mott, C. J. B. (1981). Anion and ligand exchange. *In* ''The Chemistry of Soil Processes'' (D. J. Greenland and M. H. B. Hayes, eds.), pp. 179–219. Wiley, Chichester.

Neue, H. U., and Scharpenseel, H. W. (1984). Gaseous products of the decomposition of organic matter in submerged soils. *In* "Organic Matter and Rice," in press. International Rice Research Institute, Los Baños, Philippines.

Pasricha, W. S. (1976). Exchange equilibria of ammonium in some paddy soils. *Soil Sci.* **121,** 267–271.

Patrick, W. H., Jr. (1981). The role of inorganic redox systems in controlling reduction in paddy soils. *In* "Proceedings of Symposium on Paddy Soil" (Institute of Soil Science Academia Sinica, eds.), pp. 107–117. Science Press, Beijing, China.

Patrick, W. H., Jr., and Reddy, C. N. (1978). Chemical changes in rice soils. *In* "Soils and Rice," pp. 361–379. International Rice Research Institute, Los Baños, Philippines.

Pearsall, W. H., and Mortimer, C. H. (1939). Oxidation reduction potentials in waterlogged soils, natural waters and muds. *J. Ecol.* **27,** 483–501.

Ponnamperuma, F. N. (1955). "The Chemistry of Submerged Soils in Relation to the Growth and Yield of Rice." Ph.D. Thesis, Cornell Univ., Ithaca, New York.

Ponnamperuma, F. N. (1972). The chemistry of submerged soils. *Adv. Agron.* **24,** 29–96.

Ponnamperuma, F. N. (1976). Temperature and the chemical kinetics of flooded soils. *In* "Climate and Rice," pp. 249–263. International Rice Research Institute, Los Baños, Philippines.

Ponnamperuma, F. N. (1977a). Physicochemical properties of submerged soils in relation to fertility. *IRRI Res. Paper Ser.* No. 5.

Ponnamperuma, F. N. (1977b). Behavior of minor elements in paddy soils. *IRRI Res. Paper Ser.* No. 8.

Ponnamperuma, F. N. (1978). Electrochemical changes in submerged soils and the growth of rice. *In* "Soils and Rice," pp. 421–441. International Rice Research Institute, Los Baños, Philippines.

Ponnamperuma, F. N. (1981). Some aspects of the physical chemistry of paddy soils. *In* "Proceedings of Symposium on Paddy Soil" (Institute of Soil Science Academia Sinica, eds.), pp. 59–94. Science Press, Beijing, China.

Ponnamperuma, F. N., Tianco, E. M., and Loy, T. A. (1966). Ionic strengths of the solutions of flooded soils and other natural aqueous solutions from specific conductance. *Soil Sci.* **102,** 408–413.

Ponnamperuma, F. N., Castro, R. U., and Valencia, C. M. (1969). Experimental study of the influence of the partial pressure of carbon dioxide on pH values of aqueous carbonate systems. *Soil Sci. Soc. Am. Proc.* **33,** 239–241.

Rao, D. N., and Mikkelsen, D. S. (1977). Effects of acetic, propionic, and butyric acids on rice seedling growth and nutrition. *Plant Soil* **47,** 323–334.

Reddy, K. R., and Patrick, W. H., Jr. (1975). Effect of alternate aerobic and anaerobic conditions on redox potential, organic matter decomposition and nitrogen loss in a flooded soil. *Soil Biol. Biochem.* **7,** 87–94.

Roger, P. A., and Reynaud, P. A. (1979). Ecology of blue-green algae in paddy fields. *In* "Nitrogen and Rice," pp. 287–309. International Rice Research Institute, Los Baños, Philippines.

Roger, P. A., and Watanabe, I. (1984). Algae and aquatic weeds as a source of organic matter and plant nutrients for wetland rice. *In* "Organic Matter and Rice," in press. International Rice Research Institute, Los Baños, Philippines.

Rowell, D. L. (1981). Oxidation and reduction. *In* "The Chemistry of Soil Processes" (D. J. Greenland and M. H. B. Hayes, eds.), pp. 401–461. Wiley, New York.

Russell, E. W. (1973). "Soil Conditions and Plant Growth." Longman, London.

Saito, B., and Takahashi, T. (1954). Investigation on controlling of the so-called "Akiochi" of the paddy rice plant found widely in Kyushu. I. Some chemical characteristics of irrigation waters and soil waters collected from the so-called "Akiochi rice field." *Kyushu Agric. Exp. Stn. Bull.* **2,** 273–282.

Sanchez, P. (1976). "Properties and Management of Soils in the Tropics." Wiley, New York.

Savant, N. K., and De Datta, S. K. (1983). Nitrogen transformations in wetland rice soils. *Adv. Agron.* **35,** 241–302.

Sethunathan, N., and Siddaramappa, R. (1978). Microbial degradation of pesticides in rice soils. *In* "Soils and Rice," pp. 479–497. International Rice Research Institute, Los Baños, Philippines.

Shalhevet, J., and Kamburov, J. (1976). "Irrigation and Salinity." Int. Commission on Irrigation and Drainage, New Delhi.

Sims, J. L., and Patrick, W. H., Jr. (1978). The distribution of micronutrient cations in soil under conditions of varying redox potential and pH. *Soil Sci. Soc. Am. J.* **42,** 258–262.

Soil Survey Staff (1975). "Soil Taxonomy." U.S. Dep. of Agric., Washington, D.C.

Stanford, G., Frere, M. H., and Schwaninger, D. H. (1973). Temperature coefficient of soil nitrogen mineralization. *Soil Sci.* **115,** 321–323.

Stevenson, F. J. (1979). Humus. *In* "Encyclopedia of Soil Science, Part I" (R. W. Fairbridge and C. W. Finkl, Jr., eds.), pp. 195–205. Dowden, Hutchinson & Ross, Strongdsberg, Pennsylvania.

Stevenson, F. J., and Ardakani, M. S. (1972). Organic matter reactions involving micronutrient in soils. *In* "Micronutrients in Agriculture" (J. J. Mordvedt, P. M. Giordano, and W. L. Lindsay, eds.), pp. 79–114. Soil Sci. Soc. Am., Madison, Wisconsin.

Stewart, W. D. P., Rowell, P., Ladha, J. K., and Sampaio, M. J. A. M. (1979). Blue-green algae (Cyanobacteria)—some aspects related to their role as sources of fixed nitrogen in paddy soils. *In* "Nitrogen and Rice," pp. 263–285. International Rice Research Institute, Los Baños, Philippines.

Takijima, Y. (1963). Studies on behavior of growth inhibiting substances in paddy soils with special reference to the occurrence of root damage in peaty paddy fields. *Nogyo Gijutsu Kenkyusho Hokoku B (Bull. Natl. Inst. Agric. Sci. Jpn. Ser. B)* **13,** 117–252.

Talibudeen, O. (1981). Cation exchange in soils. *In* "The Chemistry of Soil Processes" (D. J. Greenland and M. H. B. Hayes, eds.), pp. 115–117. Wiley, Chichester.

Tsutsuki, K. (1984). Volatile products and low-molecular weight phenolic products of the anaerobic decomposition of organic matter. *In* "Organic Matter and Rice," in press. International Rice Research Institute, Los Baños, Philippines.

Tsutsuki, K., and Kumada, K. (1980). Chemistry of humic acids. *Fert. Sci.* **3,** 93–171 (in Japanese).

Tsutsuki, K., and Kuwatsuka, S. (1978). Chemical studies on soil humic acids. II. Composition of oxygen-containing functional groups of humic acids. *Soil Sci. Plant Nutr. (Tokyo)* **24,** 547–560.

Van Breemen, N., and Pons, L. J. (1978). Acid sulfate soils and rice. *In* "Soils and Rice," pp. 739–761. International Rice Research Institute, Los Baños, Philippines.

Vlamis, J., and Davis, A. R. (1944). Effects of oxygen tension on certain physiological responses of rice, barley and tomato. *Plant Physiol.* **19,** 33–51.

Watanabe, I. (1978). Biological nitrogen fixation in rice soils. *In* "Soils and Rice," pp. 465–478. International Rice Research Institute, Los Baños, Philippines.

Watanabe, I. (1984). Anaerobic decomposition of organic matter in flooded rice soils. *In* "Organic Matter and Rice," in press. International Rice Research Institute, Los Baños, Philippines.

Watanabe, I., and Furusaka, C. (1980). Microbial ecology of flooded rice soils. *Adv. Microb. Ecol.* **4,** 125–168.

Wesseling, J. (1974). Crop growth in wet soils. *In* "Drainage for Agriculture" (J. van Schilfgaarde, ed.), pp. 9–37. Am. Soc. Agron., Madison, Wisconsin.

Wickham, T. H., and Singh, V. P. (1978). Water movement through wet soils. *In* "Soils and Rice," pp. 337–358. International Rice Research Institute, Los Baños, Philippines.

Yamane, I. (1978). Electrochemical changes in rice soils. *In* "Soils and Rice," pp. 381–398. International Rice Research Institute, Los Baños, Philippines.

Yoshida, T. (1978). Microbial metabolism in rice soils. *In* "Soils and Rice," pp. 445–463. International Rice Research Institute, Los Baños, Philippines.

CHAPTER 3

Effects of Flooding on Growth and Metabolism of Herbaceous Plants

MICHAEL B. JACKSON
MALCOLM C. DREW
Agricultural Research Council Letcombe Laboratory
Wantage, Oxfordshire, England

I. INTRODUCTION

Growing plants have a requirement both for rapid gaseous exchange with their environment and for sufficient water to satisfy the needs of growth and evapotranspiration. Excess of one or the other, imposed by too much or too little water, leads to stress and consequently to loss of plant productivity or even death. Overwet environments impose stress by slowing gaseous diffusion by several orders of magnitude, ultimately asphyxiating the aerobically based biology on which healthy plant life normally depends. In this chapter the conse-

ISBN 0-12-424120-4

quences of flooding on the vegetative growth of herbaceous species are examined. We review research that reveals something of the mechanisms that bring about the effects of flooding or that relate to the capacity of some species or cultivars to tolerate flooding more than others. Effects on roots are first described because they usually experience the flooded environment firsthand. Consequences for shoot development then follow, with emphasis on correlative aspects because changes in shoot behavior are largely the results of previous developments in the root system. However, account is also taken of any direct effects caused by complete or partial submergence of the shoot itself.

Flooding and its effects have a dynamic aspect that must be appreciated to form a properly integrated view. The stress intensifies with time as physiologically active gases accumulate and the waterlogged environment becomes increasingly reduced chemically. Furthermore, plants continually readjust physiologically, and early changes inevitably influence subsequent events. Some of these will be a cost to overall dry-matter accumulation (e.g., stomatal closure and premature senescence of old leaves), but in the longer term they may benefit plant performance and the prospect of survival. We call these *acclimatic* responses.

Seed germination and seedling establishment, crop yields, and methods of alleviating injury have been considered in other reviews (de Wit, 1978; Drew and Lynch, 1980; Cannell and Jackson, 1981; Krizek, 1981; Jackson, 1983).

II. ROOT GROWTH AND METABOLISM

A. Early Responses to Inundation

In well-drained soils, aerobic respiration of plant roots and soil microorganisms under temperate conditions consumes typically between 5 and 24 g oxygen for each square meter of land surface per day during the growing season (Russell, 1973). The upper value, for an agricultural crop at a soil temperature of 17°C, is equivalent to a net daily flux through the soil surface of ~17 liters O_2 m^{-2} at standard temperature and pressure, of which the roots and rhizosphere consume about half.

With flooding, the soil pore space is totally water filled, and gas exchange between soil and atmosphere is virtually eliminated (Fig. 1). Because of its low diffusivity in water, the oxygenated zone at the soil surface may be confined to a depth of only a few millimeters. Depending on soil temperature and respiration rates, the dissolved O_2 in the soil water (~1.9 liters m^{-3} soil) will be exhausted in only hours or days, and the soil becomes anaerobic or oxygen free (see Ponnamperuma, Chapter 2 in this volume).

The imposition of excess water around roots affects their development in advance of total O_2 depletion of the tissue (anoxia). These early responses are

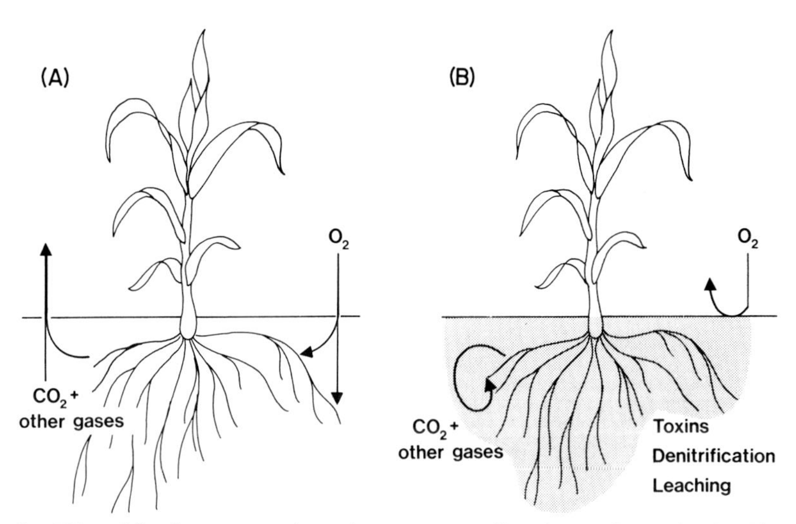

Fig. 1. Effect of flooding on gas exchange between roots, soil, and atmosphere and on leaching and toxin formation in soil. (A) Well aerated, O_2 3.5–17 liters m^{-2} day^{-1}. (B) Flooded.

principally to increases in ethylene content and to modest depressions in the internal concentration of O_2 (hypoxia). Both arise from impeded gaseous exchange caused by static water around the roots. Figure 2A,B shows some of the consequences for root extension of depletion of O_2 or increase in ethylene that can be expected shortly after inundation. Rice grows well in flooded conditions. It has a slow rate of endogenous ethylene production in the roots, and this supports only modest accumulations of the gas under water (Fig. 2C). These are likely to *stimulate* elongation because this species exhibits a marked optimal response curve to ethylene (Smith and Robertson, 1971; Konings and Jackson, 1979; Fig. 2B). In contrast, plants such as white mustard with much faster rates of ethylene synthesis accumulate larger, *inhibitory* amounts of the gas when submerged (Konings and Jackson, 1979; Fig. 2B,C). Such an accumulation probably explains reports of slow elongation under water by roots of white mustard or cress (Larqué-Saavedra *et al.*, 1975; Haberkorn and Sievers, 1977; MacDonald and Gordon, 1978). The capacity of 3,5-diiodo-4-hydroxybenzoic acid (DIHB), an inhibitor of ethylene action, to overcome the effect of submergence lends supports to this view (Larqué-Saavedra *et al.*, 1975). Aeration of the water with air or 100% O_2 (Larqué-Saavedra *et al.*, 1975; MacDonald and Gordon, 1978) also relieves the inhibition caused by submergence. This could be the result of depleting the roots of ethylene by enhancing outward diffusion to the gas stream. Equally, some effect may also be due to an enhanced supply of O_2 to relieve mild hypoxia.

Inhibition of elongation by hypoxia, even under well-stirred conditions (Fig.

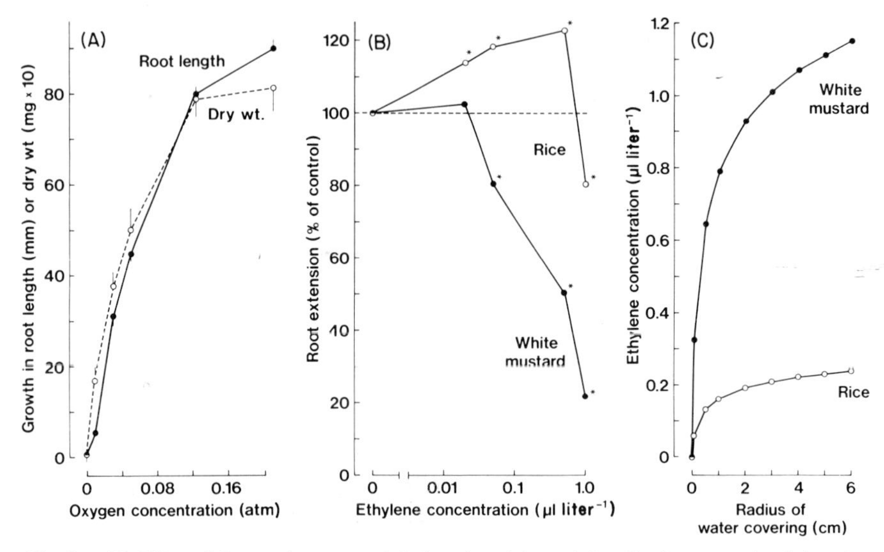

Fig. 2. (A) Effect of O_2 supply on growth in length and dry weight of barley roots after 3 days (± SE, $n = 24$). (B) Effect of increasing concentrations of ethylene on root extension in moist, flowing air after 24 hr (*significantly different from control, $p = .05$). (C) Calculated effect of a static water covering of increasing radius on ethylene at the surface of rice or white mustard roots. From Konings and Jackson (1979), and M. B. Jackson, C. M. Dobson, and B. Herman (unpublished).

2A) may arise because O_2 consumption and resistance to diffusive penetration of roots by O_2 are large enough to create anoxic conditions *within* the root (Section II,D,1). However, more subtle mechanisms may be involved because inhibition of increase in dry weight can be less than for elongation under slight O_2 depletion (0.125–0.205 atm O_2).[1] For example, oxidations other than those concerned with oxidative phosphorylation may be inhibited. The enzymes involved [e.g., polyphenol oxidase, ascorbic acid oxidase, indoleacetic acid (IAA) oxidase] are less effective scavengers of O_2 (larger K_m's) than the terminal respiratory oxidase (Section II,D,1) and are thus more susceptible to O_2 deprivation. More severe conditions of O_2 depletion (0.03–0.05 atm O_2) are known to stimulate ethylene production (Jackson, 1982), which may in turn depress root extension.

Flooding can change markedly the *direction* of root growth. This is unlikely to occur in anoxia when an extremely slow or nonexistent growth rate would preclude any redirecting of elongation. It is therefore probably a reaction to hypoxia or to an accumulation of trapped gas. Guhman (1924) found that roots of the common everlast (*Gnaphalium polycephalum*), tomato (*Lycopersicon esculentum*), and sunflower (*Helianthus annuus*) become diageotropic (horizontally growing) or negatively geotropic (upward growing) rather than positively

[1]Moist air contains 0.205 atm or 0.021 MPa of O_2 at 20°C and atmospheric pressure. Water in equilibrium with this air would contain 9.6 mg liter^{-1} O_2.

geotropic (downward growing) when they contact the water table [see also Durell (1941), Webster and Eavis (1972), Wample and Reid (1978), Ellmore (1981)]. In some circumstances when poorly drained soil is also cooler because of its greater specific heat, a more horizontal orientation may be favored in maize roots where well-drained soil temperatures exceed about 17°C (Onderdonk and Ketcheson, 1973). The mechanisms causing these changes in response to gravity are unknown but worth investigating because they may be of considerable acclimatic significance, enabling roots to escape O_2 stress by growing closer to the better-aerated soil surface.

Submergence-induced hypoxia and ethylene accumulation also stimulate formation of interconnected gas-filled channels (aerenchyma) within roots (Section II,C,5). This may be a further mechanism for avoiding anoxic stress by facilitating internal gas diffusion between roots and the well-aerated aerial shoot.

B. Mechanisms of Flooding Injury

Numerous laboratory investigations have established that deoxygenation of the rooting medium can cause root damage or death, but this does not necessarily imply that arrest of aerobic respiration is always the principal cause of injury to the roots of unadapted, nonwetland species in the flooded soil environment. The absence of O_2 at once triggers a sequence of chemical and biochemical reduction reactions in the soil. Some of these reduced components (NO_2^-, Mn^{2+}, Fe^{2+}, and S^-) as well as microbial metabolites can sometimes accumulate to concentrations that are injurious to root metabolism. Anaerobic metabolism within O_2-deficient roots may also give rise to products that are potentially injurious to metabolism if they accumulate to abnormally large concentrations. Leaching and denitrification may deplete the soil nitrogen. Few investigations have examined the relative importance of each of the proposed mechanisms of injury or whether they act simultaneously.

Information on these aspects is important to our understanding of (1) the physiological attributes of wetland plants that enable them to colonize specialized sites in nature; (2) in agriculture, the identification of features that may confer particular survival value in flooding-resistant species and varieties—these could be used in breeding programs to produce high-yielding crops resistant to temporary flooding; and (3) the means by which alleviation or prevention of flooding injury to valuable crops might be economically achieved through chemical treatments.

C. Root Growth, Structure, and Function

1. Root Growth at Controlled Oxygen Concentrations

Some higher plant organs are able to resist anaerobic conditions (nitrogen atmospheres) for periods ranging from a few hours to several months (Table I). In those instances where direct comparison is possible, the duration of root

TABLE I

Duration of Root Survival of Oxygen-Free Conditions in the Laboratory

Species	Tissue	Duration of root survival	Temperature (°C)	Reference
Maize (*Zea mays*)	Seminal root	70 hr	30	Sachs *et al.* (1980)
Rice (*Oryza sativa*)	Seminal root	96–120 hr	25	Bertani *et al.* (1980)
	Adventitious root	5–8 hr	25	Webb (1982)
	Adventitious primordium	7+ days	27	Kordan (1976)
Rye (*Triticale*)	Excised roots in sterile culture	24 hr	25	Oliveira (1977)
Deschampsia caespitosa	Seedling	8+ days	15	Barclay and Crawford (1982)
Molinia caerula	Seedling	8+ days	15	Barclay and Crawford (1982)
Scirpus maritima	Rhizomes + buds + roots	2 months	22	Crawford (1982)
Schoenoplectus lacustris	Rhizomes + buds + roots	1 month	22	Crawford (1982)
Schoenoplectus tabernaemontani	Rhizomes + buds + roots	1 month	22	Crawford (1982)
Iris pseudacorus	Rhizomes + buds + roots	1 month	22	Crawford (1982)
Iris germanica	Rhizomes + buds + roots	7 days	22	Crawford (1982)
Broad bean (*Vicia faba*)	Roots	12–48 hr	23	Williamson (1968)
Cotton (*Gossypium hirsutum*)	Roots	0.5–3 hr	30	Huck (1970)
Soybean (*Glycine max*)	Roots	0.5–5 hr	30	Huck (1970)
Tobacco (*Nicotiana tabacum*)	Roots	$<$48 hr	21–29	Williamson (1970)
Tomato (*Lycopersicon esculentum*)	Excised roots in sterile culture	3 days	—	Morisset (1973)
Pea (*Pisum sativum*)	Roots	7–9 hr	25	Webb (1982)
Pumpkin (*Cucurbita pepo*)	Roots	13–20 hr	25	Webb (1982)

survival approximately corresponds to the period over which mitochondrial structure undergoes no irreversible structural degeneration (Morisset, 1973; Oliveira, 1977; Vartapetian *et al.*, 1977, 1978b). Growth does not continue in higher plants under anoxia, except in some organs of highly specialized species that are adapted to elongation while submerged in anaerobic aquatic environments. In coleoptiles of rice (*Oryza sativa*) and barnyard grass (*Echinochloa crus-galli*), a weed of paddy rice fields, cell expansion of the coleoptile is able to continue slowly (Opik, 1973; Kordan, 1976; Vartapetian *et al.*, 1978a,b; Kennedy *et al.*, 1980; Rumpho and Kennedy, 1981; Atwell *et al.*, 1982) although cell division is either greatly slowed or arrested (Opik, 1973; VanderZee and Kennedy, 1981; Atwell *et al.*, 1982). Barclay and Crawford (1982) demonstrated conclusively that a number of rhizomatous wetland species can continue to produce healthy shoots when maintained in strictly O_2-free atmospheres in the dark for 7 days or more. Survival of seedlings of *Deschampsia caespitosa* and *Molinia caerula,* both wetland grass species, was up to 8 days. For *Scirpus maritima* and other rhizomatous species adapted to overwintering in anaerobic mud (Table I), survival without growth can be for several months (Crawford, 1982). Because of their very much greater survival time under anaerobic conditions, such species were described by Crawford (1982) as anoxia tolerant. The possible mechanisms by which plant organs can generate sufficient energy for growth under anaerobic conditions are discussed in Section II,D,2. In all of these investigations root growth was not observed in the absence of O_2.

Roots vary in their tolerance to anaerobic conditions when tested in the laboratory (Table I). Soil perfused with nitrogen gas kills taproot apices of cotton and soybean after 3 and 5 hr, respectively, although as little as 30 min can be fatal to a proportion of root tips (Huck, 1970). In contrast, maize seminal roots survive for up to 70 hr (Sachs *et al.*, 1980). The resistance of rice adventitious roots (8 hr) is *less* than that of the nonwetland species pea and pumpkin (Webb, 1982), but with rice, seminal roots viability is retained for 96 hr or more (Bertani *et al.*, 1980). Root primordia may have a greater resistance to anoxia than do the apices of roots extending at the onset of the anaerobic treatment because primordia only need energy for cell maintenance. Adventitious root primordia in rice seedlings survive for ~7 days and grow normally when O_2 is then introduced (Kordan, 1976). For excised roots of tomato in aseptic culture, after prolonged anoxia (duration not specified, but >3 days) only lateral root primordia are capable of resuming growth (Morisset, 1973). However, once lateral roots emerge from the main root, their sensitivity to anaerobic conditions is similar to that of the terminal meristem (Williamson, 1968).

There is abundant information that the *initial* root system of nonwetland species degenerates soon after the commencement of flooding under laboratory conditions. Existing roots cease extending, dry-matter accumulation is arrested, and degeneration of tissues, particularly the apical meristems, becomes apparent

(Burrows and Carr, 1969; Purvis and Williamson, 1972; Drew and Sisworo, 1979; Trought and Drew, 1980a). The capacity of unadapted roots to survive in O_2-free media in the laboratory therefore seems to be greatly restricted and accords broadly with sensitivity to flooding in the field (Section II,B,2).

Roots can be subjected to less extreme O_2 deficiency in solution culture by bubbling with mixtures of O_2 and nitrogen or in nonaerated cultures in which O_2 is partially depleted by root respiration. The threshold concentration at which root extension begins to be slowed is commonly about half that in air (Fig. 2A; Hopkins *et al.*, 1950; Lopez-Saez *et al.*, 1969; Turner *et al.*, 1981; Bertani and Brambilla, 1982a,b). Apart from reducing the length of the main roots, root morphology can be simultaneously modified by low concentrations of O_2 through stimulation of the number and extension of laterals (Geisler, 1965), perhaps a reflection of the general response of roots to a temporary reduction of "apical dominance." The lower concentration of O_2 at which root extension can be expected to cease cannot be predicted with precision, because of the influence of stirring rate on unstirred layers at the root surface, the effectiveness with which air contamination of the medium is prevented, the extent to which O_2 diffuses from the shoot within the gas spaces of the root, and the radial leakage of this O_2 from the root to the medium. In practice, when roots are deprived of a supply of O_2 from the shoot, solutions with O_2 partial pressures of 0.01–0.03 atm at 20°C provide insufficient O_2 for further root extension.

2. *Root Growth and Distribution in the Field*

High water tables cause plants of wetland and nonwetland habitats to root more superficially. For dryland crop species, raising the water table even transiently causes the death of deeper roots and often the proliferation of surface ones. The potential problems arising from this change in root distribution are numerous: cultivations can result in root damage; plants are more sensitive to subsequent drought and temperature fluctuations in the surface soil when the water recedes; and there may be greater irrigation or fertilizer requirements than for more deeply rooted plants that can exploit a larger soil volume (Nicholson and Firth, 1958; Van Hoorn, 1958; Sieben, 1964; Armstrong, 1978).

High water tables for extended periods through lack of field drainage can greatly inhibit the growth of roots of winter wheat during the spring, in association with adverse effects on nutrient uptake and crop yields. Bragg *et al.* (1983) found that in a clay soil, the water table in winter and early spring was high and frequently approached the soil surface in undrained plots. Before January, drainage had no discernible effect on root distribution, but between January and April, when soil and air temperatures increased greatly, rooting density on drained plots increased approximately fivefold at all depths between 15 and 75 cm but failed to increase on the undrained plots. In the upper 15 cm, rooting density was unaffected by drainage, perhaps because of the occasional entry of O_2 with water-table fluctuations or because of the growth of aerenchymatous adventitious roots. The overall effect of high water tables was thus to produce a smaller, more superficial root system.

The severity of the effect of transient flooding on root systems may depend on the growth stage of the plant. Soybean (*Glycine max*) grown in a rhizotron located in a field plot was subjected to raised water tables for 7 days, after which the soil was allowed to drain (Stanley *et al.*, 1980). Those roots that became submerged during the vegetative (preflowering) stage immediately ceased extension but resumed growth at their earlier rate once the soil was drained. With flooding immediately after flowering, before pod set, roots below the water table ceased extension and decomposed when the soil was drained. New roots compensated for this by growing from roots located in the upper part of the soil, but they did not reach the earlier water-table level. Flooding after pod set caused virtual cessation of root growth both above and below the water table.

For wetland species, bog sites saturated with moving, partly oxygenated water provide more favorable habitats than stagnant ones. In the moving water healthy roots can develop, but in adjacent, more reducing habitats roots and rhizomes of many species form a mat just above the water table, failing to penetrate much below it when conditions are highly reducing (Emerson, 1921; Armstrong and Boatman, 1967; Sheikh and Rutter, 1969). On stagnant, O_2-free areas, Armstrong and Boatman found that *Molinia caerulea* main roots were restricted to the upper 5 cm in the summer, with upward-growing laterals. Roots of *Menyanthes trifoliata* and *Narthecium* appeared healthy but penetrated only 10 cm. Roots of these species, as well as those of *Phragmites, Myrica gale,* and *Potamogeton polygonifolius,* had deposits of iron oxide. On an adjacent site, highly reducing conditions released H_2S and only *Menyanthes* was present. The main roots that had developed recently in the highly reducing layer appeared stunted and dead, but numerous laterals from them extended near the surface. Aerenchymatous roots of *Molinia caerulea* in the valley bog studied by Sheikh and Rutter (1969) extended no more than ~20 cm below the water-table depth (located 7–9 cm from the soil surface), but about 90% of the total root weight was in the upper 19 cm. Some species seem tolerant of continuous submergence of their subterranean organs and extend their rhizomes and roots well below the water table. In the bog studied by Emerson (1921), rooting depths of 20–50 cm below the water table were recorded for *Typha latifolia, Sagittaria latifolia, Scirpus validus,* and *Eriophorum.*

Few ecological studies have followed in detail the development of the root system during the growing season or have estimated the proportion of the root system remaining alive. In bogs, an early-season flush of water may temporarily provide a well-oxygenated environment for root growth so that when conditions become stagnant and anaerobic, little further growth occurs. Plant survival will then depend on maintenance in a functional state of some of the roots that developed earlier in the season.

The root habit of paddy rice shows many similarities to that of native wetland species. Paddy-rice soils typically have a surface layer of water that is kept well oxygenated by algae. The surface soil to a depth of several millimeters is oxygenated, but below this it abruptly becomes anaerobic, although surface water moving through channels in the soil can maintain occasional pockets of aerobic soil (Aomine, 1962). During early growth, adventitious, aerenchymatous roots extend into the anaerobic layer, creating oxidized sheaths of soil around them. At the end of tillering a surface mat of horizontally growing roots with many fine laterals develops in the surface oxidized layer (Alberda, 1953), when impedance to O_2 diffusion from the elongated rice stem into the older roots seems to increase. In degraded, acid soils, less well suited to paddy, root extension into the anaerobic, highly reduced layer is much restricted, probably through injurious concentrations of iron, H_2S, or organic acids (Aomine, 1962; Hollis and Rodriguez-Kabana, 1967; Strickland, 1968; Allan and Hollis, 1972; Hollis *et al.*, 1975).

3. *Root Nodules and Mycorrhizae*

Nitrogen accumulation by shoots of nodulated peas and other legumes is slowed by soil waterlogging (Minchin and Pate, 1975). This is partly the consequence of reduced nodulation of the roots by nitrogen-fixing bacteroids (*Rhizobium* spp). The specific activity of the enzyme (nitrogenase) involved in converting nitrogen gas to reduced forms useful to the host plant is also suppressed (Sprent, 1969; Minchin and Pate, 1975). The effect on nitrogenase arises from decreased synthesis by the bacteroids of one of the two protein components of the enzyme (Bisseling *et al.*, 1980). Lack of O_2 may cause this because a small amount of the gas (0.0016 atm) is needed for optimal nitrogen fixation (Keister and Rao, 1977). Paradoxically, greater amounts of the gas can be inhibitory, but leghemoglobin contrives to protect the bacteroids from this oxygen poisoning. *Rhizobium* is thus "microaerobic." The success of the nitrogen-fixing endophyte also depends on the host tissue of the nodule. This supplies intermediates from aerobic pathways of carbohydrate metabolism and also ATP (Sprent, 1971). These services are curtailed in waterlogged, O_2-deficient conditions and a considerable amount of ethanol (14 mol m^{-3}) is generated, by both bacteroids and root cells of the nodule (Sprent and Gallacher, 1976). This may be sufficient to damage the nodule further. Any ethylene formed and trapped within nodules may also be detrimental, both to nodulation (Grobbelaar *et al.*, 1970; Goodlass and Smith, 1979) and to nitrogen fixation as measured by ^{15}N incorporation (Grobbelaar *et al.*, 1971) or acetylene reduction (Goodlass and Smith, 1979). As little as 0.4 μl $liter^{-1}$ ethylene is sufficient to diminish nodulation by 90%, and

10 μl liter^{-1} can inhibit nitrogenase activity by 75%. Carbon dioxide (3%, v/v) is also known to depress nodulation (Grobbelaar *et al.*, 1971).

Three distinct types of acclimatic response by nodules of the more flood tolerant nodulated plants have been reported. The most generally applicable of these is the expansion and dissolution of cortical cells to produce a swollen, more loosely packed nodule, especially at the sites of lenticels (Fig. 3A). The effect was reported in soybean (Pankhurst and Sprent, 1975), the shrubby *Myrica gale* (Sprent, 1971; Torrey and Callaham, 1978), the tropical, floating aquatic legume *Neptunia* (Fig. 3B), and cowpea (Minchin and Summerfield, 1976). In the latter species, nitrogen-fixing ability of the nodules was reduced by only 18% after 8 days of flooding, in association with enlargement of the nodule cortex and lenticels to form interconnected gas-filled spaces. This open infrastructure is thought to facilitate gaseous exchange and thus favor symbiotic nitrogen fixation. In *Neptunia* this also increases buoyancy, enabling the nodules to float. Hydrostatic pressure maintained by a link between the vascular trace and the lenticels is involved in expansion of the cells (Ralston and Imsande, 1982). Nodules that develop this open structure are more active nitrogen-fixers in a poorly aerated environment than the more compact nodules that developed under better drained conditions (Sprent, 1971; Criswell *et al.*, 1976).

A second structural adaptation, found in the actinomycete-induced nodules of the nonlegume *Myrica gale,* is the formation of negatively geotropic (upward growing) roots. The environmental factors causing this reversal of the normal direction of root growth are unknown. The vertical roots are likely to improve aeration of the nodules in the boggy conditions in which *Myrica gale* often grows (Bond, 1952). Tjepkema (1977) has provided experimental evidence for such a role.

A third acclimatic response is the dissimilatory reduction of nitrate to nitrous oxide and nitrogen gas (denitrification); this has been observed in bacteroid cultures of *Rhizobium japonicum* and cowpea rhizobia (Zablatowicz *et al.*, 1978) under anaerobic conditions. These organisms are therefore capable of using nitrate as an electron acceptor alternative to oxygen for ATP generation. This capability may enable the bacteroids to grow and survive anaerobic soil conditions, a contention supported by the observation that additions of nitrate to anaerobic bacteroids increase nitrogen fixation under anoxia (Zablotowicz and Focht, 1979). Interestingly, cowpea, which is tolerant of waterlogging (Minchin and Summerfield, 1976), possesses denitrifying rhizobia, whereas the *Rhizobium* species typical of the flood-intolerant garden pea (*Pisum sativum*) is less able to denitrify (Zablotowicz *et al.*, 1978).

Vesicular–arbuscular (VA) mycorrhizae frequently occur in herbaceous species in natural habitats and in some agricultural crops, where they improve the nutrient status of the plant for phosphate and possibly other soil-immobile nutrients. Numerous reviews have drawn attention to the failure of mycorrhizae to

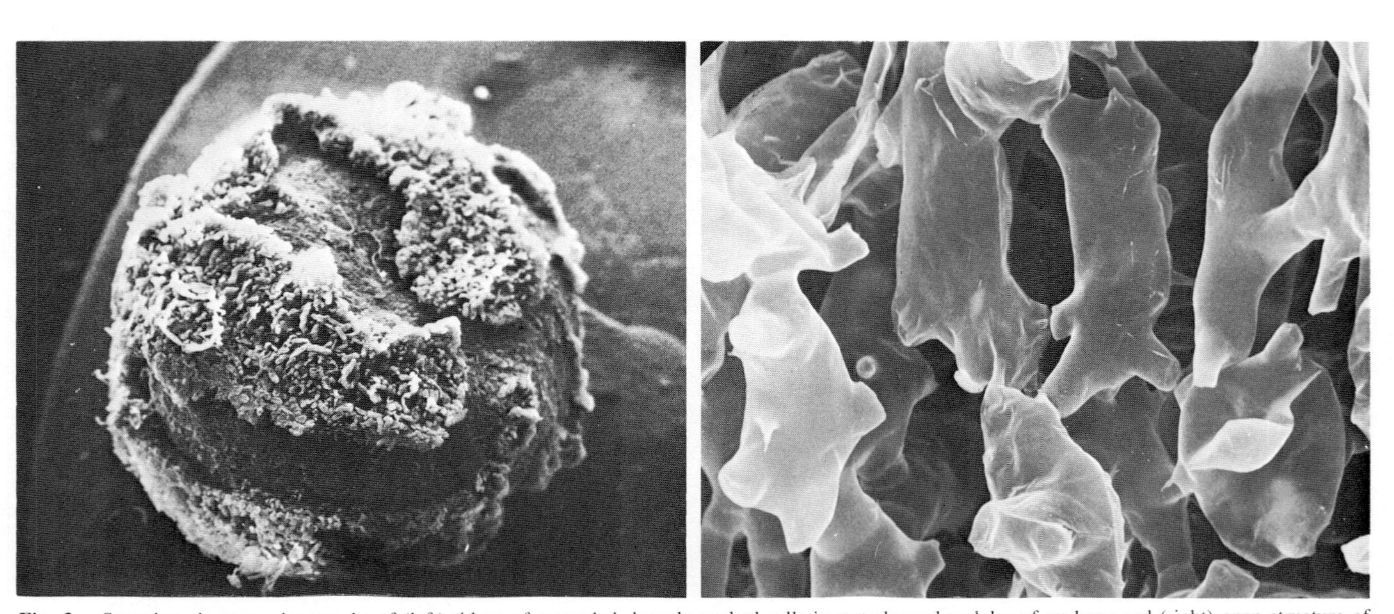

Fig. 3. Scanning electron micrographs of (left) ridges of expanded, loosely packed cells in waterlogged nodules of soybean and (right) open structure of nodules of *Neptunia*, an aquatic tropical legume. From C. E. Pankhurst and J. I. Sprent (unpublished).

develop in plants in wetland habitats (Gerdemann, 1968; Harley, 1972; Bristow, 1974). Members of the Cyperaceae and Juncaceae, frequently found in flooded environments, are usually nonmycorrhizal (Gerdemann, 1968). But some wetland species such as rice that are nonmycorrhizal in flooded soil are mycorrhizal in well-drained soil (Gerdemann, 1968). Similarly, *Ipomoea carnea,* which is mycorrhizal in dry habitats, becomes nonmycorrhizal when temporarily flooded during the monsoon (Khan, 1974). Little or no mycorrhizal infection was found in 6 marshland species in England (Read *et al.*, 1976). None of the 17 species of monocotyledons and dicotyledons growing in ponds, swamps, and ditches in Pakistan were found to be mycorrhizal and the occurrence of spores in the soil was rare, in contrast to their abundance in adjacent, drier sites (Khan, 1974). This suggests that spore survival in the soil and the frequency of root infection are greatly diminished with flooding. Additionally, mycorrhizal plants can show a greater sensitivity to poor aeration than nonmycorrhizal plants. When *Eupatorium odoratum* was grown in phosphate-deficient soil continuously perfused with controlled gas mixtures (Saif, 1981), shoot and root growth of mycorrhizal plants declined when O_2 was less than 0.12 atm. In contrast, the slower-growing, phosphate-limited, nonmycorrhizal plants did not decline appreciably until O_2 was less than 0.04 atm. The percentage infection of roots was little affected by O_2 concentration, but at 0.04 atm O_2 and below, the development of vesicles and arbuscles was greatly depressed, suggesting that the infection process was less sensitive to O_2 deficiency than the physiological activity of the fungus within the host.

Despite many observations that excess water depresses mycorrhizal association, aquatic species sometimes develop VA mycorrhizae. Root samples collected from nutrient-poor sediments at 0.3- to 0.8-m depths in Danish freshwater lakes were mycorrhizal for 5 of 7 species that had little root-hair development (Søndergaard and Laegaard, 1977). Whether mycorrhizae would improve the supply of nutrients to the plant in the aquatic environment depends on the relative uptake of nutrients by roots and submerged leaves and stem (Bristow, 1974).

4. *Oxygen Requirements for Root Meristematic Activity*

Fast O_2 consumption by root apices, combined with lack of intercellular spaces between the densely packed cells in some species, might be expected to render meristemic activity highly sensitive to lack of O_2. In anaerobic nutrient-mist culture (nitrogen atmosphere), the proportion of root-tip cells apparently undergoing mitosis (determined by counts of cells in prophase and anaphase) in *Vicia faba* was detectably reduced in 15–30 min and ceased completely in 24 hr, at which time some root apices were dead (Williamson, 1968). Inhibition of cell division caused by exposure for 12 hr or less was reversible when roots recovered in air.

Anaerobic conditions, like root excision or treatment with cyanide or dinitrophenol (DNP), restrict entry of cells into mitosis but do not inhibit those already dividing (Amoore, 1961a; Webster and Van't Hof, 1969). Undoubtedly the common factor in these treatments is an inhibition of aerobic respiration either at the level of the respiratory chain or through lack of substrates. By carefully controlling the O_2 partial pressure around pea root tips, Amoore (1961b) determined the sensitivity to O_2 deficiency of different stages in the cell cycle. A very low concentration, between 5×10^{-4} and 20×10^{-4} atm, was the threshold at which the progression of cells from G_2 into mitotic division was halted, whereas cells that had begun mitotic division completed that stage even at 0.1×10^{-4} atm O_2. Only at half the latter concentration was mitosis completely arrested. Amoore (1962a,b) pointed out that much lower concentrations of O_2 were required to arrest mitosis than to inhibit severely aerobic respiration. Amoore suggested that mitosis was stopped at these low concentrations of O_2, not by the energy status of the cells, but by inhibition of oxidation of a nonrespiratory complex essential to mitosis (Amoore, 1963). Alternatively, perhaps sufficient ATP is generated anaerobically to energize mitosis. It would be interesting to have direct information on the energy status of cells (e.g., adenylate nucleotide ratios) under these conditions, using reliable techniques for extraction and estimation of adenine nucleotides.

The conclusion from these studies must be that anaerobic metabolism is unable to provide sufficient energy for initiation of S phase (DNA-synthesis phase) or mitosis in the root meristem. However, in view of the likely inhibition of phloem transport of assimilates to the root tip under anoxia and cessation of cell division in the absence of a source of easily respired substrate, it is questionable for intact roots in O_2-deficient media whether carbohydrate starvation or inhibition of aerobic respiration is the primary cause of inhibition of the cell cycle.

Root-extension rates depend on rates of cell division and elongation. The relative sensitivity of these components was examined with intact roots of *Allium cepa* over a range of O_2 concentrations from 0.2 atm (air) to 0.02 atm, which reduced root-extension rate from 21.3 to 8.8 mm day^{-1} (Lopez-Saez *et al.*, 1969). The principal component retarding growth at the lower O_2 concentrations was a slower production of new cells; the duration of the cell cycle was increased from 13.5 to 56 hr mainly because of a longer cell interphase (G_1, S, G_2). Cell extension was relatively insensitive to O_2 supply in this range, and the final cell length was unaffected.

5. *Structural Changes in Roots with Low Oxygen Supply*

a. Root Anatomy. In many wetland species, the cross-sectional area of the root available for gaseous diffusion is appreciably increased by development of aerenchyma tissue, which arises by cell separation (schizogenously) or by partial breakdown of the cortex (lysigenously), giving rise to continuous gas-filled lacunae. The formation of such aerenchyma seems to contribute to the survival of many wetland plant species (Kawase, 1981a). The occurrence of aerenchyma in roots of wetland plants like rice (*Oryza sativa*) and wild rice (*Zizania aquatica*) even under well-oxygenated conditions suggests that its development is under genetic control in these species. Restricted aeration can encourage still further the breakdown of the cortex to increase the internal porosity in the roots (Armstrong, 1971; Das and Jat, 1977) and in the culm of rice and other wetland species (Arashi and Nitta, 1955; Arikado, 1959).

Lysigenous aerenchyma can also develop in the newly formed adventitious roots that emerge from the base of the shoot in many nonwetland herbaceous species, including crops, in response to waterlogging or low O_2 concentration (Table II). These specialized roots are thus "adapted" anatomically to O_2-deficient media. It is widely assumed (Section III,C,2) that they functionally replace the original root system, which is inhibited or killed by lack of O_2, because their development coincides with a partial recovery from waterlogging injury to shoots (Kramer, 1951; Arikado, 1955a,b, 1959, 1961; Jackson, 1955; Yu *et al.*, 1969). However, there have been few studies that have directly examined whether such structurally modified roots can continue to function as absorbers of nutrient ions (Section II,C,6).

The formation of lysigenous aerenchyma in response to a shortage of O_2 takes place in the cortex, external to the endodermis, and becomes well developed in the zone 3–4 cm behind the root tip in elongating roots of maize (Drew *et al.*, 1979a), rice (Armstrong, 1971), and wheat (Trought and Drew, 1980b), where cell extension growth is complete (Fig. 4). The formation of such gas spaces has been attributed, since McPherson (1939), to the death and dissolution of cells caused by anoxia. There is now evidence that the mechanism of cell lysis, at least in maize, involves endogenous ethylene. Its concentration increases in root tissues in nonaerated media, probably through stimulation of ethylene biosynthesis by small concentrations of O_2 (Jackson, 1982) and entrapment of the gas by unstirred layers of water. As little as 0.1 $\mu l \ liter^{-1}$ of exogenous ethylene in well-oxygenated solution stimulates the formation of aerenchyma in which the pattern of cell breakdown appears to be identical with that induced by O_2 shortage (Drew *et al.*, 1979a, 1981). Aerenchyma formation, whether induced by nonaeration or exogenous ethylene in air, can be blocked by low concentrations of Ag^+ (Drew *et al.*, 1981), which inhibits ethylene action. Furthermore, compounds that inhibit ethylene synthesis also prevent aerenchyma formation in nonaerated solution (Konings, 1982). Changes in the ultrastructure of cortical cells during lysis induced by poor aeration, examined by transmission electron microscopy (Campbell and Drew, 1983), suggest that the pattern of organelle degeneration is more compatible with that induced by ethylene than by anoxia. The question arises as to the mechanism by which O_2 shortage stimulates eth-

TABLE II

Induction of Gas Spaces (Aerenchyma) in Roots and Stem Base of Nonwetland Herbaceous Plants in Response to Waterlogging or Oxygen-Deficient Solution

Species and tissue	Method of induction[a]	Method of gas-space measurement or observation[b]	Reference
Maize (*Zea mays*) roots	1 3	LM	McPherson (1939)
	1	R	Arikado (1959)
	2	P	Yu *et al.* (1969)
	1	P	Das and Jat (1972)
	3	LM	Drew *et al.* (1979a)
	3	TEM	Campbell and Drew (1983)
Wheat (*Triticum vulgare* and *T. aestivum*) roots	1	R,P	Arikado (1961)
	2	P	Yu *et al.* (1969)
	4	P	Varade *et al.* (1970)
	4	SEM	Trought and Drew (1980b)
Barley (*Hordeum vulgare*) roots	3	LM	Bryant (1934)
	2	R	Arikado and Adachi (1955); Arikado (1959)
	2	LM	Drew and Sisworo (1979)
	4	P	Benjamin and Greenway (1979)
Tomato (*Lycopersicon esculentum*) roots	2	P	Yu *et al.* (1969)
Sunflower (*Helianthus annuus*) roots	2	P	Yu *et al.* (1969)
Stem base	2	LM	Kawase and Whitmoyer (1980)

[a] 1, Soil waterlogged in the field; 2, soil waterlogged in glasshouse or controlled-environment room; 3, solution culture, nonaerated; 4, solution culture, bubbled with low O_2 concentration or nitrogen gas.

[b] LM, Light microscopy; SEM, scanning electron microscopy; TEM, transmission electron microscopy; R, resistance to flow of gas under a pressure gradient; P, measurement of gas-filled porosity.

ylene accumulation in the root tip. There is a possibility that the ethylene precursor 1-aminocyclopropane-1-carboxylic acid (ACC) is produced in the hypoxic root tip, ACC being converted to ethylene in the subapical zone in the presence of O_2 diffusing from the aerial tissues. The restriction of cell lysis to the cortical cells suggests that the cortex contains "target cells" that, unlike those of the other tissues in the root, are responsive to elevated concentrations of the gas.

We do not know whether ethylene stimulates the formation of other types of aerenchyma found in some wetland and semiaquatic species. In *Ludwigia*

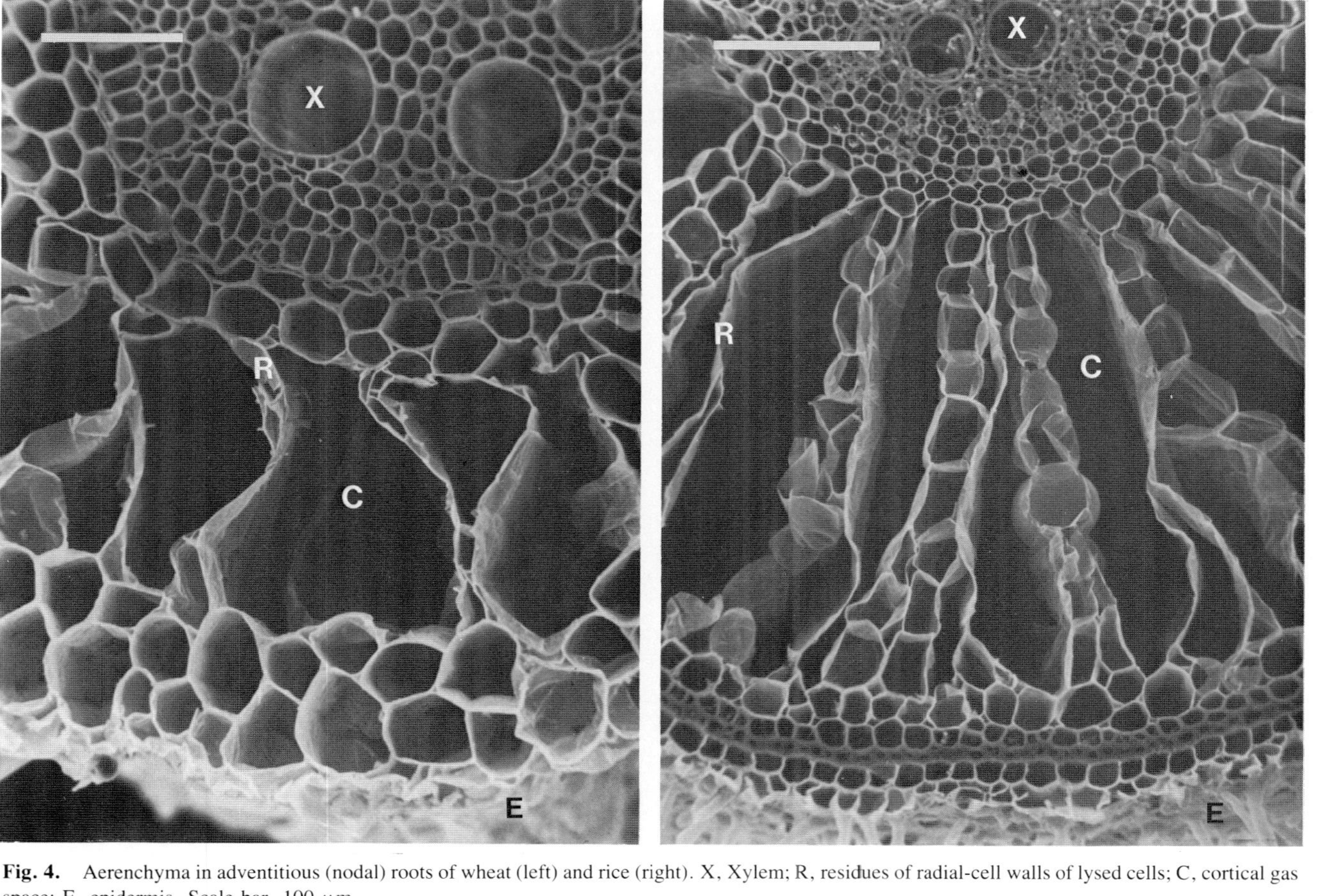

Fig. 4. Aerenchyma in adventitious (nodal) roots of wheat (left) and rice (right). X, Xylem; R, residues of radial-cell walls of lysed cells; C, cortical gas space; E, epidermis. Scale bar, 100 μm.

peploides some roots produced at the nodes on the submerged stem grow upward and develop a spongy, aerenchymatous-like tissue in the cortex through expansion of living cells, giving a swollen appearance to the root. Extraction of gas from these roots by reduced pressure showed no detectable ethylene in the internal atmosphere (Ellmore, 1981), but it is not clear whether these roots were still extending and developing gas spaces at the time of sampling.

Through loss of the cortex, aerenchymatous roots are vulnerable to radial hydrostatic pressure when submerged. Additional wall structures, presumed to be of lignin or suberin, occur in the epidermis or cortex of some wetland species (*Aster subulatus* and *Phragmites communis*) and in the aerenchymatous roots of nonwetland plants (wheat, barley, maize, and sorghum) in waterlogged soil (Arikado, 1959). In rice (Fig. 4) these involve suberization of the hypodermis immediately beneath the epidermis and prominent lignification of the adjacent cortical cells beneath the hypodermis (Juliano and Aldana, 1937). According to Arikado (1959), such thickenings might have a role not only in preventing collapse of the root but in simultaneously excluding toxic components that may occur in the soil water. An additional function might be to reduce radial leakage of O_2 from the root to the outer medium (Armstrong, 1971). The properties of the cell-wall thickenings in aerenchymatous roots have had little direct investigation. Measurements on ''hypodermal sleeves'' prepared from aerenchymatous roots of *Carex arenaria,* comprising epidermis and two or three layers of cortical cells with multilayered suberin lamellae, revealed very low permeability to water or phosphate ions (Robards *et al.,* 1979).

b. Mitochondrial Structure. When the growing roots of nonwetland species are excised and subjected to anaerobic conditions in the laboratory, their mitochondria show abnormal ultrastructure (irregular shape and dilated cristae) after several hours, followed by generalized degeneration, usually within 24–48 hr (Oliveira, 1977; Vartapetian *et al.,* 1977, 1978b). No such changes are observed in mitochondria in well-oxygenated roots. The degeneration of mitochondrial structure in the roots of wetland species was at least as rapid under anoxia as that of nonwetland species [rice (Vartapetian *et al.,* 1977, 1978b) and *Carex leporina* (Vartapetian, 1982)]. Provided the roots are returned to air before the cristae degenerate, the initial structure is restored and respiratory activity partially regained (Oliveira, 1977). Mitochondria in lateral-root meristems may show less structural degeneration than those in apical meristems under anaerobic conditions and quickly regain normal appearance on reintroduction of O_2 (Morisset, 1973).

In *excised* roots, coleoptiles, and leaves of rice, mitochondrial degeneration is complete in 1–2 days of anoxia. However, with additions of glucose to the medium, or in the *intact* seedling, an appreciable increase in mitochondrial survival is accompanied in all three organs of the plant by marked changes in ultrastructure, the long axis increasing severalfold with dense packings of parallel cristae (Vartapetian *et al.*, 1977, 1978b). With apices of excised pumpkin roots, an increase in duration of mitochondrial integrity from 12 to 96 hr with glucose additions was associated with changes in mitochondrial structure similar to those in rice but with sometimes highly branched (or fused) mitochondria forming complex shapes or even networks (Vartapetian *et al.*, 1977). Morisset (1973) also noted the occurrence of parallel cristae in elongated mitochondria in differentiating cells in tomato roots surviving under anoxia. Vartapetian and colleagues (1977) argued plausibly that mitochondrial and cell survival of anoxia depend on an adequate supply of respirable substrates to maintain rapid glycolysis and sufficient energy for maintenance. They speculated that the close proximity of elongated mitochondria with regions of the endoplasmic reticulum that are densely packed with ribosomes could be of special significance under anaerobic conditions. Potential energy generated in the cytoplasm during glycolysis might be directed along mitochondrial membranes toward the endoplasmic reticulum, there to be transformed into ATP to energize protein synthesis. However, it is difficult to assign a role to the striking changes observed in mitochondrial ultrastructure; no such changes were found in mitochondria in the coleoptiles of anaerobically germinated rice (Vartapetian *et al.*, 1978b) in which the energetic status of cells is clearly maintained by fermentation for many days. Furthermore, increase in mitochondrial survival in the presence of glucose is of limited duration only, despite an apparently plentiful supply of respiratory substrate. The ultimate cause of loss of mitochondrial structure and the failure of cell metabolism thus remains uncertain. Is it a gradual decline in ATP and energy charge? Do products of anaerobic metabolism accumulate to damaging concentrations? Is membrane instability the result of failure of O_2-dependent synthesis of unsaturated lipids?

c. Membrane Properties. Because of the essential role of O_2 in biosynthesis of unsaturated fatty acids, membrane structure and function in plants cells would be expected to be sensitive to O_2 supply. There is no evidence that the "anaerobic" pathway of synthesis present in anaerobic microorganisms occurs in higher plant or animal cells. Instead, the creation of double bonds by desaturase activity depends on O_2 together with a reductant (NADH or NADPH) and ferredoxin or some other electron carrier (Hitchcock and Nichols, 1971; Stumpf, 1980). Changes in membrane lipid composition (e.g., degree of saturation of fatty acyl residues, chain length, and presence of sterols) can be expected to modify membrane fluidity and alter permeability and membrane transport processes. In mitochondria extracted from yeast cells, degeneration under anoxia is closely associated with interference in synthesis of unsaturated fatty acids, leading to increased permeability of the inner mitochondrial membrane and uncoupling of ADP phosphorylation from respiration (Quinn and Chapman, 1980).

The existence of anoxia-tolerant and -intolerant plant organs (Table I) suggests that differences between them might originate either in the synthesis of membrane lipids or in avoidance of their degradation under anoxia. But in anoxia-tolerant coleoptiles of rice germinating anaerobically, the lipid composition including the unsaturated fatty acids was similar to that of aerobically grown controls. The mitochondria that were assembled anaerobically had functionally intact membranes (Vartapetian *et al.*, 1978a). However, labeling with [^{14}C]acetate revealed that no *de novo* synthesis of unsaturated fatty acids using this precursor was occurring under anoxia. The authors concluded that this puzzling result must indicate either the existence of an unusual pathway for lipid synthesis in the coleoptile or, equally implausibly, that lipids are transported from the seed to the coleoptile without prior conversion to glucose by the glyoxylate cycle and gluconeogenesis.

Enigmatic results concerning lipid composition were also obtained by Hetherington *et al.* (1982) with rhizomes of *Iris pseudacorus* (anoxia tolerant) and *I. germanica* (anoxia intolerant). The lipid composition of the two species was similar in aerated controls, and when rhizomes were subjected to 14 days of anaerobic treatment, little change could be detected in the composition of the anoxia-intolerant *I. germanica* lipids. However, in the anoxia-tolerant *I. pseudacorus* there was a marked decline in content of the saturated fatty acids (palmitic and stearic) but not, as might have been expected, the unsaturated ones. A more extensive survey of the lipid composition of a range of anoxia-tolerant and -intolerant species is required to assess whether this difference indicates an adaptive mechanism. However, an increase in the ratio of unsaturated to saturated fatty acids confers increased fluidity to biological membranes, counteracting the proposed reduction in fluidity resulting from the presence of ethanol (Ingram, 1976), and is a feature associated in microorganisms with resistance to the penetration of ethanol at high concentrations (Ingram, 1976; Thomas *et al.*, 1978; Thomas and Rose, 1979).

These results contrast sharply with the response of anoxia-intolerant plant cells. With *Acer pseudoplatanus* cells in suspension culture (Rebeille *et al.*, 1980), the synthesis of polyunsaturated fatty acids was progressively suppressed when the O_2 partial pressure was at or below 0.031 atm. There was a simultanteous rise in the concentration of lipids with less saturated carbon chains. Chirkova *et al.* (1981) found a comparable decline in total unsaturation in the phospholipids extracted from homogenates or mitochondria of wheat and rice roots that had been anoxic for several days. Some differences between species may be revealed under the combined effects of anoxia and ethanol. Although the lipids of wheat roots were degraded under anaerobic conditions, especially in the presence of 2 *M* ethanol, the lipids of rice roots were much less affected by these conditions (Chirkova *et al.*, 1980).

However, the significance of changes in unsaturation of lipids alone is difficult to assess, particularly from gross analyses where no information is obtained for the individual phospholipid components and for positional effects concerning the two acyl chains. Membrane fluidity and function can be modified by other components, for example, interactions between the polar head groups (Horvath *et al.*, 1981) or the presence of sterols (Thomas and Rose, 1979). Furthermore, it has been suggested that the occurrence of polyunsaturated phospholipids in membranes is concerned more with protection of membrane proteins from free-radical attack than with maintenance of fluidity (Quinn and Williams, 1978). The possible mechanisms by which the superoxide radical is disposed of when aerotolerant anaerobes are exposed to O_2 have been explored (Quinn and Williams, 1978), but little is known about the means by which roots contend with the potentially damaging effects of reintroducing O_2 following a period of anoxia.

6. *Ion Transport*

Numerous studies with soil-grown plants have shown that flooding depresses the concentrations of nitrogen, phosphorus, and potassium in the shoots (Letey *et al.*, 1962, 1965; Lal and Taylor, 1970; Leyshon and Sheard, 1974; Elkins and Hoveland, 1977; Jackson, 1979a; Drew and Sisworo, 1979; Trought and Drew, 1980b,c), ion accumulation being inhibited to a greater extent than photosynthesis and dry-matter production by the shoot.

Anaerobic conditions, like the presence of uncoupling agents, almost immediately inhibit uptake from solution and radial transport of nutrient ions by roots of nonwetland species (Lüttge and Pitman, 1976; Pitman, 1976; Rao and Rains, 1976; Cheeseman and Hanson, 1979; Cheeseman *et al.*, 1980), indicating that anaerobic metabolism provides energy insufficient to maintain the activity of ion pumps. Additionally, disruption of cell-membrane properties would cause a dissipation of proton-concentration gradients across the plasmamembrane that, together with wholesale leakage of accumulated ions and cell metabolites, would be expected to curtail any further energy-dependent ion transport. Measurements of ion transport in roots of nonwetland species at different O_2 concentrations indicate a progressive decline at partial pressure below 0.05–0.02 atm in the aerating gases (Vlamis and Davis, 1944; Hopkins *et al.*, 1950; Hopkins, 1956; Letey *et al.*, 1962). Some transfer of ions from the rooting medium to the shoots may then take place passively, ions moving with little selectivity across damaged cell membranes by mass flow with the transpiration stream (Trought and Drew, 1980b). The provision of greater concentrations of ions in the outer solution can thus partially offset the anoxia-induced inhibition of transport to the shoot (Trought and Drew, 1981).

With intact roots of wetland species, ion uptake can be almost independent of the O_2 concentration of the solution because of the internal supply of O_2 from the shoot (Vlamis and Davis, 1944; John *et al.*, 1974), but the pathways for ion transport across such roots remain in doubt because of the structural modifications that occur. There has been little work to determine if the marked changes to their structure impede the uptake of ions and water. In maize, gas-filled lacunae can occupy some 80% of the midcortex in aerenchymatous roots, but potassium transport to the stele remains unaffected (Drew *et al.*, 1980). Despite the loss of much of the root cortex, the occasional bridges of intact cells and residual walls of lysed cortical cells appear to maintain low-resistance pathways for movement of ions between the epidermis and endodermis (Fig. 4).

In the roots of wetland species, extensive lignification and suberization of the hypodermis, features that may help retain internal O_2, might be expected to act as potent barriers to movement of ions across the root. In *Carex arenaria*, "hypodermal sleeves" were highly impermeable to labeled water and phosphate and calcium ions (Robards *et al.*, 1979). Uptake by intact roots of potassium and phosphate was also very slow once the hypodermis became suberized (2–5 cm from the root tip), but this inhibition was reversible if the hypodermis was removed surgically (Robards *et al.*, 1979). Clearly, the hypodermis in *Carex* can act as a potent barrier to ion penetration, effectively sealing off the root from its environment, other than in the apical zone. How, therefore, do water and ions reach the stele, with suberization occurring to within at least 2 cm of the root tip and with intact roots, and ions absorbed by the tip region transported only to the root apex? Robards *et al.* suggested that the fine, more ephemeral lateral roots might supply the nutrient requirements of the remainder of the plant. With its typical wetland structure, it is possible that rice behaves similarly.

In contrast to most other ions, the concentration of sodium in shoots of cereal species can *increase* during flooding (Letey *et al.*, 1962, 1965) or in anaerobic solution culture (Hopkins *et al.*, 1950). In salt-sensitive species, such increases in sodium content can give rise to economically important depression of growth and yield. Mechanisms of sodium exclusion from the shoot, which depend on sodium extraction from the xylem stream by adjacent xylem parenchyma cells or by outwardly directed sodium pumps in the root cortex (Kramer *et al.*, 1977; Yeo *et al.*, 1977; Jeschke and Jambor, 1981), presumably break down under anoxia.

D. Respiratory Metabolism and Oxygen Deficiency

1. Oxygen Requirements of Roots

With the rare exception of roots that are "adapted" in some way to O_2-deficient media, root growth and function are dependent on aerobic respiration, using as terminal electron acceptor O_2 supplied predominately by the rooting environment. Cannon (1925) recognized the importance of measuring the O_2 concentration in the rooting medium and defining the concentration at which root respiration would first begin to suffer from O_2 shortage—the critical O_2 pressure (COP).

If the O_2 requirement of plant cells is considered at the cellular or subcellular level, extremely small concentrations should be adequate to maintain at least the fraction of the total O_2 consumption associated with oxidative phosphorylation of adenine nucleotides (i.e., with cytochrome oxidase and the remainder of the electron-transport chain in mitochondria). The K_m for the binding of O_2 to cytochrome oxidase may be as low as 0.1 or 0.01 μM at the active site (Griffin, 1968; Bonner, 1973), but it may be more relevant to consider the O_2 requirements of intact mitochondria. With mitochondria from pigeon heart, half-maximum reduction of cytochrome *c* occurs between 0.27 and 0.03 μM O_2, depending on metabolic activity. In rapidly respiring (state 3) mitochondria supplied with 0.27 μM O_2, O_2 uptake was 30% of that of the fully aerobic controls (Sugano *et al.*, 1974). These minute concentrations should be compared with that of O_2 in air-saturated water at 15°C and 760 cm Hg total pressure (which is 314 μM; Hitchman, 1978). This concentration would be depressed ("salted out") by only about 10% in the cytoplasm if the concentration of dissolved substances gives an osmotic potential of ~5 atm.

Although the practice of defining the COP with respect to the outer solution is widely accepted, it provides an inaccurate measure of the O_2 concentration within root cells. It has long been recognized that diffusion of O_2 across the root in the aqueous phase provides a major resistance (Berry and Norris, 1949; James, 1953; Griffin, 1968). Berry and Norris (1949) hypothesized that at O_2 concentrations below the COP, O_2-uptake rates are limited by diffusion to an outer sleeve of fully aerobic cells extending a variable distance into the root and surrounding a cylindrical core of anaerobic cells. Calculations based on a diffusion model suggest that even in well-aerated media, radial O_2 diffusion in the aqueous phase would be sufficiently slow to produce an anaerobic core (Griffin, 1968). Further experimental support for the notion of an anaerobic core comes from the work of Bertani and Brambilla (1982a,b), who found a progressive increase in activity of alcohol dehydrogenase in root tips when the O_2 partial pressure was at or below 0.1 atm (wheat) or 0.05 atm (rice). At these same concentrations the pattern of polypeptide synthesis gradually changed to that associated with anaerobic conditions (Section II,D,2).

Thus, with high temperatures and rapid respiration, especially in tissues with few intercellular spaces for gaseous diffusion such as apical meristems, respiration and root metabolism can be limited by the diffusion of O_2 through the tissue. This limitation is exacerbated in excised tissues that artificially become water filled.

A further difficulty in defining the COP with respect to the rooting medium, whether soil or nutrient solution, is the influence of the unstirred layer of water at the root surface. Stationary water films of unknown thickness exert a large resistance to the diffusion of dissolved gases, so that the O_2 concentration at the outer wall of the epidermis may differ from that measured in the soil air or in the bulk nutrient solution. In unsaturated soil, the thickness of such water films depends on soil water potential ψ. Thus, with vigorously stirred nutrient solution, which ensure thin films of static water around the root, much lower COP's were found (with excised roots of mustard, *Sinapis alba*), O_2 uptake being half-maximal at a partial pressure of 0.006 atm (Greenwood and Goodman, 1971).

Limiting O_2 concentrations equivalent to the COP can also be defined in relation to a range of root physiological properties such as extension, ion transport, and water uptake. Rates of root fresh-weight increase close to maximal have been recorded at O_2 partial pressures of 0.015–0.05 atm with tomato, tobacco, and soybean (Hopkins *et al.*, 1950) in well-stirred solution. Uptake of ions and water (Hopkins *et al.*, 1950; Willey, 1970) were also near maximal at these concentrations. A close correspondence between respiratory COP's and effects of limiting O_2 concentrations on root physiology or metabolism would not necessarily be expected because the proportion of total O_2 consumption linked to phosphorylation of ADP is uncertain (Section II,D,3).

Alternatively, COP may be defined with respect to the internal environment—the gas concentration in the intercellular spaces. Using cylindrical platinum electrodes located around the root tip to measure radial O_2 efflux polarographically, Armstrong and Gaynard (1976) derived values for the internal O_2 concentration of roots of intact plants transferred from waterlogged soil to anaerobic agar gel. The supply of O_2 to the root tip was varied by changing the partial pressure of O_2 in the vessel enclosing the leafy parts of the plant. They concluded that for certain wetland species (rice and cotton grass, *Eriophorum angustifolium*), respiration would be little affected in the relatively porous cortex by partial pressures as low as 0.001 atm. For the apical meristem and stelar tissues, of much smaller gas-filled porosity than the cortex and therefore of greater diffusional resistance, COP's were in the range 0.02–0.026 atm. In roots of pea, a nonwetland species, the internal COP was between 0.002 and 0.0207 atm (Webb, 1982; Webb and Armstrong, 1983), values at least as low as in the wetland species. Differences between these species in their resistance to flooding are thus not readily explained in terms of differences in ability to respire when O_2 concentrations within the tissue are low.

In the roots of nonwetland herbaceous species, the interconnected system of intercellular spaces, occupying up to 10% of the root volume, is gas filled and allows limited but measurable diffusion of O_2 from the shoot into the roots (Fig. 5). By supplying $^{15}O_2$ to the leaves (Barber *et al.*, 1962; Jensen *et al.*, 1967), the radioisotope was rapidly detected in the roots of barley (*Hordeum vulgare*) and corn (*Zea mays*) in air, its velocity being that expected for gaseous diffusion. In seedlings of 12 herbaceous crop species, O_2 moved from the aerial environment

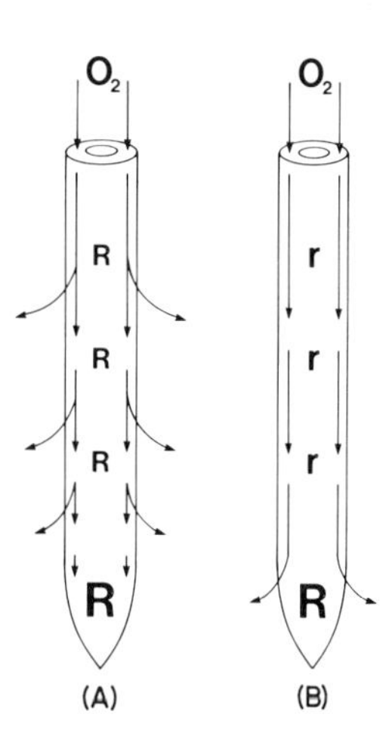

Fig. 5. Oxygen diffusion and respiration in roots representative of nonwetland and wetland species. (A) Nonwetland. Fast radial leakage, impeded longitudinal diffusion, rapid respiration. Oxygen concentration in apical meristem becomes limiting. (B) Wetland. Slow radial leakage, small resistance to longitudinal diffusion in aerenchyma, slower respiration. Oxygen concentration in apical meristem is adequate. Radial leakage near tip oxidizes rhizosphere. R, Highest respiration rate (per unit volume of tissue) in apical meristem; R, rapid respiration typical of nonwetland, subapical tissue; r, slow respiration typical of aerenchymatous roots of wetland species

through the roots into O_2-free nutrient agar or solution, where it was detected polarographically (Greenwood, 1967a,b; Greenwood and Goodman, 1971). Because roots were able to extend a short distance into the anaerobic media and respire using O_2 diffusing from the shoot, Greenwood (1967) concluded that "when the plants were grown under anaerobic conditions such diffusion could satisfy all the roots' requirements for oxygen [p. 337]." The demonstration of O_2 diffusion encouraged the speculation that root injury in flooded soil is attributable more to the accumulation of soil toxins than to lack of O_2 for respiration. However, Greenwood's measurements were confined to short seedling roots with diffusion distances of only 0.5–4.2 cm.

Diffusion rates of O_2 over greater path lengths within "unadapted" roots in anaerobic media were found to be insufficient for satisfying their O_2 requirements. Depletion of O_2 from solution by root respiration of 25-day-old cotton was unaffected by removal of the leaves or entire aerial parts of the plant, indicating that the amount of O_2 diffusing to the roots of the intact plant was insignificant (Vartapetian *et al.*, 1978b). This was confirmed using chemiluminescence (ultraweak glow at 450–650 nm attributed to peroxide oxidation of membrane lipids) as a monitor of the aeration status of the root tissue; chemiluminescence was extinguished, indicating anoxia, after 2 hr of bubbling O_2-free nitrogen gas through the culture solution, with either intact plants with the shoots in air or with excised roots. An alternative measure of the oxygenation or respiratory activity of tissue is given by the nucleotide phosphate ratios (energy charge) when the roots are in anaerobic media and the shoots in air (Raymond *et al.*, 1978). In maize, energy charge declined from the fully oxygenated values to those associated with anaerobic tissue.

Nonwetland species are characterized by a high radial permeability to internally diffusing O_2, and in stirred anaerobic solution, steep concentration gradients at the root surface can cause O_2 leakage from the basal zones near the junction with the shoot (Fig. 5). This effect can be mitigated experimentally using unstirred deoxygenated solution or an anaerobic agar gel so that a virtual "jacketing" effect is produced by the slow diffusion of O_2 in the static aqueous medium. In O_2-free gel of stiff agar (3% w/v), roots of pea seedlings lacking aerenchyma were able to extend to lengths greater than 20 cm (Healy and Armstrong, 1972). In a semiliquid O_2-free agar gel (0.05% w/v), root extension at 23°C was limited to ~9 cm, presumably because of a greater radial leakage of O_2 to the medium from the basal zones close to the shoot.

The following experimental evidence indicates that the occurrence of aerenchyma improves the internal supply of O_2 in roots of herbaceous wetland and nonwetland species:

1. Increased gas-space formation in rice roots is associated with faster radial leakage of O_2 (detected polarographically) in the tip region (Armstrong, 1971). Abundant precipitation of iron (as Fe^{3+} compounds) from Fe^{2+} around the epidermis and in the root cortex in rice must also be indicative of appreciable movement of O_2 through the roots (Green and Etherington, 1977). Diffusion of $^{15}O_2$ from shoots to a point 15 cm along a root was readily detected in aerenchymatous rice roots, but not in nonaerenchymatous barley roots (Barber *et al.*, 1962). However, the smaller radial leakage of O_2 from the basal zones of rice roots compared with the nonwetland cereal could have contributed to this difference.

2. The composition of the gases in internal cavities in plant organs in anaerobic or near-anaerobic media indicates that the lower but appreciable concentrations of O_2 in them are likely to have been maintained by diffusion from the aerial parts [see Arikado, 1961 (barley and wheat, as well as flood-resistant species like *Phalaris arundinacea, Alopeculus fulvus, Veronica anagallis,* and *Ranunculus sceleratus*); Barber, 1961 (*Equisetum limosium*); Teal and Kanwisher, 1966 (*Spartina alterniflora*); Laing, 1940 (*Nuphar advenum*); Conway, 1937 (*Cladium mariscus*); Ellmore, 1981 (*Ludwigia peploides*)].

3. Growth of roots an appreciable distance into strictly anaerobic media is associated with the presence of aerenchyma in rice (Armstrong, 1979) and in wheat (Trought and Drew, 1980b). In maize, chemical inhibition of aerenchyma formation using silver ions slows root extension by 45% in nonaerated culture (Drew *et al.*, 1981).

4. The presumptive pathway for the diffusion of O_2 can be inferred from anatomical examination of sectioned plant material. Nontortuous gas-filled spaces can be traced from the root, through the root–shoot junction, into the leaf base or the culm (Arashi and Nitta, 1955; Arikado, 1955a,b, 1959, 1961). Such observations have been coupled by Arikado with measurements of enhanced gas flow (i.e., viscous mass flow) under a pressure gradient. The entry of O_2 in

herbaceous plants is largely through the stomata (Armstrong, 1979), but in *Puccinellia peisonis,* a flood-tolerant grass, appreciable passage of O_2 into aerenchymatous roots takes place when the upper root zone near the junction with the shoot is aerated (Stelzer and Läuchli, 1980).

Theoretical modeling of diffusion, together with the use of an electrical analog, allows quantitative assessment of the importance of O_2 diffusion within roots for the respiratory requirements of tissues, especially of the apical meristem, which is remote from the O_2 source. The concentration of O_2 reaching the apical meristem depends on the diffusion path length from the aerial parts, the gas-filled porosity of the root, the respiratory use of O_2, and radial leakage (Fig. 5). The latter is modified by "wall" permeability of the root and the rate of soil respiration, which affects the concentration gradient from the root surface outward (Armstrong, 1979). Calculated values for the internal concentration of O_2 at the root apex, derived from an electrical analog, are given as a function of root length (Fig. 6A–D), using boundary conditions representative of the nonwetland root. It is assumed that root growth ceases when the internal O_2 partial pressure declines to 0.02 atm. Under all conditions, increasing the root gas space from 1.5 to 15% would be expected to increase root length in an anaerobic medium before the O_2 concentration in the root apex became limiting. Roots are appreciably shorter when radial leakage of O_2 occurs (Fig. 6, compare A with B), and this effect is accentuated by faster soil respiration (Fig. 6, compare B with C). Finer roots extend a shorter distance (Fig. 6D), particularly when fast soil respiration causes a steep concentration gradient from the root surface outward. At the limit, when internal O_2 is conserved to the maximum, extension is independent of radius (Fig. 6A) and roots should attain a length of ~22 cm. Although the predictions developed in this approach have not yet been fully tested experimentally, Armstrong (1979) pointed out that such a length is only a few centimeters in excess of the deepest root penetration, recorded by Yu *et al.* (1969), for nonwetland species in waterlogged soil. This compares with the deepest penetration of aerenchymatous roots of wheat at 14°C in waterlogged soil (20 cm; Trought and Drew, 1980a) and anaerobic stirred solution (12 cm; Trought and Drew, 1980b). Although diffusion of O_2 along more bulky structures (e.g., rhizomes) has not been modeled, the trend with increasing diameter is to conserve oxygen (Fig. 6D). Coult and Vallance (1958) recorded O_2 (0.08 atm partial pressure) in *Menyanthes trifoliata* stolons in anaerobic cultures at a distance of 75 cm from the shoot exposed to air.

When the soil first becomes water saturated, particularly when temperatures are low and soil respiration slowed, the concentration of O_2 in the water may decline gradually. During winter waterlogging of cereals in a clay soil, Blackwell and Ayling (1981) found that 13 days elapsed before the concentration of dissolved soil O_2 at 20-cm depth had declined to 0.02 atm. The same change

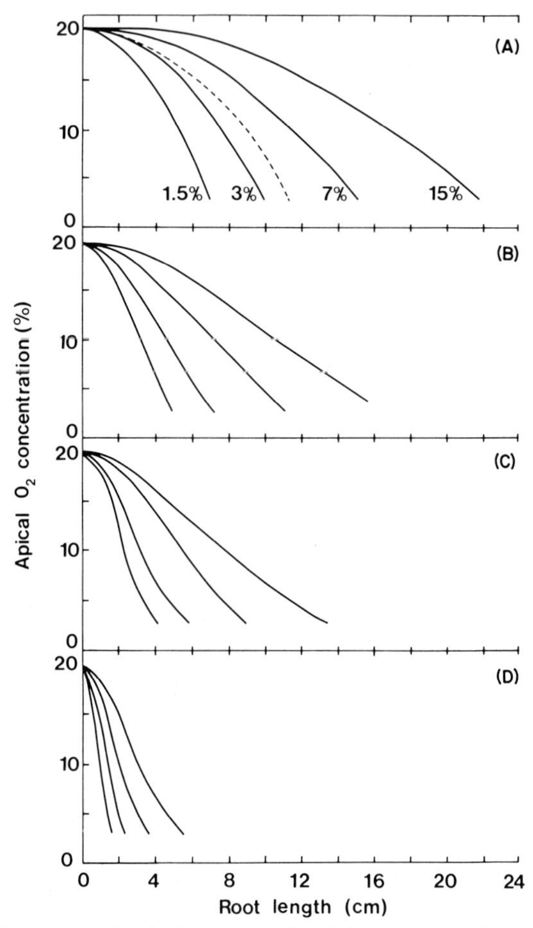

Fig. 6. Expected O_2 concentration in the root apex in relation to root length, root porosity, radius (cm: A–C, 0.05; D, 0.01), and soil O_2 consumption. Values are derived from an electrical analog. Boundary conditions: (1) root respiration is uniform along the root (30 ng O_2 cm^{-3} sec^{-1}); (2) porosity is uniform along the root (1.5, 3, 7 or 15%); (3) no radial O_2 leakage in (A), therefore no soil O_2 consumption. Uniform O_2 consumption in rhizosphere soil is (B) 5.3 and (C, D) 53 ng cm^{-3} sec^{-1}; (4) root wall permeability to O_2 (B–D) is 100% at apex and declines to a 60% minimum at 6 cm from the apex; (5) extension ceases at an O_2 partial pressure of 0.02 atm in the gas spaces in the apex. In (A), the dotted line indicates the effect of increasing root respiration to 120 ng cm^{-3} sec^{-1} when root porosity is 15%. Redrawn from Armstrong (1979, part of Fig. 17).

required only 3.5 days in spring. Although appreciable concentrations of O_2 remain in the soil water, the respiration of unacclimatized roots is controlled by the diffusion of O_2 through the soil to their surfaces. Armstrong and Wright (1975, 1976a,b) and Blackwell (1983) have measured O_2 flux to platinum microelectrodes in water-saturated soil, taking great care to use voltages at which only the flux of O_2 to the electrode surface contributed to the measured current. While the soil-O_2 concentration slowly declined (Blackwell and Wells, 1983), frequent measurements were made of the extension of the seminal roots of oats (observed through inclined transparent container walls) and of O_2 flux to electrodes located in the soil (10°C) near the extending roots. The time at which root extension first slowed corresponded to an O_2 flux to the electrode of 56 ng cm^{-2} min^{-1}, a flux that an independent calculation indicated would be just sufficient to sustain an uninhibited O_2-consumption rate of the root apex. Root extension ceased when O_2 flux to the electrode was effectively zero. Such observations suggest that in unstirred water-saturated media, the electrode method can provide a reliable means for assessing the maximal potential supply of O_2 to a root surface. Oxygen fluxes in the topsoil of arable land rarely exceed 56 ng cm^{-2} min^{-1} during waterlogging in autumn and spring, when soil temperatures are around 10°C (Blackwell and Ayling, 1981), suggesting that some restriction to root growth would soon occur after waterlogging.

2. *Pathways for Carbon Metabolism in Relation to Theories of Metabolic Adaptation to Anoxia*

Under fully aerobic conditions, the oxidation of 1 mol of sugar (glucose) to CO_2 and water yields 38 mol of ATP. In the production of 2 mol of pyruvate by glycolysis, synthesis of 4 mol of ATP occurs through substrate-linked phosphorylation of ADP (Fig. 7), but as 2 mol of ATP are required for phosphorylations of glucose and of fructose 6-phosphate, the net yield is 2 mol.

Under anoxia, the transfer of electrons from cytochrome oxidase to O_2 cannot take place, and synthesis of ATP via the electron-transport system is therefore blocked. Furthermore, NADH and reduced flavoproteins produced during the TCA cycle are no longer oxidized by the electron-transport system, and the cycle stops. However, glycolysis can continue, provided a means is available for using the NADH it generates. If the respiratory substrate is derived from starch, starch phosphorylase in the presence of inorganic phosphate degrades 1–4 linkages to release glucose 1-phosphate, which is converted to glucose 6-phosphate to enter the glycolytic pathway. This would allow a net synthesis of 3 mol of ATP. However, there is evidence that starch in roots is not readily mobilized for respiration (Saglio and Pradet, 1980; Saglio *et al.*, 1980; Massimino *et al.*, 1981; Bertani *et al.*, 1981), so a dependence on the current supply of assimilates from the phloem is paramount.

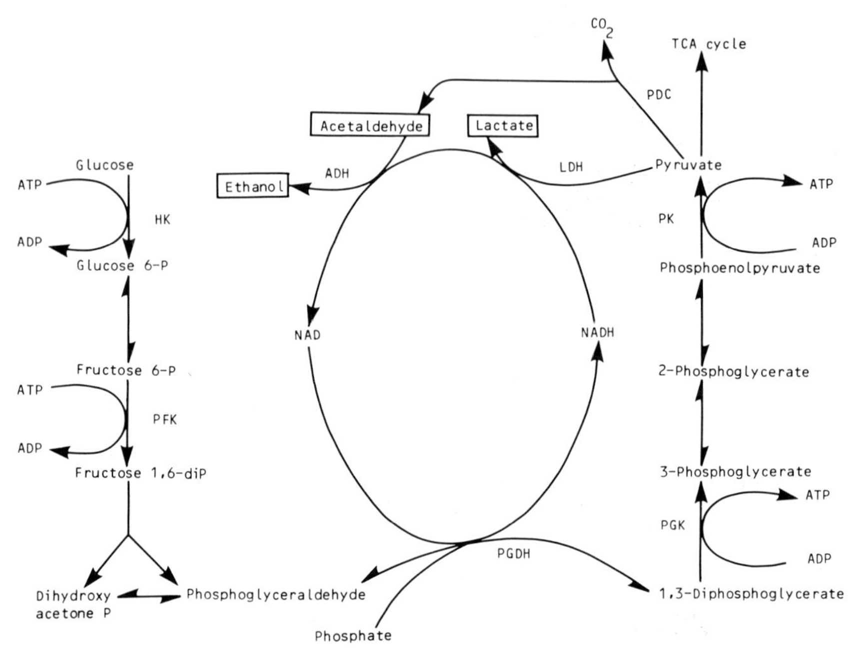

Fig. 7. Glycolytic (Embden–Meyerhof–Parnas) pathway for carbohydrate oxidation showing end products of anaerobic metabolism, ATP synthesis, and NAD regeneration. HK, Hexokinase; PFK, phosphofructokinase; PGDH, phosphoglyceraldehyde dehydrogenase; PGK, phosphoglycerate kinase; PK, pyruvate kinase; LDH, lactate dehydrogenase; PDC, pyruvate decarboxylase; ADH, alcohol dehydrogenase.

In considering the effectiveness of anaerobic respiration in higher-plant cells, it is important to establish, first, whether sufficient energy can be derived to supply the requirements of the cell for maintenance and growth; second, whether the end products of incomplete oxidation are compatible with cell metabolism; third, whether the supply of assimilates is adequate to maintain respiration; and fourth, whether the supply of intermediates normally supplied by the tricarboxylic acid (TCA) cycle is maintained. Growth under anaerobic conditions is rare (Section II,C,1), but the possibility remains that sufficient energy could be generated anaerobically to maintain cell integrity and thus aid survival of anoxia-tolerant organs.

In excised barley roots transferred from an aerated to an O_2-free medium, the concentration of ATP declined in 15 min by 88% (Jacoby and Rudich, 1980). This was sufficient to destroy the metabolically generated pH gradient across cell membranes and the associated chloride influx. Such changes accord with observations of a loss of membrane integrity in roots exposed to a nitrogen atmosphere, detected by loss of cell constituents to the outer solution (Hiatt and

Lowe, 1967). However, the possibility cannot be ruled out that when tissue becomes gradually O_2 deficient over an extended period in natural conditions, metabolic adaptations take place that are not revealed after the sudden imposition of anoxia carried out in the laboratory. As yet there is no convincing information on this last point.

Whether the flooding tolerance of wetland species derives from a specialized metabolism, allowing plants to survive or even grow in the absence of O_2, has long been a subject of interest. Crawford (1966, 1967) proposed that flood tolerance of a number of herbaceous species is related to the capacity to avoid the generation of potentially toxic ethanol at the end of the glycolytic pathway (Fig. 7). Also, in flood-intolerant species, ethanol was thought to be formed in larger amounts during flooding as a result of increased alcohol dehydrogenase (ADH) activity and/or a Pasteur effect. These proposals were later developed into a generalized theory of flooding tolerance (Crawford, 1967, 1978; McManmon and Crawford, 1971), summarized in Fig. 8. In flood-susceptible species, accumulation of ethanol derived from glycolysis via phosphoenolpyruvate (PEP) was assumed to cause death of cells in roots (Crawford, 1967). The occurrence of ethanol in the transpiration stream (Fulton and Erikson, 1964) had already been identified as a possible causative factor in flooding damage to shoots (but see Section II,E,2). In contrast, in flood-tolerant species there was evidence that

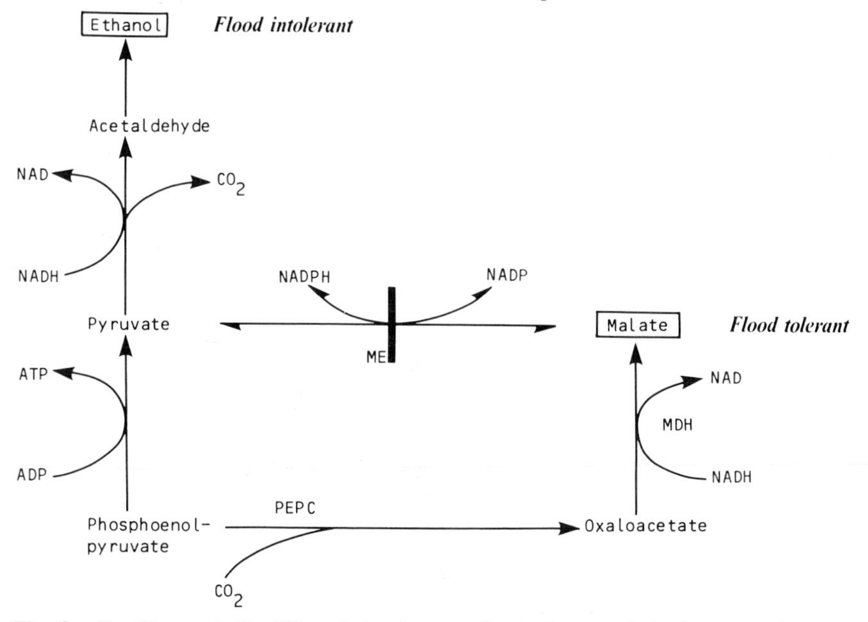

Fig. 8. Possible metabolic differentiation between flood-tolerant and -intolerant species. *Note proposed absence of malic enzyme activity in flood-tolerant plants.* PEPC, Phosphoenolpyruvate carboxylase; MDH, malate dehydrogenase. From Crawford (1978).

PEP was converted instead to oxaloacetate by fixation of CO_2 through the action of PEP carboxylase. Malate dehydrogenase then reduced oxaloacetate to malate, thus allowing the regeneration of NAD, to maintain glycolysis (Fig. 8). Malate was assumed to be less toxic than ethanol. It was suggested that any further conversion of malate to pyruvate and on to ethanol was prevented in flood-tolerant species either by constitutive lack of the appropriate enzyme (malic enzyme) or inhibition of its action (McManmon and Crawford, 1971). Support for the hypothesis came from observations of 19 marsh species where the relative increase in activity of ADH, induced by O_2 deficiency, was correlated with increased sensitivity to flooding (McManmon and Crawford, 1971). Similar results were obtained by Francis *et al.* (1974) for subterranean clover varieties.

More recently, attention has been directed to the possible combinations of metabolic pathways in roots and rhizomes of flood-tolerant plants by which NADH could be reoxidized while avoiding ethanol accumulation. Evidence of the occurrence of some of these end products of anaerobic metabolism, including glycerol, shikimate, lactate, and the amino acids alanine and glutamic acid, have been reviewed (Crawford, 1978). Although reducing power is utilized in producing these metabolites, only the formation of lactate and alanine (like ethanol) would give a net ATP synthesis (2 mol ATP per mole hexose respired). Several investigations revealed marked increases in the concentration of alanine in the root tissues of rice (Bertani *et al.*, 1981; Bertani and Brambilla, 1982a) and other wetland species (Smith and ap Rees, 1979a) as well as in *Pisum sativum* (Smith and ap Rees, 1979b) and *Zea mays* (Kohl *et al.*, 1978; Saglio *et al.*, 1980) under anaerobic or near-anaerobic conditions. These changes suggest increased availability of pyruvate for transaminations when the TCA cycle is suppressed. Whether alanine synthesis is sufficiently great to give an appreciable yield of ATP is doubtful. Information on the rate of production of end products compared with utilization of substrates suggests that synthesis of ethanol, and to a much lesser extent lactate, predominates by far (Davies *et al.*, 1974; Avadhani *et al.*, 1978; Smith and ap Rees, 1979a,b; Bertani *et al.*, 1980; Saglio *et al.*, 1980).

Some have suggested that in the absence of O_2, nitrate could act as a terminal electron acceptor in respiration (Arnon, 1937; Garcia-Novo and Crawford, 1973). Nitrate respiration or dissimilatory nitrate reduction occurs in some anaerobic microorganisms, where it has been thoroughly characterized (Hamilton, 1979). The evidence for its occurrence in higher plants is not properly established (see Drew and Lynch, 1980, and following discussion).

However, despite the widespread attention that the metabolic theory of flood tolerance has attracted, in our view the evidence is *incomplete* for the following reasons:

1. *Ethanol production.* Plant cells under anaerobic conditions commonly produce ethanol (James, 1953), but it is doubtful whether this response in roots

accords with susceptibility to flooding. Rice and relatively flood-tolerant winter wheat and rye readily synthesize appreciable amounts of ethanol under anaerobic conditions (Phillips, 1947; John *et al.*, 1974; John and Greenway, 1976; Beletskya, 1977; Avadhani *et al.*, 1978; Bertani *et al.*, 1980), as do roots of flood-sensitive pea (Smith and ap Rees, 1979a). In barley varieties and rice, flood tolerance was associated with *increased* ADH activity (Wignarajah *et al.*, 1976) rather than the opposite.Similar accelerated ethanol production takes place in the flood-tolerant roots of trees (Hook and Scholtens, 1978; Keeley and Franz, 1979; see also Hook, Chapter 8 in this volume).

2. *Ethanol toxicity*. There is no evidence that ethanol accumulates in affected tissues to toxic concentrations. Indeed, Jackson *et al.* (1982) showed that applied ethanol displayed little toxicity to roots or even to cell protoplasts at concentrations up to 0.1 *M*. To halve protoplast viability within 3.5 hr required as much as 1 *M* ethanol.

3. *ADH activity*. Although there is indication of increased ADH activity with O_2 deficiency, it seems likely that the synthesis of ethanol is limited more by the activity of pyruvate decarboxylase than by ADH (John and Greenway, 1976). It is possible that ADH is of greater significance in catalyzing the reverse reaction in the presence of O_2, that is, that of oxidizing ethanol to acetaldehyde and thus redirecting carbon to pyruvate and the TCA cycle in tissue (Phillips, 1947; Cossins, 1978). That root tips generate small amounts of ethanol even in fully oxygenated media, due perhaps to an anoxic "core" of the tissue, indicates a role for ADH in the aerobic "sheath." Additionally, ADH could serve a similar function when O_2 reenters the root environment.

4. *Malate and ethanol*. In a laboratory examination of three of the flood-tolerant species investigated by Crawford (*Ranunculus sceleratus, Glyceria maxima,* and *Senecio aquaticus*) Smith and ap Rees (1979b) obtained the following results that seem difficult to reconcile with Crawford's metabolic hypothesis of flood tolerance. Malate failed to accumulate under anaerobic conditions (4 hr), and appreciable activity of malic enzyme was found in the roots of these flood-tolerant plants, in agreement with Davies *et al.* (1974). Nonaeration caused large increases in ADH activity and smaller increases in lactate dehydrogenase. In all three species carbohydrate was fermented predominantly to ethanol, showing essentially the same pattern as in roots of flood-susceptible pea (Smith and ap Rees, 1979a).

5. *ATP production*. No net ATP production would occur in the formation of malate via oxaloacetate if PEP carboxylase catalyzes the step from PEP to oxaloacetate (Fig. 8). If starch were respired to the level of malate, the net yield would be 1 mol ATP per mole hexose. The action of PEP carboxykinase would yield 2 mol ATP, but no PEP carboxykinase activity could be detected in the roots of nonaerated marsh plants (Smith and ap Rees, 1979a) or in seedlings of *Echinochloa crus-galli* (Rumpho and Kennedy, 1981). The significance there-

fore of a respiratory pathway giving no net yield of ATP is highly questionable. As yet, no reliable measurements have been presented for the ATP content or energy status of roots in which it is known that growth or survival is being sustained by anaerobic metabolism, other than by ethanol production.

6. *Gas spaces.* In studies of metabolism in flood-tolerant and -intolerant species, roots were frequently sampled after 1–4 weeks of flooding treatment. In view of the widespread formation of aerenchymatous roots in flood-tolerant species, it seems likely that a large proportion of the roots sampled had improved internal aeration. One might, therefore, expect less fermentation in such roots. This contrasts with the roots of susceptible species that lack aerenchyma. There fermentation is symptomatic of incomplete respiratory metabolism and degeneration under O_2 stress.

7. *Substrates for respiration.* The metabolic theory proposes that glycolysis should be stimulated under anaerobic conditions (Pasteur effect) in *flood-intolerant* species, causing ethanol to be accumulated more rapidly than in the flood-tolerant ones. However, there is abundant evidence to the contrary. Survival of rice seedlings under anoxia is closely associated with rapid production of ethanol (Phillips, 1947; Bertani *et al.*, 1980) and an *I/N* ratio greater than 0.33, which may be indicative of faster utilization of substrates. Furthermore, increasing the supply of respirable substrate by addition of glucose to anaerobic media *prolongs* the survival of roots of wetland and nonwetland species rather than decreasing it. In the absence of exogenous glucose the average survival time of the apical zone in rice and pea was, respectively, 7 and 8 hr, increasing to 30 and 24 hr with exogenous glucose (Webb, 1982). Because rice and pea are known to undergo a mainly alcoholic fermentation under anoxia, it suggests that a lack of substrates for anaerobic respiration restricts phosphorylation to the point that cell maintenance collapses. Similar conclusions were reached by Saglio *et al.* (1980) with excised maize root tips. The adenylate energy charge in air (0.9; Section II,3,a) declined rapidly after transfer to a nitrogen atmosphere to values ranging from 0.6 to 0.15, these values as well as the soluble sugar content becoming lower with increasing duration of aging in air before transfer to nitrogen. The presence of 0.2 *M* glucose restored the energy charge to 0.6, irrespective of the earlier period of aging. Of the sugars consumed by respiration under anoxia, 70% could be accounted for in terms of ethanol and lactate formation, and alanine accounted for 5%. In view of the inhibition of phloem transport in roots under anaerobic conditions (Vartapetian *et al.*, 1978b; Drew and Sisworo, 1979), it is clear that a lack of respirable substrates could soon become critical to cell survival.

8. *Nitrate respiration.* Support for the possibility that nitrate could act as a terminal electron acceptor in plant respiration comes from two inconclusive sets of observations. First, Arnon (1937) and others noted that cereal species grow better in nonaerated solution when supplied with nitrate compared with ammonium. Second, increases in the *in vivo* activity of nitrate reductase with

flooding have been recorded for the leaves and roots of waterlogging-tolerant *Senecio* species (Garcia-Novo and Crawford, 1973; Lambers, 1976; Lambers *et al.*, 1978). However, the first of these effects might be the result of nitrate, present at high concentration around the roots entering the plant and simply acting as a substrate for shoot growth (Trought and Drew, 1981). Lee (1978) did not find any evidence of nitrate respiration in sterile excised barley roots when placed in a nitrogen atmosphere; there was no increase in nitrate reductase activity and no direct suppression of ethanol production by nitrate. The previous nitrogen nutrition did influence ethanol production indirectly, but this was entirely explicable by an associated increase in the amount of substrates available for fermentation.

In conclusion, the significance of malate in relation to the metabolic hypothesis of flood tolerance remains in doubt until further experimental evidence is available to demonstrate its unambiguous contribution to cell survival. Even in organs in which there is little growth (e.g., during winter submergence of rhizomes or roots), some net ATP synthesis would be necessary for cell maintenance. On the other hand, a body of detailed evidence indicates that fermentation reactions generate sufficient ATP to permit cell survival in at least some plant organs for several days. Many that display an ability to survive anoxia readily produce ethanol and rapidly dispose of it to the outer solution (Bertani *et al.*, 1980) or into the transpiration stream (Fulton and Erikson, 1964; Jackson *et al.*, 1982). The biochemistry of cell survival of prolonged anoxia in rhizomes and associated organs remains to be examined in comparable detail.

3. *Energy Metabolism*

a. Adenine Nucleotide Ratios. Although short-term changes in the ATP concentration in tissues can provide a useful index of the consequences of O_2 deficiency for respiration, the energy status of cells cannot be estimated reliably by ATP level alone; changes in cell metabolism causing alteration in the concentration of the total pool of adenine nucleotides (Adn) would also modify the ATP concentration without necessarily affecting the energy status of the cell. A convenient means of comparing energy status is to express ATP in terms of the ratio of Adn in the various equilibrating components of the pool maintained by adenylate kinase. One such ratio, the adenylate energy charge (AEC) is given by (Atkinson, 1969; Pradet and Bomsel, 1978; Pradet and Raymond, 1983)

$$[ATP] + 0.5[ADP]/[ATP] + [ADP] + [AMP]$$

A value of 1.0 is obtained when all the Adn has been converted to ATP, and zero when all is AMP. Because of equilibria between ATP-regenerating and -utilizing reactions, AEC tends to stabilize between 0.8 and 0.95. Other ratios, such as ATP/ADP or ATP/(ADP + AMP + P_i), some of which are correlated with the

AEC, are of interest (Erecinska *et al.*, 1977; Davies, 1980). The AEC and other adenylate ratios take on special significance in relation to the allosteric control of the activity of enzymes involved in the respiratory pathways. In some cases a low proportion of ATP acts as a positive effector in enzyme activity, and the consequent acceleration of respiration increases the production of ATP. Conversely, some ATP-utilizing reactions are accelerated by a high proportion of ATP (Turner and Turner, 1980; Pradet and Raymond, 1983), although it is now a matter of controversy as to whether AEC exerts the precise regulation of plant enzymes in the manner initially proposed by Atkinson.

Pradet and Bomsel (1978) outlined two broad adaptations for plant survival of anoxia in relation to AEC. In the first, high values are maintained and correlate with sustained metabolic activity. In embryos and coleoptiles of rice the AEC was maintained at 0.75–0.85 for days, while synthesis of DNA, RNA, and protein remained active (Mocquot *et al.*, 1977, 1981). When maize root tips were exposed to a nitrogen atmosphere, the rapid decline in AEC (to 0.2) was partially reversed by exogenous glucose (Saglio *et al.*, 1980). Because low values of AEC are associated in general in cells in vegetative organs with a low level of metabolic activity, the glucose-stimulated rise in fermentation and AEC (from 0.2 to 0.6) in this work agrees well with the observations of Webb (1982) of increased duration of survival of cells and organelles under anoxia in the presence of glucose. It is noticeable that in maize roots, as in other tissues that do not survive anoxia for extended periods (pea seeds and wheat leaves), the energy charge shows a gradual decline in the absence of exogenous glucose, leading to cell degeneration.

The alternative adaptation, exemplified by germinating lettuce seeds (Raymond and Pradet, 1980), is maintenance of a low AEC and low metabolic activity under anoxia. Whether induced ''dormancy'' of this type might occur in roots or rhizomes of flood-tolerant species is unknown.

b. The Alternative Pathway. At least two terminal oxidases present in plant mitochondria require O_2. Cytochrome oxidase, sensitive to cyanide, is ubiquitous and, together with the remainder of the cytochromes, closely involved in ADP phosphorylation. The nature of the alternative oxidase, resistant to cyanide, is less well understood. Electrons initially share the same pathway as that preceding the cytochromes, allowing a single phosphorylating step. Branching occurs at or near coenzyme Q, where the alternative nonphosphorylating bypass of the cytochromes begins (Solomos, 1977; Laties, 1982).

The extent to which the alternative path is engaged *in vivo* is important to the energy metabolism of the cell, especially when O_2 is in short supply because (1) despite a somewhat lower affinity for oxygen than that of cytochrome oxidase, it may still significantly compete with it for oxygen and (2) the energy derived from it is much less than from the cytochrome pathway. The alternative pathway, unlike the cytochrome pathway, can be blocked by substituted hydroxamic acids such as salicylhydroxamic acid (SHAM) at appropriate concentrations (1 m*M* or less). However, the use of cyanide to inhibit the cytochrome pathway often induces a compensatory flow of electrons through the cyanide-resistant alternative pathway. Thus the potential capacity of the alternative pathway is not necessarily a measure of the extent to which it is engaged *in vivo*.

The possibility that flood-tolerant and -intolerant *Senecio* species might differ in their respiratory metabolism, depending on the affinity of their terminal oxidases for oxygen, has been suggested (Lambers, 1976). However, in *S. aquaticus* (flood tolerant) and *S. jacobaea* (flood intolerant), sensitivity to flooding did not correlate with a greater activity of the alternative pathway, which appeared to contribute approximately the same extent to total respiration in both species (Lambers and Smakman, 1978). On the basis of SHAM inhibition of O_2 uptake, the alternative pathway accounted for 70% of total root respiration and did not seem to be involved specifically in growth respiration or maintenance respiration. However, in view of the high concentration of SHAM (25 m*M*) used in this and other studies by Lambers and co-workers and the lack of specificity that would most likely occur (Day *et al.*, 1980; Laties, 1982), the conclusions are open to question. For the roots of pea, the response to the presence of respiratory inhibitors (Webb, 1982) indicated that the alternative pathway was unlikely to have been operative. The flooding sensitivity of this species compared with others cannot therefore be explained in terms of the operation of a low-affinity terminal oxidase. In flood-tolerant *S. aquaticus* and *S. congestus* in nitrogen-bubbled solution, the internal transport of O_2 maintained O_2 consumption at only half its fully aerobic rate (Lambers *et al.*, 1978), but this appeared to be sufficient to ensure that oxidative phosphorylation was unimpaired.

Presumably only the oxidases that contribute nothing to phosphorylation would become inhibited at, and just below, the critical O_2 pressure because of their low affinity for O_2. Energy metabolism might therefore be expected to continue at a similar level to the fully aerobic state even when O_2 concentration within cells is very small. The O_2 requirements of the other terminal oxidases, and the influence of their inhibition on metabolism and growth, are less easily predicted. Concerning the alternative oxidase, some view its role in roots as purely regulatory, an energetically inefficient means of respiring carbohydrate in excess of that required for growth, energy, or osmoregulation (Lambers *et al.*, 1978; Lambers, 1980). On the other hand, engagement of the alternative pathway may be a means of producing certain TCA intermediates for biosyntheses and organic acid accumulation, where these exceed the requirement for ATP, making a smaller generation of ATP advantageous (Laties, 1982). *In vivo*, the alternative path becomes involved when electron flow through the cytochrome pathway is already saturated, suggesting that the energy requirements of the cell are first met. Consumption of O_2 by pathways insensitive to both cyanide and

SHAM may be as little as 10% of the total aerobic consumption (Lambers and Smakman, 1978). But it is possible that the control of the redox state of cell metabolites and the control of oxidative metabolism and biosyntheses by the "direct" oxidases (Butt, 1980), together with processes dependent on the alternative pathway, could be critical for some aspect of root growth and differentiation.

4. *Stress Proteins*

Under a wide range of adverse environmental conditions, animal and plant cells synthesize increased amounts of specific polypeptides distinguishable by gel electrophoresis (Webster, 1980). More than 100 polypeptides were identified in the primary roots of maize under aerobic conditions. Under anaerobic conditions protein synthesis was much inhibited, but RNA-dependent synthesis of a small group of polypeptides (the anaerobic polypeptides, ANPs) that were virtually absent from aerobically treated roots became apparent after ~2 hr. The ANPs accounted for about 70% of the incorporation of [^{3}H]leucine into protein under anaerobic conditions (Sachs *et al.*, 1980). Prominent among the 20 ANPs were two alcohol dehydrogenase isozymes, but the possible functions of the remainder of the polypeptides were unidentified. Despite a profound change in the pattern of protein synthesis in maize roots, cell death occurred after ~72 hr. The significance of the ANPs may thus seem open to question, but it should be borne in mind that any biochemical feature improving cell survival of anoxia could be valuable to the plant, no matter how short term. Changes in ANP production have been studied by Bertani and co-workers as a function of the O_2 partial pressure in the aerating gases. With wheat, even at 0.10 atm O_2, the pattern of *de novo* protein synthesis began to change to that characteristic of severe O_2 deficiency, becoming most pronounced at 0.01 atm, approximately in parallel with increases in ADH activity (Bertani and Brambilla, 1982b). However, when conditions were strictly anaerobic, roots failed to incorporate labeled amino acids into protein and the viability of the root tips was only a "few hours."

Under anaerobic conditions the behavior of maize and wheat roots contrasts markedly with that of rice. In germinating rice embryos (Mocquot *et al.*, 1981), only a small number of polypeptides were synthesized during the first few hours of anoxia when the AEC declined from 0.9 (aerobic) to 0.5–0.6. But after 1–2 days the AEC was partly restored (to 0.8) and numerous ANPs were synthesized, a capability that remained for up to 9 days under anoxia. Bertani and Brambilla (1982b) found that ANP production in rice began at 0.05 atm O_2 in the aerating gas stream, a lower concentration than with wheat. With rice there was appreciable ANP production under strictly anaerobic conditions, again unlike wheat, and root tips survived for 96–120 hr but without growth (Bertani and Brambilla,

1982a). The ANP pattern could develop rapidly (0.5–2.5 hr) following exposure to anaerobic conditions and revert equally rapidly to the fully aerobic pattern on exposure to air (Bertani *et al.*, 1981).

Clearly, in rice some mechanism for ATP synthesis was maintained under anoxia and there seems little doubt that the ANPs included enzymes catalyzing production of ethanol and perhaps other end products of respiratory metabolism. One difference between embryos and roots of rice and the roots of maize and wheat may reside in the greater ability of the former to produce ATP and maintain a favorable AEC by fermentation reactions.

III. SHOOT GROWTH AND DEVELOPMENT

Flooding modifies almost every aspect of shoot behavior (Whitlow and Harris, 1979). The occurrence and extent of any particular response depends on many interrelated factors such as the species or cultivar, its age or stage of development, the duration, depth and timing of flooding, the soil type, and the conditions of temperature and light during or even before flooding.

The theme of our discussion is causality. In this way principles can be identified that bind conceptually many apparently disparate effects of flooding on shoot systems. At the simplest level are direct responses of shoot tissue to flooding of its external environment (e.g., underwater growth). More complicated are phenomena induced indirectly in the shoot when only the root environment is waterlogged and becomes O_2 deficient. Here, flooding effects on the shoot arise from modifications to the *internal* flow of substances between root and shoot. We distinguish three sorts of internally transmitted chemical messages: (1) increases in supply of substances from the flooded roots or soil to the shoot (positive messages), (2) decreased supply of substances to the shoot (negative messages), and (3) accumulation in the shoot of substances usually transported down to the roots (accumulation messages; (Cannell and Jackson, 1981; Bradford and Hsiao, 1982; Jackson and Kowalewska, 1983). Water, photosynthate, inorganic nutrients, hormones or their precursors, and toxins are the substances most involved. We indicate which of their effects appear conducive to survival of the plant (acclimatization) or which constitute injury that may prejudice recovery and set lower limits on postflooding performance. Some acclimatic responses are among the first reactions to inundation (i.e., within minutes or hours), modifying the plant's ability to grow and survive if the stress is extended in time.

A. Inhibition of Growth

Many of the responses of shoots to flooding, such as hypertrophy, adventitious rooting, epinasty, and abscission, discussed subsequently, involve an element of growth stimulation, but they usually occur within a context of an overall decrease in shoot growth. This decrease is undoubtedly the consequence of the gamut of changes in the movement of minerals, water, hormones, toxins, and assimilates that anoxic incapacitation of the roots can bring about. Because it is difficult to ascribe particular causal mechanisms, only selected aspects of growth inhibition are considered here.

1. Inhibition of Leaf Growth

Leaf growth is extremely sensitive to flooding and root anoxia (Trought and Drew, 1980a). Sojka *et al.* (1975) found an 83% decrease in leaf area of wheat after 25 days of root anoxia at 21°C, and a 76% decrease at 9°C (see also Purvis and Williamson, 1972), both attributable to inhibitions of leaf expansion and tiller production. Leaf expansion can slow within 20–40 min (Wenkert *et al.*, 1981), an effect ascribed by Brouwer (1977) to water stress because the inhibition is restricted to the photoperiod during the first few days. However, Wenkert *et al.* did not find any link with water stress. In common with Trought and Drew (1981), they believed deficiencies in nitrogen and other major nutrients are a principal cause, especially in the medium term (up to 7 days). Ethylene may also contribute, because the basal growing zone of graminaceous leaves is especially vulnerable to ethylene from the soil or trapped in the base of the plant by water (Jackson *et al.*, 1981). This may favor development of a smaller shoot that is commensurate with the reduced size of the functional root system.

2. Inhibition of Stem Extension

Inhibition of stem extension is almost always observed when soil is waterlogged (see reviews cited in Section I). In tomato, its partial recovery when new adventitious roots emerge (Reid and Crozier, 1971) or its prevention for at least 72 hr if such roots are induced prior to flooding (Jackson and Campbell, 1979) suggests that negative messages may be involved, at least to begin with (e.g., lack of gibberellin hormones). But persistent slow extension will have multifarious causes. The importance of hormones will diminish as the effects of toxins, deficiencies in water, inorganic nutrients, or assimilates begin to take effect. An involvement of ethanol passing from anaerobically respiring roots to shoots (Fulton and Erickson, 1964) has not been substantiated by more recent research (Jackson *et al.*, 1982; Harberd and Edwards, 1982; but see Smucker and Adler, 1980).

3. Inhibition of Photosynthesis

Anoxic root environments depress increase in dry weight (e.g., Sojka *et al.*, 1975; Jackson and Campbell, 1979; Trought and Drew, 1980b). Surprisingly, this may be preceded by an increased dry-weight accumulation (Trought and Drew, 1980a,b), possibly arising from inhibition of dark respiration (Wiedenroth and Poskuta, 1981) and depressed export of photosynthate to the roots (Nuritdinov and Vartapetian, 1980), presumably caused by a lack of sink capacity and injured phloem (Quereshi and Spanner, 1973).

Waterlogging undoubtedly depresses net carbon fixation per unit of leaf area when calculated as a net assimilation rate (Trought and Drew, 1980a) or estimated by $^{14}CO_2$ assimilation (Cannell *et al.*, 1977). Bradford (1982) deduced that inhibition of ribulose-bisphosphate carboxylase/oxygenase production is partly responsible, together with stomatal closure. Moldau (1973) came to a similar conclusion. A negative message in the form of cytokinin deficiency was thought by Bradford to be a possible cause. Indeed, foliar application of cytokinin can increase the gain in shoot dry weight of flooded tomato plants (Jackson and Campbell, 1979), although the effect is not large.

B. Promotion of Extension Growth

Although flooding or submergence is generally thought to be inhibitory to growth, some species are stimulated to elongate their stems or petioles more quickly under these conditions. Most, but not all, are aquatic plants or marsh dwellers. Ridge and Amarsinghe (1981) tested 20 such species. All but 2 extended faster under water than above it and these included *Ranunculus repens* and *Plantago major,* common weeds of both wet *and* well-drained sites.

The reaction time to submergence can be less than 20 min and is reversible by refloating (Jackson, 1971). Musgrave *et al.* (1972) showed that small concentrations of ethylene (>0.01 μl $liter^{-1}$) applied to floating plants of the starwort (*Callitriche platycarpa*) promoted stem extension in a manner exactly similar to submergence. The ethylene response, like that of submergence, involves cell extension rather than division and is also reversible by removing the gas (Jackson, 1982). Furthermore, naturally produced ethylene in the gas phase of submerged plants exceeds the maximum concentration needed for fastest growth (1 μl $liter^{-1}$). Figure 9 summarizes some of these findings. Depth-controlled elongation in the starwort therefore seems to be mediated by endogenous ethylene trapped in the submerged tissue, the accumulation occurring because radial diffusive losses of the gas to the surroundings are slowed 8000- to 1000-fold by the water covering. The fast growth results in the shoot apex reaching the water surface quickly. On arrival the shoot can easily ventilate to the aerial environment, thus depleting the store of growth-promoting ethylene and slowing extension as a consequence. The system appears to confer competitive and survival advantages, permitting photosynthesis to proceed in the highest light intensity, and allowing flowers to be pollinated. It is an example of a direct effect of excess water around the shoot mediated by an entrapped gas, in this case ethylene.

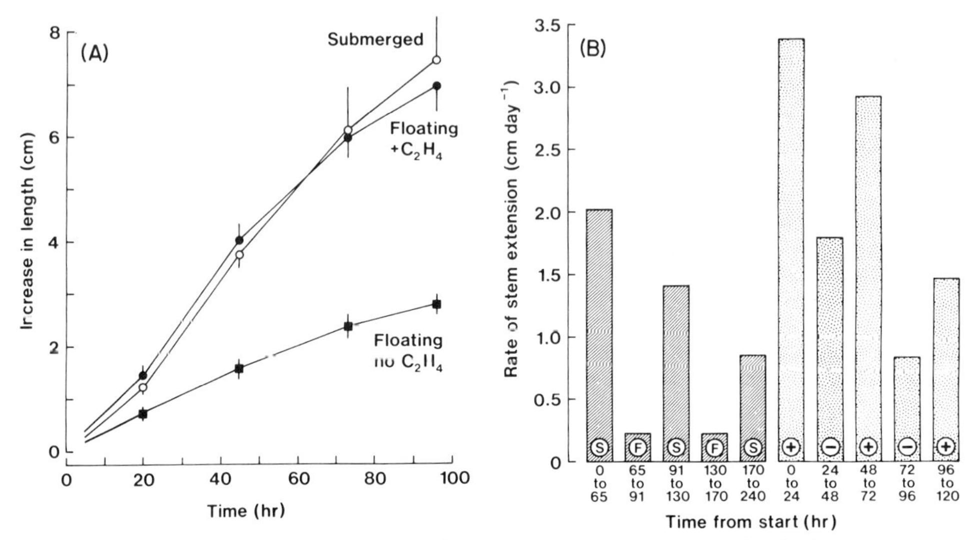

Fig. 9. Effect of ethylene (1.5 μl $liter^{-1}$) or submergence on stem extension in the starwort *Callitriche platycarpa*. (A) Continuous treatment. (B) Intermittent treatment. S, Submerged; F, floating; +, C_2H_4 supplied; −, C_2H_4 withheld. From Jackson (1982).

Unlike many other responses to flooding, the exclusion of O_2 is not a necessary feature because the rapid extension can take place equally in well-aerated or stagnant water (McComb, 1965). The possibility that a reduction in light intensity plays a part in submergence responses has yet to be examined.

This hypothesis based on work with *Callitriche platycarpa* has been confirmed and extended to the petioles of *Ranunculus sceleratus* (Musgrave and Walters, 1973; Horton and Samarakoon, 1982), the rachis of the semiaquatic fern *Regnellidium diphyllum* (Musgrave and Walters, 1974), the water lily *Nymphoides peltata* (Malone and Ridge, 1983), internodes of floating or deep-water rice (Metraux *et al.*, 1982), petioles of *Ranunculus repens* and other species (Ridge and Amarsinghe, 1981), and rice coleoptiles (Ku *et al.*, 1970; Ishizawa and Esashi, 1983). Knowledge of these reactions may help in the chemical control of troublesome aquatic weeds of reservoirs and watercourses (Boyd, 1971; Jackson, 1983).

In *Callitriche platycarpa* the presence of endogenous gibberellins seems necessary for ethylene action in the submergence response to ethylene (Musgrave *et al.*, 1972), whereas in others (e.g., *Regnellidium diphyllum*) auxin, perhaps transported from the leaf lamina (Horton and Samarakoon, 1982), is a prerequisite together with tension forces created by buoyancy under water (Musgrave and Walters, 1974; Cookson and Osborne, 1979). One likely mode of action for ethylene is a decrease in cell ψ brought about by loosening of the cell wall

(Cookson and Osborne, 1979) through active acidification (Malone and Ridge, 1983). Inhibitors of ethylene production or action when applied in nontoxic amounts prevent the growth response to submergence or to ethylene (Cookson and Osborne, 1978; Jackson, 1982). This seems to confirm the involvement of the gas in the "drowning avoidance" mechanism.

The intriguing effect of submergence on transforming the shape of new leaves (heterophylly) to a characteristic underwater form remains largely unexplained. In *Callitriche platycarpa* ethylene does *not* reproduce the effect of submergence on leaf shape. Allsopp (1961) believed that in *Marsilea drummondii*, differences in carbohydrate content may be responsible.

Flooding and submergence have been reported to promote growth in other kinds of plants. For example, Rozema and Blom (1977) measured an 84% stimulation of stolon length when *Agrostis stolonifera*, a flood-tolerant grass, was submerged in 5 cm of water for 6 weeks, and Stelzer and Läuchli (1977) found that *Puccinellia peisonis*, another flood-tolerant grass, grew 70% larger in nonaerated than in aerated solution cultures. These effects are unexplained.

C. Hypertrophy

Swelling of the stem base or hypocotyl is a common occurrence in flooded herbaceous species. It is a result of cell enlargement in the cortex, often accompanied by collapse of some cells to form gas-filled spaces (Kawase, 1981a). Hypertrophic expansion of lenticels and the emergence of adventitious roots are also sometimes seen on hypertrophied stems (e.g., *Ipomoea crassicaulis;* Mishra, 1973). Hypertrophy has also been reported in tomato (Kramer, 1951), sunflower (Phillips, 1964a), maize (Kuznetsova *et al.*, 1981), and alsike clover (Arikado, 1955a). A hypertrophic hypocotyl of sunflower flooded to the height of the cotyledonary node is illustrated in Fig. 10, together with a close-up of swollen cells of a hypocotyl of flooded soybean. In his first paper on this subject, Kawase (1974) classified hypertrophy as "flooding injury" but later (Kawase, 1979) considered that because of the associated air space it may "enhance the air diffusion system of mesophytic plants from leaves to roots [p. 183]" and thus be an acclimatic phenomenon, increasing tolerance to waterlogging. In Yushin, a high-yielding cultivar of Korean rice, a paucity of air space at submerged internodes and resultant lack of root aeration probably explains low flood tolerance [International Rice Research Institute (IRRI), 1979].

Both Kawase (1974) and Wample and Reid (1975) examined the causes of hypertrophy in sunflower. The latter proposed that the presence of excess water, independent of any effect of partial or complete O_2 deficiency of the roots or stem, induced hypocotyl hypertrophy. They found that flooding to the cotyledonary node with either stagnant or aerated water promoted hypocotyl expansion equally well, commencing within 24 hr of inundation (Wample and Reid, 1975).

Fig. 10. (Left) Swollen, hypertrophic hypocotyl of sunflower flooded to the cotyledonary node for 2 weeks. From M. B. Jackson and J. H. Palmer (unpublished). (Right) Scanning electron micrograph of swollen cells of hypertrophic hypocotyl of flooded soybean. From C. E. Pankhurst and J. I. Sprent (unpublished).

Unpublished experiments by M. B. Jackson and J. H. Palmer confirmed that aerated water around the hypocotyl induces hypertrophy. Indeed, hypertrophy is probably oxygen requiring because aeration of the surrounding water can intensify its development (Kawase, 1978). Both Kawase (1974) and Wample and Reid (1979) detected enhanced concentrations of ethylene in the gas vacuum extracted from the swelling tissue. Kawase reported that Ethephon, an ethylene-releasing chemical, applied to sunflower plants (albeit at a high concentration of 400 mg liter^{-1}) promotes hypertrophy, suggesting that ethylene trapped within the water-covered hypocotyl of flooded plants may cause this reaction as well as the associated development of air space. It has since been found that ethylene per se at small, physiological concentrations of 1–5 μl liter^{-1} can indeed promote swelling of the hypocotyl (Kawase, 1981b), as can auxin (Jackson and Palmer, 1981), another hormone that accumulates at the stem base of flooded sunflowers (Wample and Reid, 1979). Auxin and ethylene may stimulate the activity of cellulase enzymes that weaken cell walls, favoring cell expansion followed by cellular collapse (Kawase, 1979, 1981b). The stem pithiness observed in flooded tomatoes (Aloni and Rosenshtein (1982) could have a similar cause.

The available evidence, therefore, suggests that hypertrophic growth in sunflower, as with extension growth of water plants, results principally from the trapping in of ethylene by water, with the possibility of a supplement of ethylene moving in from the soil. The roles of CO_2 and hypoxia as promoters of hypertrophy remain to be determined.

D. Adventitious Rooting

1. *Occurrence*

Accelerated rooting of this type, from the submerged part of the stem or hypocotyl, has been reported in many monocotyledons and dicotyledons in both field and laboratory conditions. Examples include barley and alsike clover (Arikado, 1955a,b), wheat (Karishnev, 1958), *Zea mays* (Jat *et al.*, 1975; Jackson *et al.*, 1981), rice (Kar *et al.*, 1974), the amphibious *Ludwigia peploides* (Ellmore, 1981), the grass *Holcus lanatus* (Watt and Haggar, 1980), tomato, and sunflower (Kramer, 1951). The response in sunflower after several weeks of flooding is shown in Fig. 11. Bergman (1920) observed adventitious rooting in non-wetland mesophytes such as *Pelargonium, Phaseolus,* and *Coleus* after ~10 days of flooding and noted its association with death of the original root system below. In the hydrophytes *Ranunculus sceleratus, R. abortivus,* and *Cyperus alternifolius,* the original roots survived flooding and *no* adventitious roots emerged. Thus flood-induced rooting from the stem base is probably best considered as regenerative, occurring when the original roots succumb to anoxia.

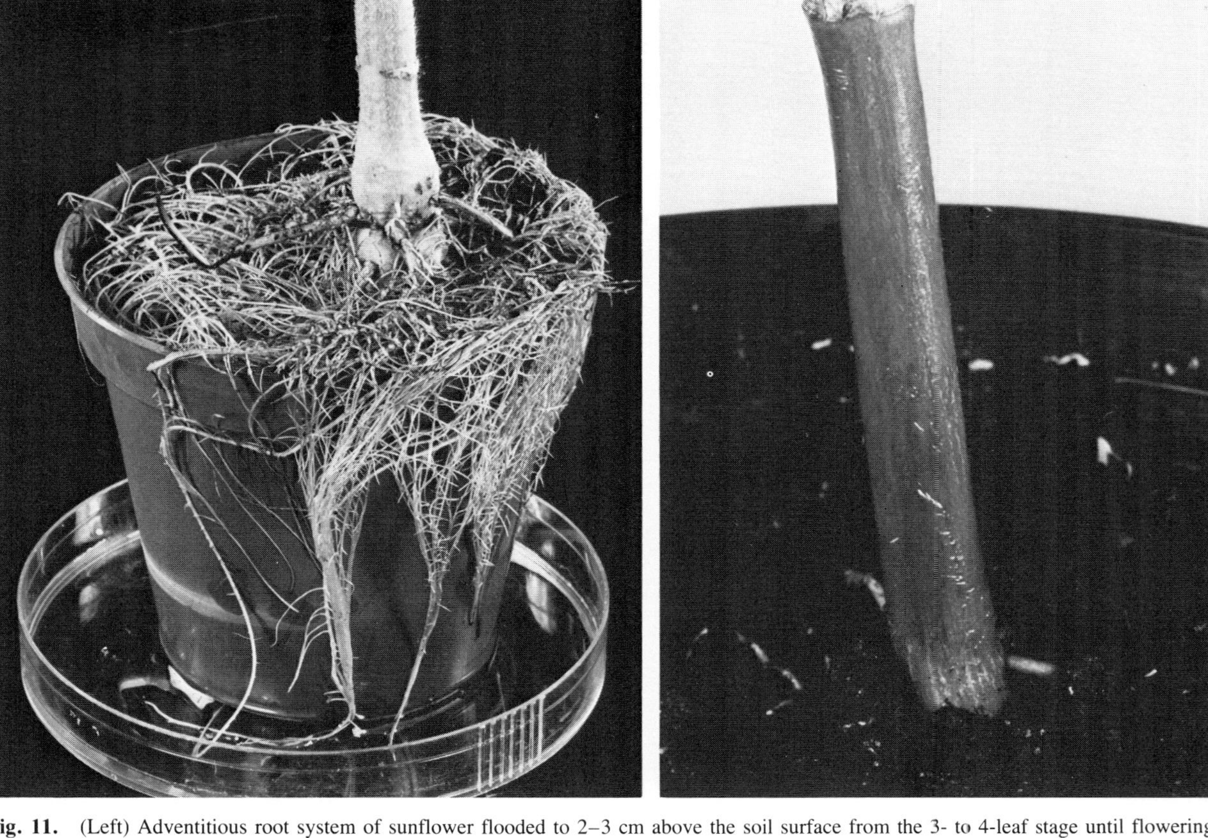

Fig. 11. (Left) Adventitious root system of sunflower flooded to 2–3 cm above the soil surface from the 3- to 4-leaf stage until flowering. (Right) Hypocotyl of plant grown in well drained soil. From M. B. Jackson and J. H. Palmer (unpublished).

2. *Acclimatic Significance*

Most authors consider that the new roots replace many functions of the original, defunct root system and thus aid survival and partial recovery. Circumstantial evidence for this is substantial. Bergman (1920) and Kramer (1951) observed shoot regrowth or new leaf emergence in flooded mesophytes *after* adventitious roots had formed at the stem base. Sartoris and Belcher (1949) found that flood-tolerant cultivars of sugarcane produced the most copious adventitious roots, and a similar relationship is known for *Zea mays* (Jat *et al.*, 1975). In tomatoes, adventitious root emergence coincides with recovery from leaf curling (M. B. Jackson, unpublished), resumption of leaf growth (Aloni and Rosenshtein, 1982), and upward flow of gibberellins into the shoots (Reid and Crozier, 1971). In flooded maize, severe reductions in transpiration and leaf ψ are relieved when adventitious roots emerge (Wenkert *et al.*, 1981). Kramer (1951) found that those tomato plants with the most adventitious roots recovered best from 9 days of flooding, a treatment that almost killed tobacco (a species with only limited capacity for replacement rooting). Intolerance of flooding by peas (Jackson, 1979a), *Lolium perenne* (Watt and Haggar, 1980), and genge (Arikado, 1955a,b) is also associated with failure to form adventitious roots.

The evidence just cited lacks direct demonstration of the survival value of adventitious roots; cause and effect are not distinguished—new roots may form because the plant is surviving or the plant survives because adventitious roots have developed. Some evidence also runs contrary to that expected. For example, Wample and Reid (1978) found that 25 days of flooding can be fatal to sunflower despite the formation of many adventitious roots, and Kuo and Chen (1980) reported that the *least* flood tolerant of 20 cultivars to tomato produced the most roots. In contrast, Jackson and Palmer (1981, and unpublished) found that sunflowers grown in potting compost survived many weeks of flooding and flowered successfully. Daily removal of emerging adventitious roots, however, almost killed the flooded plants. In tomato, the inhibiting effect of flooding for 72 hr on shoot extension can be offset by inducing an adventitious root system to grow into a well-aerated medium before and during flooding (Jackson and Campbell, 1979). A similar result was previously obtained by W. T. Jackson (1955).

In reaching a conclusion concerning the acclimatic significance of adventitious rooting, it is important to realize that it is least likely to releive effects such as epinasty, stomatal closure, and certain deficiencies in nutrients (Sections III,D,1 and III,E,2,a,b) that take place during the earliest stages of flooding. This is because the source of the message (anoxic roots or soil) delivers the damaging effect before the new roots have emerged.

On balance, evidence favors a vitally important role for adventitious roots in the survival of many mesophyte species by replacing the functions of the original roots, such as supplying water, nutrient, and hormones and acting as a sink for shoot metabolites. Complete replacement of function, however, is improbable because of the inevitable delay in the formation of the new root system and its limited size (Trought and Drew, 1980a), its restriction often to regions that may be nutrient poor (e.g., in rice; Vamos, 1957), and its vulnerability to subsequent drying following drainage (Wenkert *et al.*, 1981).

3. *Acclimatic Features*

Adventitious roots probably survive while the original roots die because the new roots emerge from the stem close to the water surface where water and O_2 are available and where anaerobically generated toxins are absent. In many cases, their physical connections with hypertrophic, aerenchymatous stem tissue (see Kawase and Whitmoyer, 1980) may also aid their aeration because gas space in the stem facilitates internal aeration. Often, the adventitious roots themselves are also permeated by aerenchyma [e.g., maize (Drew *et al.*, 1979a), wheat (Trought and Drew, 1980b), sunflower (Kawase and Whitmoyer, 1980), rice (John, 1977), and *Spartina alterniflora* (Teal and Kanwisher, 1966)] that could interconnect with that of the stem, although information about such connections is often lacking (Section II,C,5). Despite their hollow internal structure (Fig. 4), aerenchymatous roots are able to absorb nutrient ions or their analogs (Drew *et al.*, 1980) and to increase the activity of adenylate kinase (Rodionova *et al.*, 1977), a key enzyme regulating the production of ATP from other adenine nucleotides.

Other morphological features contributing to root survival include a capacity to float (Yu *et al.*, 1969) and grow horizontally (diageotropism). Both attributes maintain the roots close to the air–water interface where sufficient dissolved O_2 is present to support growth and functions such as nutrient uptake and hormone production. The environmental factors that favor diageotropic growth are largely unknown (Jackson and Barlow, 1981) but may be related to other disturbances to normal root orientation brought about by flooding (Section II,A). In *Ludwigia peploides* both upward- and downward-growing roots extend from submerged nodes (Ellmore, 1981). The former are aerenchymatous, and their high O_2 content suggests an aeration function.

4. *Causes of Adventitious Rooting*

In most instances flood-induced rooting occurs at or just below water level. Here the stem or hypocotyl is subject to direct effects of a water covering *and* to the consequences of anoxia on the original roots below. It is thus a zone of rapidly changing biochemistry (Grineva and Nechiporenko, 1977), where anabolic processes necessary for primordium formation and outgrowth may proceed rapidly because O_2 is readily available (Zimmerman, 1930; Kordan, 1976).

Adventitious rooting has been studied most closely in sunflower hypocotyls, in which primordia are initiated within 24 hr of waterlogging (Wample and Reid, 1978). However, its causes remain unclear. Kawase (1974) found increases in ethylene below water in taproot and hypertrophied hypocotyl while adventitious roots were emerging (Fig. 12). The ethylene found within the hypocotyl may have been trapped there by the floodwater. He concluded that "the high ethylene concentration results in . . . new root formation in hypocotyls in the early stages of flooding [p. 36]." Unfortunately, Kawase did not test the activity of ethylene in root formation, although he showed that Ethephon, the ethylene-releasing chemical, could induce other symptoms of flooding such as hypertrophy.

Because aeration of the floodwater does little to reduce rooting or ethylene content (Kawase, 1978) it seems that a direct effect of an aerated water covering is sufficient. This has been confirmed independently by Wample and Reid (1978, 1979). However, no one has yet demonstrated that ethylene can induce rooting from the hypocotyl of intact sunflowers. Wample and Reid (1979) failed using Ethephon, and 1–5 ppm ethylene gas applied for 5 days also was without a promoting effect (M. B. Jackson and J. H. Palmer, unpublished). Similar lack of activity by ethylene was also reported in mung bean (Batten and Mullins, 1978) and *Eucalyptus* spp. (Clemens *et al.*, 1978). In some circumstances ethylene can even inhibit root formation (Mullins, 1972; Coleman *et al.*, 1980).

Zea mays differs from sunflower by having *preformed* root primordia at stem nodes. These emerge soon after flooding commences in association with more ethylene within the plant. In this species ethylene *can* speed up the process

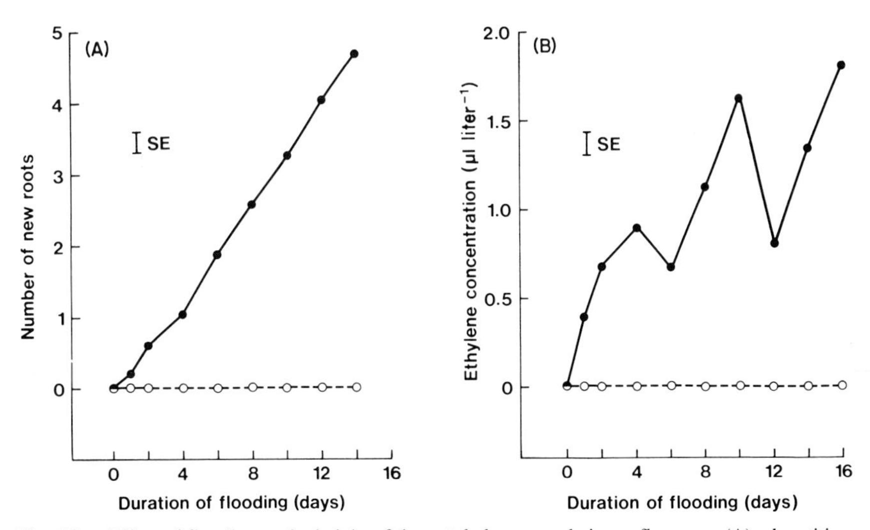

Fig. 12. Effect of flooding to the height of the cotyledonary node in sunflower on (A) adventitious roots formed from the hypocotyl above the soil and (B) ethylene concentration in gas extracted by partial vacuum from hypocotyl and taproot. ○, Control; ●, flooded. Redrawn from Kawase (1974).

(Jackson *et al.*, 1981). A similar situation probably applies in tomato (Jackson and Campbell, 1975a), although Zimmerman and Hitchcock (1933) claimed that ethylene stimulates root initiation in maize in addition to simply encouraging outgrowth.

In most flooding situations, the original roots are incapacitated by anoxia. Influences other than just a water covering of the stem base will then come to bear on the morphogenesis of this zone. For example, in sunflower, auxin transport from shoots to roots stops and the hormone accumulates in the shoot system (Phillips, 1964a; Wample and Reid, 1979). This transport is thought to occur principally in the phloem (Morris and Kadir, 1972). Auxin has a reputation for promoting new root formation (e.g., Jackson and Harney, 1970); unfortunately, such activity was not found in hypocotyls of intact sunflowers when a wide concentration range of the auxin IAA was applied (M. B. Jackson and J. H. Palmer, unpublished). The importance of auxin is therefore as uncertain as that of ethylene.

A second accumulation message that should stimulate rooting is a buildup of carbohydrates (Grineva and Nechiporenko, 1977; Trought and Drew, 1980b), probably brought about during the early stages of flooding by an imbalance in assimilate source and sink capacities (Wiedenroth, 1981). Active phloem transport (Kursanov, 1963) may also be stopped in roots by anaerobic soil conditions (Mason and Phillis, 1936), thus favoring a buildup of carbohydrates at the stem base. A similar effect is obtained by steam-girdling the stem (Stoltz and Hess, 1966). Sugars and high concentrations of internal carbohydrates are known to favor adventitious rooting (Gregory and Samantari, 1950; Stoltz and Hess, 1966).

Negative messages that influence adventitious rooting in flooded plants have been little studied. Fabijan *et al.* (1981) concluded from work with hypocotyls of 6-day-old sunflower seedlings under well-aerated conditions that the original roots normally supply substances that inhibit root formation in the hypocotyl. Flooding might well be expected to reduce the amount of such an inhibitor passing into the hypocotyl. Cytokinins and gibberellins are candidates for the inhibitor, and their application of flooded tomato plants diminished the rooting response (Selman and Sandanam, 1972; Railton and Reid, 1973; Jackson and Campbell, 1979).

E. Reorientation of Growth

This seemingly obscure class of responses warrants discussion because it contains several visually conspicuous and early reactions to flooding that may have acclimatic significance. Epinastic curvature of leaf petioles in tomato is probably the best-known effect of this type. It is certainly one of the most completely understood responses of shoots to soil waterlogging.

1. *Epinastic Curvature of Leaf Petioles*

a. Occurrence. The direction of nastic curvatures such as epinasty is determined predominantly by internal asymmetry of structure or physiology in the potentially bending organ. Gravity also plays some part in epinasty, although this is not well understood (Ball, 1969; Kang, 1979). In tomato (Crocker *et al.*, 1932) and sunflower (Palmer, 1975), petiole epinasty comprises an acceleration of growth in the adaxial half of the petiole causing the leaves to swing downward. Shoot geometry is thus much altered (Fig. 13B,C). Epinasty is a turgor-requiring growth process, quite distinct from the wilting that flooding can occasionally induce (Section III,G). Flooding induces epinasty in potato, tomato (Turkova, 1944), sunflower (Phillips, 1964a), *Mercurialis perennis* (Martin, 1968), and probably many other species, although testing has been limited. In tomato, growing at 18–25°C, curvatures develop most rapidly between 24 and 48 hr of flooding (Jackson and Campbell, 1976). The bending takes place at the base of the petiole in older leaves but can involve the whole of young growing leaves, causing them to curl.

b. Causes of Epinasty. The early physiological work was carried out with ethylene as an atmospheric pollutant (Harvey, 1913; Doubt, 1915). Crocker *et al.* (1932) identified 89 species or cultivars showing ethylene-induced epinasty. The effect on tomato is shown in Fig. 13B. A link between these studies and unfavorable soil conditions was made by Neilson-Jones (1935), who extracted gases from infertile organic soil that promoted epinasty when supplied to tomatoes. Neilson-Jones speculated that any ethylene present within the soil could influence plants growing in it. Similarly, Turkova (1944) believed that the epinasty seen in flooded field crops in the Alma-Ata region of U.S.S.R. could be ascribed to ethylene. W. T. Jackson (1956) also mentioned this possibility. Support for these ideas was forthcoming after gas chromatographic methods sensitive to the extremely small concentrations of ethylene needed to affect plants (>0.001 μl liter^{-1}) were developed. The technique enabled Smith and Russell (1969) to find up to 10.6 μl liter^{-1} ethylene in flooded soil and Musgrave *et al.* (1972) and Kawase (1972) to detect increases of similar magnitude in submerged parts of plants. Kawase (1972) also found slightly elevated ethylene concentrations immediately above the waterline in stems of flooded tomatoes, suggesting that gas trapped within submerged tissue could move internally to aerial parts. This was later shown to be possible by experiments with radioactive ethylene (Jackson and Campbell, 1975a). Sufficient quantities of the gas may move from roots to leaves to promote epinastic curvature when more than 2 μl liter^{-1} are present around the roots. In this situation ethylene from flooded soil would be a positive message, giving an obvious epinastic reaction ~3 hr after arrival at the target tissue.

Fig. 13. Reorientation responses in tomato to ethylene or anaerobic treatments to the roots. (A–C) Epinasty in cultivar Ailsa Craig. (D–F) Reorientation of shoot extension in the diageotropic mutant. (A,D) Controls. (B,E) Ethylene treated. (C,F) Roots anaerobic. From Jackson (1979b).

Supporting evidence of ethylene's involvement in epinasty of flooded plants includes the temporal and spatial relationship between epinasty and increases in ethylene in leaves (Fig. 14) and the ability of inhibitors of ethylene action such as CO_2, benzothiadiazole, DIHB, silver nitrate, benzyladenine, and gibberellic acid to inhibit flood-induced epinasty (Jackson and Campbell, 1976, 1979; Bradford and Dilley, 1978; Wilkins *et al.*, 1978).

Movement of ethylene per se from flooded soil to the leaves is *not* the only or indeed the principal mechanism by which leaves of flooded tomato plants become enriched with ethylene and develop epinasty. Ethylene produced by the leaves themselves is also involved. Increased synthesis of the gas occurs in leaves when the roots are deprived of O_2, not only in flooded soil but also in solution cultures containing less than 0.03 atm O_2 or in soil flushed with nitrogen gas to exclude O_2 and preclude ethylene accumulation around the roots (Jackson and Campbell, 1976; Bradford and Dilley, 1978). The presence of O_2-deficient roots is essential for the increases in ethylene and epinasty in the leaves to take place.

What then is the causal link between the anaerobic roots and the elevated rates of foliar ethylene production? Anoxic roots behave as if they are sources of an

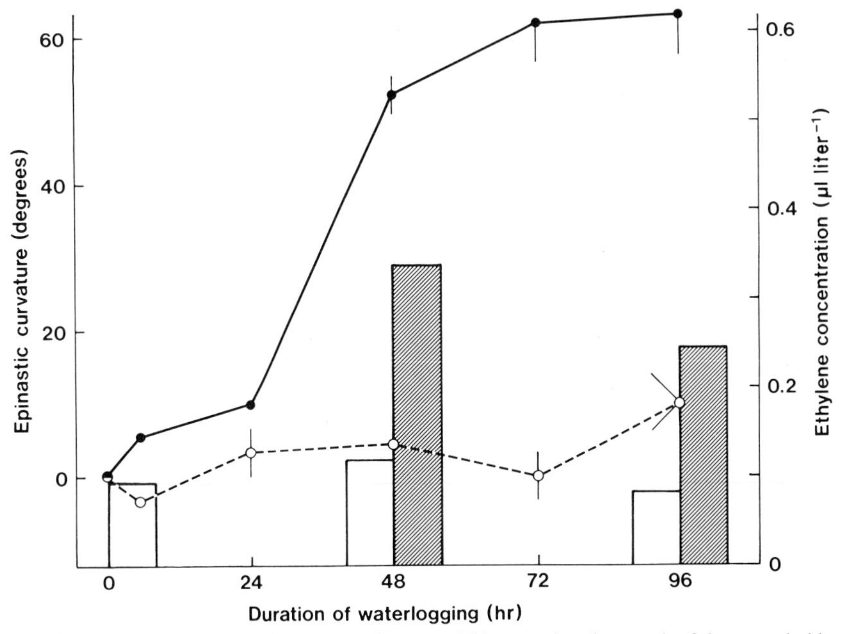

Fig. 14. Effect of flooding tomato plants for up to 96 hr on epinastic growth of the second oldest leaf (●, flooded; ○, control) and the concentration of ethylene in the oldest three leaves (□, control; ▨, flooded). Adapted from Jackson and Campbell (1976).

ethylene precursor (Jackson and Campbell, 1975b, 1976; Jackson *et al.*, 1978). By analogy with earlier work with apples (Burg and Thimann, 1959), Jackson and Campbell proposed that because anoxia halts ethylene production in the roots themselves, a precursor of ethylene biosynthesis probably accumulates at an oxygen-requiring step. The characteristic leakiness of anoxic roots would then allow the precursor to be drawn into the transpiration stream and up to the shoot, where readily available O_2 would permit rapid conversion to ethylene. Split-root experiments (Jackson, 1980) indicate the importance of interconnecting xylem as the pathway for movement. A rapid closing of stomata (Jackson *et al.*, 1978; Aloni and Rosenshtein, 1982) may help conserve the ethylene generated.

These findings were placed on a biochemical basis by Bradford and Yang (1980a). They identified the precursor as ACC and found that its transport in xylem sap from roots to shoots preceded increased ethylene production and epinasty in leaves. This was shown for flooded tomato plants and also those grown in anaerobic solution cultures (Amrhein *et al.*, 1982). The ACC measured in the xylem sap (73 μl liter^{-1}) was sufficient to stimulate ethylene biosynthesis and the subsequent epinasty. These workers also suggested that O_2 deprivation may even stimulate ACC production in roots in a manner analogous to effects of this kind reported for other forms of chemical and physical wounding.

Epinasty is not of course entirely under ethylene control. The requirements in tomato for a leaf lamina (Needham, 1969) and in sunflower for a shoot apex (Phillips, 1964a,b) indicate that other inputs are needed from elsewhere in the plant. Phillips's work points to an auxin requirement and also to an epinasty-inhibiting role of gibberellins that would normally come from the roots. The capacity of gibberellic acid and a synthetic cytokinin supplied to leaves to decrease epinasty in flooded or ethylene-treated plants also suggests this (Selman and Sandanam, 1972; Railton and Reid, 1973; Jackson and Campbell, 1975b, 1979). In contrast, experiments with plants with two horizontally separate root systems have not supported the view that cytokinins and gibberellins originating from the roots play a significant role in suppressing epinasty (Jackson and Campbell, 1979).

c. Acclimatic Features of Epinasty. In the first hours or days of soil flooding, plants face an increased risk from foliar water stress caused by a temporary increase in the resistance to water flow through roots to shoots (Section III,G). The acclimatic significance of epinasty may be to minimize the risk by decreasing the radiation load and thus lowering evaporative demand at this critical time. Oppenheimer (1960) calls such leaf postures "sun's-rays evading," and there is limited experimental evidence that the amount of incident radiation is decreased, especially around midday (Waggoner, 1966; Begg, 1980). However, we are unaware of any published demonstration of an effect of epinasty in flooded plants on the interception of radiation or its consequences for plant water relationships, despite claims for its acclimatic value (e.g., Smith and Jackson, 1974; Bradford and Yang, 1980a). The appropriate tests are overdue.

2. *Other Examples of Shoot Reorientation Involving Ethylene*

The diageotropic mutant of tomato usually elongates horizontally, but applying ethylene (Zobel, 1973) induces upright growth (Fig. 13E). The effect is reversible by withdrawing ethylene (Jackson and Campbell, 1975b). Exposing the roots to anoxic conditions typical of flooded soil also gives rise to an upright posture (Fig. 13F) in association with an abnormally high ethylene content in the leaves (Jackson, 1979b). Bradford and Yang (1980b) found that ACC transported from the roots was probably responsible for the ethylene enrichment. The mutant is a particularly efficient converter of ACC into ethylene.

A response of greater acclimatic significance is seen in the stolons of the flood-tolerant strawberry clover (*Trifolium fragiferum*). These plants are naturally prostrate in well-drained conditions, but when covered with water they rapidly grow erect (Bendixen and Peterson, 1962). This feature is absent from Ladino clover, which is killed by overwet habitats. Applying ethylene (50 μl liter^{-1}) to prostrate plants reproduces the effect of submergence within 24 hr (Hansen and Bendixen, 1974). In submerged stolons, ethylene presumably accumulates because of the "water jacket" effect and in this species a marked alteration to gravity responsiveness is brought about by the gas. Measurements of ethylene in submerged stolons and experiments with inhibitors of ethylene production or action have, however, not been carried out. As with many other reactions to ethylene connected with rapid responses to flooding (Jackson, 1982), reorientation of *T. fragiferum* stolons involves a stimulation of cell extension (Hansen and Bendixen, 1974).

F. Senescence

1. *Occurrence*

Senescence is a blanket term describing the deterioration that gives rise to the death of part or all of a plant (Wang and Woolhouse, 1982). In leaves, yellowing followed by necrosis are the usual visual symptoms. There are many reports of accelerated senescence in flooded plants [e.g., *Impatiens*, beans, and *Pelargonium* (Bergman, 1920), tobacco and tomato (Kramer, 1951), sunflower (Kawase, 1974), carrot (Olymbios and Schwabe, 1977), barley (Drew and Sisworo, 1977), peas (Jackson, 1979a), wheat (Trought and Drew, 1980a), and maize (Wenkert *et al.*, 1981)]. But the effect is not always seen. Gilbert and Shive (1942) did not observe chlorosis in soybean with roots exposed to O_2 deficiency even after 23 days, and in *Phaseolus vulgaris*, Hiron and Wright

(1973) found that leaves were a darker green after 4 days of flooding. Unlike many of the responses of shoots described previously, senescence probably has no one overriding cause. Thomas and Stoddart (1980) remarked succinctly that "the diversity of living things is matched by the variety of deaths that await them [p. 104]."

The biochemical and ultrastructural basis of ordinary, aging-related senescence has been thoroughly described (Gahan, 1982). It comprises a regulated dismantling of cellular macromolecules and membranes under genetic control from nuclear, chloroplast, and mitochondrial genomes that is keyed into the development of the plant as a whole (Thomas and Stoddart, 1980). There is, however, no certainty that the internal processes of deterioration are the same as this in flooded plants, but merely accelerated. In some instances the syndrome may be induced by a different pathway dominated by external influences such as anaerobically generated toxins. In addition to positive messages of this kind, negative and accumulative messages are also involved. The possibilities are discussed here.

2. *Senescence and Negative Messages*

a. Nitrogen Deficiency. Four to five days of soil waterlogging gives a noticeable loss of chlorophyll in the oldest leaves of barley plants (Leyshon and Sheard, 1974). The yellowing starts at the tip of the leaf and is *preceded* by reduction in the concentrations of free nitrate and total nitrogen in the tissue (Table III). A halt to nitrogen uptake by the shoot after 2 days of flooding (Drew and Sisworo, 1977) and an apparent movement of nitrogen out of the oldest leaves to younger parts seem to cause the chlorosis. The same symptoms are reproduced by starving plants of nitrogen by other means, such as root excision or transferring plants to nitrogen-deficient nutrient solution. Furthermore, supplying nitrate fertilizer to the surface of flooded soil, where a few roots remain alive to absorb ions, prevents chlorosis and associated nitrogen deficiency (Drew *et al.*, 1979b). Thus an important role for nitrogen deficiency in causing chlorosis in leaves of flooded barley has been demonstrated.

Studies with wheat show that the almost complete absence of dissolved O_2 in soil water after 1 day of flooding is mainly responsible for inhibition of nitrogen absorption (Trought and Drew, 1980a). This occurs because anoxia disrupts cell structure and production of chemical energy required for the transport of nitrate and ammonium from the soil and across roots to the stele (Section II,C,6). Drew and co-workers were unable to demonstrate a significant contribution from the soil of likely senescence-promoting substances such as CO_2, ethylene, lower aliphatic acids, manganese, and iron. However, when soil is particularly rich in such substances, they must inevitably play some part, but this is least likely in cases of temporary short-term flooding typical of temperate agriculture (cf. rice

TABLE III

Effect of Flooding on Chlorophyll and Nitrogen Concentration in Leaves of 13-Day-Old Barley Plants[a]

Time (days)		Chlorophyll concentration (abs. mg^{-1} fresh wt) Leaf 2	Leaf 3	Nitrate concentration per shoot ($\mu mol\ g^{-1}$ dry wt)	Total nitrogen concentration per shoot ($\mu mol\ g^{-1}$ dry wt)
0		31	29	790	3470
2	Control	26	26	774	3750
	Flooded	26	26	409[b]	3190[b]
4	Control	31	30	668	3170
	Flooded	22[b]	24[b]	148[b]	1768[b]
8	Control	25	27	644	3074
	Flooded	3[b]	13[b]	38[b]	1188[b]

[a]From Drew and Sisworo (1977).
[b]Significantly different from controls; $p < .05$.

culture). Potassium and phosphorus concentrations also decline in flooded wheat plants, but this is less important than for nitrate in controlling senescence. Supplementary feeding of potassium and phosphate via a single aerobic root does not prevent chlorosis, in contrast to nitrate (Trought and Drew, 1981). Loss of nitrogen-uptake capacity is thus a dominant negative message inducing early chlorosis of older leaves in flooded wheat and barley plants and possibly also maize (Wenkert *et al.*, 1981). A major unsolved question is the mechanism by which nitrogen is remobilized from older to young leaves, thus precipitating senescence in the lower leaves and ensuring that they fail first. Such leaves are likely to be the least efficient photosynthetically. Their loss perhaps provides a means of supplying nitrogen to young, more efficient parts of the shoot. Any loss of nitrogen from the soil by anaerobic denitrification by bacteria (Broadbent and Clark, 1965), by slowing of oxidative mineralization of soil organic matter (Wesseling, 1974, p. 7) or by leaching (Cooke, 1976), will also contribute to the nitrogen starvation.

In nodulated plants such as legumes, nitrogen deficiency can also be the result of inhibited nitrogen fixation and nodulation by the endophytic symbiont (Section II,C,3).

Other nutrient deficiencies that may promote senescence have been less studied. Flooding can induce temporary boron deficiency in sunflowers, causing necrosis in the leaves expanding during the first few days of the stress (Fig. 15). Supplements of boric acid to the soil completely eliminate the symptom (Jackson, 1983). Leaves emerging on flooded plants after adventitious roots are

Fig. 15. Leaf necrosis in flooded sunflower caused by boron deficiency. From M. B. Jackson and J. H. Palmer (unpublished).

formed (Section III,C) are not necrotic, presumably because the new roots are effective absorbers of the micronutrient.

b. Hormone Deficiency. Bioassays of hormones in xylem sap during the first 3 days of flooding show a marked decline in cytokinin- and gibberellin-like substances (Carr and Reid, 1968; Reid *et al.*, 1969). This is probably caused by depressed metabolic activity or death of root apices, the presumptive sites for synthesis of these hormones (Jones and Phillips, 1966; Burrows and Carr, 1969). Under less-severe conditions of O_2 starvation, nitrogen and phosphate deficiency may also favor a decrease in cytokinin production by roots (El-D; Salama and Wareing, 1979). These observations have significance for senescence because applications of xylem sap (Kulaeva, 1962; Kende, 1964), emergence of aerobic roots (Wheeler, 1971; Reid and Crozier, 1971; Engelbrecht, 1972) or application of gibberellins and cytokinins can retard leaf chlorosis in a variety of species and circumstances (e.g., Richmond and Lang, 1957; Fletcher and Osborne, 1966), including flooded tomato plants (Railton and Reid, 1973). Thus a deficiency in root-sourced gibberellins and cytokinins may accelerate leaf senescence in some plants. However, their role has not been reexamined with modern analytical techniques, and there is some skepticism concerning the importance of hormones in the transpiration stream in controlling shoot performance (King, 1976). This is because shoots are capable of producing cytokinins and gibberellins themselves and perhaps do not rely on roots to any great extent. It is notable that in flooded barley (Drew *et al.*, 1979b), spraying with a cytokinin gives only a small and short-lived retardation of leaf chlorosis.

Stomatal closure may form a link between leaf senescence and reduction in cytokinin flux from roots to shoots. Closure occurs soon after flooding or anaerobic treatment to roots [beans (Wright, 1972), wheat (Sojka *et al.*, 1975), tomato (Jackson *et al.*, 1978), and sunflower, cotton, and the xerophyte *Simmondsia chinensis* (Sojka and Stolzy, 1980); pea (Jackson and Kowalewska, 1983)]. Thimann (1980) argued cogently that leaf senescence is much accelerated by stomatal closure or at least varies in parallel with it. The reasons why this should be so are unclear (for a review see Wardle, 1982), but it is possible that closing of stomata in flooded plants accelerates senescence. Because applications of cytokinins keep stomata open (Livne and Vaadia, 1965; Meidner, 1967), even in

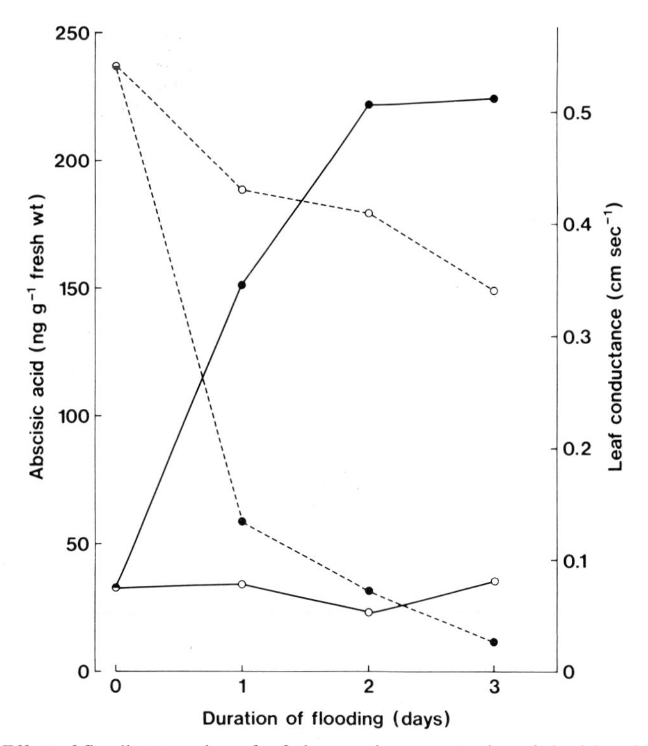

Fig. 16. Effect of flooding pea plants for 3 days on the concentration of abscisic acid (———) in the leaves, and leaf conductance (- - -; used as an estimate of stomatal opening). ○, Control; ●, flooded. Flooding commenced 1 hr after the start of the 18-hr photoperiod. From M. B. Jackson, K. C. Hall, and A. K. B. Kowalewska (unpublished).

flooded tomato plants (Jackson and Campbell, 1979), it is possible that deficiencies in root-sourced cytokinin may favor closure of stomata, thus accelerating senescence indirectly. Closure can certainly be induced by removing roots or by detaching leaves from the plants and incubating them in vials of water. (Xu and Lou, 1980; Bradford and Hsiao, 1982; Jackson and Kowalewska, 1983). This is a sure way of depriving the leaves of hormones such as cytokinins that are produced or recycled from roots. Stomatal closure in flooded peas is associated with a rise in methanol-extractable abscisic acid (ABA; Fig. 16), a hormone capable of inducing closure quickly. It is feasible that lack of cytokinin from roots may favor increased ABA production, and this in turn could promote senescence via stomatal closure.

3. Senescence and Positive Messages

Discussion of senescence in peas will serve to illustrate the possible involvement of substances that actively promote senescence after transport from roots. In this plant, 4 days of soil flooding promoted extensive senescence and desiccation (Jackson, 1979a). It is the result of membrane injury in the leaves causing a loss of semipermeability and competence to retain water osmotically. The leaves consequently desiccate despite previous water conservation resulting from closure of the stomata (Jackson and Kowalewska, 1983). The injury *cannot* be reproduced by detaching leaves from healthy plants and incubating them in vials of water. Thus deficiencies in substances normally supplied by roots cannot alone be responsible. An alternative explanation for the damage is for substances injurious to membranes to pass into the shoot from the anoxic soil or root system. Two candidates for such a positive message, ethanol and ethylene, increase in shoots of flooded peas. However, detailed testing of these substances shows them to be insufficiently active (Jackson *et al.*, 1982; Jackson and Kowalewska, 1983), even at concentrations well in excess of those measured in flooded plants or soil. This is despite extensive claims in the literature that ethylene stimulates senescence (Abeles, 1973; El-Beltagy and Hall, 1974) and that ethanol is a potent phytotoxin (Hook and Crawford, 1978).

Surprisingly, phosphorus concentration increases abruptly in leaves of flooded peas, and this may offer an explanation for flooding injury to leaves (Jackson and Kowalewska, 1983). High concentrations of phosphorus are known to be injurious, especially at the leaf margins (Clarkson and Scattergood, 1982). There are several other reports of abnormally high phosphorus levels in flooded plants (Harris and van Bavel, 1957; Rozema and Blom, 1977; Singh and Ghildyal, 1980).

Where soil is flooded for long periods, anaerobic soil microorganisms may produce toxins. These toxins are in the form of chemically reduced alternatives to oxygen as acceptors of electrons at the final step in the respiratory generation of ATP. Divalent manganese and iron and hydrogen sulfide are examples (see Ponnamperuma, Chapter 2 in this volume). For these to be significant, plants must be capable of surviving long enough (up to several weeks) in anoxic soil to allow time for these toxins to accumulate. In continually overwet soil their presence may exclude certain species in natural communities. Grable (1966, pp. 96–97) discussed the complexity of possible interactions between these chemically reduced toxins and absorption of nutrients.

Jones and Etherington (1970) related chlorosis and reddening of the foliage of flooded *Erica cinerea* to a fivefold increase in the concentration of foliar iron (from 200 to 1000 $\mu g\ g^{-1}$ dry weight). This amount was sufficient to damage the leaves when iron was supplied to cuttings in the transpiration stream. Further support for iron as a positive senescence-promoting message in *E. cinerea* when flooded in iron-rich soil comes from experiments with antitranspirant silicone oil sprayed onto the foliage. This inhibits the uptake of iron while prolonging the life of flooded plants (Jones, 1971). Direct toxicity by iron also occurs in rice (Howeler, 1973). Divalent iron is a strong candidate for excluding *Mercurialis perennis* from waterlogged boggy soil in Cambridgeshire, England (Martin, 1968). There it appeared to act by killing the roots and thus indirectly causing chlorosis and death. As little as 2–4 ppm ferrous sulfate was fatal to this species. The high sensitivity was attributed by Martin to the absence of aerenchyma and thus of internally diffusing O_2 that could oxidize ferrous iron to the insoluble nontoxic ferric form in the roots or at their surface. *Primula elatior,* a flood-tolerant species, possesses extensive aerenchyma and tolerates up to 30 ppm ferrous sulfate (Martin, 1968).

Flax (*Linum usitatissimum*) is susceptible to manganese toxicity (Olomu and Racz, 1974). In four of six Manitoba (Canada) soils, waterlogging caused interveinal chlorosis and necrosis in association with high concentrations of manganese in the leaves (299 ppm maximum). The concentration of iron, however, was much lower in the flooded plants. Chlorosis occurred whenever the ratio of manganese to iron exceeded 4. However, the effects of manganese:iron ratios were not tested directly.

In the chemically reduced soils of paddy fields, overcast weather can favor the formation and penetration of hydrogen sulfide into rice plants. This may reach the shoots and cause "browning disease" (Vámos and Köves, 1972). The symptoms derive from an accumulation of oxidized phenols and the exudation onto the leaf surface of sugars and amino acids that would otherwise have been translocated internally. This encourages fungal attack. The principal mode of action of H_2S is to inhibit the action of metalloenzymes such as cytochrome and polyphenol oxidases.

A final example of a positive message that promotes leaf senescence is sodium chloride. Under saline conditions its uptake can be greatly increased by stagnant flooding (Section II,C,6). This is thought to cause senescence in *Phaseolus vulgaris* (West and Taylor, 1980). It may be generally true that roots injured by anoxia allow toxins such as NaCl present in the soil to enter the plant more readily.

G. Abscission

Abscission of leaves and other appendages occurs commonly in many dicotyledons and some monocotyledons. It is a consequence of enzymatically promoted lysis of cells and intercellular material that physically weakens the connection at precisely located abscission zones, comprising only a few tiers of cells (Webster, 1973). In *Phaseolus*, separation occurs between a single tier of growing cells and more distal nongrowing cells (Wright and Osborne, 1974). These events proceed most rapidly in the presence of ethylene (Jackson and Osborne, 1970). The additional ethylene present in flooded plants may thus be an explanation for the premature abscission of leaves, flowers, and young fruits reported in several species exposed to overwet soil [beans (Bergman, 1920), peas (Cannell *et al.*, 1979), cotton (Lloyd, 1920; Albert and Armstrong, 1931), *Vicia faba* (El-Beltagy and Hall, 1974)]. Leaf abscission in *droughted* cotton is promoted for this reason (McMichael *et al.*, 1972; Jordon *et al.*, 1972). In cotton, flooding destroys pollen viability and this in turn leads to ovary abortion, senescence, and abscission (Lloyd, 1920). Senescence may be a necessary preliminary to abscission for two reasons: (1) senescent tissue often produces abnormally large amounts of ethylene, and this may accelerate separation (Jackson and Osborne, 1970); and (2) senescent tissue produces and exports only small amounts of auxin (Böttger, 1970). This hormone powerfully inhibits ethylene action in abscission zones (Jackson *et al.*, 1973). A senescence-linked decline in auxin therefore is probably needed to sensitize the abscission zone to any ethylene present.

The hormone ABA was named in 1967 at an international conference on plant hormones in the belief that it was a potent abscission-promoting substance in plants. Subsequently, its activity was disappointing in species other than cotton (Osborne *et al.*, 1972). Flood-induced increases in ABA (Section III,E) may therefore be of little consequence in the timing of separation.

There is little published evidence of any acclimatic value of abscission for flooded plants. Intuitively, it seems possible that this form of self-pruning may reduce the transpirational load and contribute to maintenance of a typical shoot:root ratio. It may also prevent spread of microbial infection from dying appendages.

H. Wilting

Flooding can cause rapid wilting (''flopping'') in a wide range of species, in both field and laboratory conditions, sometimes within a few hours [e.g., *Impa-*

tiens and *Pelargonium* (Bergman, 1920), sunflower (slightly) and tobacco (Kramer, 1951; Campbell and Seaborn, 1972); tomato (Jackson, 1956), maize (Szlovak, 1975), alfalfa (Van't Woudt and Hagan, 1957); and broad bean (El-Beltagy and Hall, 1974)]. The effect is accentuated by conditions conducive to fast transpiration. Anoxic solution cultures also induce wilting or some loss of turgor (Kramer and Jackson, 1954; Willey, 1970; James, 1974), suggesting that lack of O_2 in flooded soil is the principal cause. However, some controversy exists concerning a role also for CO_2. Applying 100% CO_2 to roots of intact plants can cause more rapid wilting than the exclusion of O_2 alone (Kramer, 1940; Chang and Loomis, 1945; Kramer and Jackson, 1954). Furthermore, in very short term (3 min) experiments, Glinka and Reinhold (1962) found that 100% CO_2 severely inhibited water influx into pea roots, whereas anoxia in the absence of CO_2 had no effect. It is, of course, impossible for waterlogged soil in nature to be saturated with 100% CO_2, and smaller amounts are much less effective. Willey (1970) failed to detect an influence of 21% CO_2 on tobacco over and above that effected by anoxia. Similarly, Whitney (1942) found that 20% CO_2 did no further damage to cotton, tobacco, or maize growing in anoxic solution cultures, although in *Coleus* some further injury was sustained. Hunt *et al.* (1981) are perhaps alone in believing that ethylene from flooded soil is a cause of wilting and low leaf ψ's.

Rapid wilting seems to be a consequence of higher resistances to the mass flow of water through the roots of flooded plants. Waterlogging, CO_2, or anoxic solution cultures can increase this resistance, as measured by forcing water at known pressures or suctions through detached root systems (Kramer, 1940; Kramer and Jackson, 1954; Mees and Weatherley, 1957; Parsons and Kramer, 1974; Bradford and Hsiao, 1982) or calculated from leaf ψ and transpiration data (Janes, 1974). Moldau (1973) demonstrated relief from the water stress when roots (the sites of high resistance) were removed from flooded bean plants.

Chemical inhibitors of respiration raise the resistance of roots to water flow, presumably by suppressing formation of ATP required to keep resistances down (Brouwer, 1954; Mees and Weatherley, 1957), perhaps by maintaining particular structural conformations in membranes. Anoxia is presumed to have a similar effect.

Anoxia-induced resistance to water transport in roots is short lived. Once roots are killed, resistance falls to very low levels (Kramer, 1933; Mees and Wetherley, 1957; Bradford and Hsio, 1982). The temporary nature of the high-resistance phase may partly explain why wilting is not detected or does not always occur when plants are flooded [e.g., tomato (Whitney, 1942; Kramer, 1951; Jackson *et al.*, 1978; Bradford and Hsio, 1982), maize (Szlovak, 1975; Wenkert *et al.*, 1981), wheat (Sojka *et al.*, 1975; Trought and Drew, 1980a), and peas (Jackson and Kowalewska, 1983; but see Cannell *et al.*, 1979)]. Also, evapotranspirational demand may sometimes be insufficient to stress the plants even when root resistance has increased. Wilting is almost certainly prevented in many instances by the stomatal closure that commonly takes place within a few hours of flooding (Section III,E). Indeed, the leaves of flooded plants often become less water stressed than control plants, even in the complete absence of any preceding water deficit (Jackson *et al.*, 1978; Bradford and Hsiao, 1982; Jackson and Kowalewska, 1983). The occurrence of wilting is thus determined by a combination of circumstances, including the duration and intensity of the initial phase of high resistance to water movement through roots, the temperature and humidity of the air that determine evaporative demand, and the timing of stomatal closure.

A possibility that at later stages of flooding plugging of the xylem with material from degenerating roots may impose a second, later phase of high resistance to water flow to the shoots has been raised (Kramer and Jackson, 1954), but it has not examined experimentally.

IV. CONCLUSIONS

Although much remains to be learned about responses of herbaceous plants to flooding, a great deal is now known concerning the factors in the flooded environment that affect growth, certain biochemical lesions that preclude root growth or that lead to death of roots, morphological changes in roots and shoots that appear to favor plant survival, and the correlative links between flooded root systems and the responding shoot. The information is sufficiently detailed to assist in management of water tables in agricultural and more natural environments. It may also allow more precise selection of flood-tolerant crop types and identification of ways to manipulate with chemical treatments the development of plants affected by flooding.

Topics that warrant close attention in the future include: (1) endogenous regulation of water relations and avoidance of wilting, (2) mechanisms of adventitious rooting and reasons why some flood-susceptible species do not produce these roots, and (3) the biochemical attributes that enable a small number of species to survive without O_2 for a considerable time. Discovering how glycolysis and fermentation pathways of anaerobic respiration are regulated in these plants to provide sufficient chemical energy for maintenance and even growth is a priority task. Information on this point may enable appropriate biochemical features to be sought in other species or even introduced into plants by advanced methods of genetic engineering.

ACKNOWLEDGMENTS

We thank Drs. M. G. T. Shone and B. Atwell for their critical reading of the manuscript, Mrs. Jean M. Nash and Mrs. Meg J. Strange for typing the chapter, and Mrs. Lynda J. Hall, Mrs. Clare L. Pinkney, and Miss Alina K. B. Kowalewska for help with the figures and references.

REFERENCES

Abeles, F. B. (1973). "Ethylene in Plant Biology." Academic Press, New York and London.

Alberda, T. (1953). Growth and root development of lowland rice and its relation to oxygen supply. *Plant Soil* **5,** 1–28.

Albert, W. B., and Armstrong, G. M. (1931). Effects of high soil moisture and lack of soil aeration upon fruiting behaviour of young cotton plants. *Plant Physiol.* **6,** 585–591.

Allan, A. I., and Hollis, J. P. (1972). Sulfide inhibition of oxidases in rice roots. *Phytopathology* **62,** 634–639.

Allsopp, A. (1961). Morphogenesis in *Marsilea. J. Linn. Soc. London Bot.* **58,** 417–427.

Aloni, B., and Rosenshtein, G. (1982). Effects of flooding on tomato cultivars: the relationship between proline accumulation and other morphological and physiological changes. *Physiol. Plant.* **56,** 513–517.

Amoore, J. E. (1961a). Arrest of mitosis in roots by oxygen-lack or cyanide. *Proc. R. Soc. London B* **154,** 95–108.

Amoore, J. E. (1961b). Dependence of mitosis and respiration in roots upon oxygen tension. *Proc. R. Soc. London B* **154,** 109–129.

Amoore, J. E. (1962a). Oxygen tension and the rates of mitosis and interphase in roots. *J. Cell Biol.* **13,** 365–371.

Amoore, J. E. (1962b). Participation of a non-respiratory ferrous complex during mitosis in roots. *J. Cell Biol.* **13,** 373–381.

Amoore, J. E. (1963). Action spectrum of mitotic ferrous complex. *Nature (London)* **199,** 38–40.

Amrhein, N., Breuing, F., Eberle, J., Skorupka, H., and Tophof, S. (1982). The metabolism of 1-aminocyclopropane-1-carboxylic acid. *In* "Plant Growth Substances 1982" (P. F. Wareing, ed.), pp. 249–258. Academic Press, New York and London.

Aomine, S. (1962). A review of research on redox potentials of paddy soils in Japan. *Soil Sci.* **94,** 6–13.

Arashi, K., and Nitta, H. (1955). Studies on the lysigenous intercellular space as the ventilating system in the culm of rice and some other graminaceous plants. *Nippon Sakumotsu Gakkai Kiji (Proc. Crop Sci. Soc. Jpn.)* **24,** 78–81.

Arikado, H. (1955a). Anatomical and ecological responses of barley and some forage crops to the flooding treatment. *Bull. Fac. Agric. Mie Univ.* **11,** 1–29.

Arikado, H. (1955b). Studies on the development of the ventilating system in relation to the tolerance against excess-moisture injury in various crop plants. VI. Ecological and anatomical responses of barley and some forage plants to flooding treatment. *Nippon Sakumotsu Gakkai Kiji (Proc. Crop Sci. Soc. Jpn.)* **24,** 53–58. (In Japanese with extended summary and figure captions in English.)

Arikado, H. (1959). Supplementary studies on the development of the ventilating system in various plants growing on lowland and on upland. *Bull. Fac. Agric. Mie Univ.* **20,** 1–24.

Arikado, H. (1961). Comparative studies on the gas content and oxygen concentration in the roots of lowland and upland plants. *Bull. Fac. Agric. Mie Univ.* **24,** 17–22.

Arikado, H., and Adachi, Y. (1955). Anatomical and ecological responses of barley and some forage crops to the flooding treatment. *Bull. Fac. Agric. Mie Univ.* **11,** 1–29.

Armstrong, A. C. (1978). The effect of drainage treatments on cereal yields: results from experiments on clay lands. *J. Agric. Sci.* **91,** 229–235.

Armstrong, W. (1971). Radial oxygen losses from intact rice roots as affected by distance from the apex, respiration and waterlogging. *Physiol. Plant.* **25,** 192–197.

Armstrong, W. (1979). Aeration in higher plants. *Adv. Bot. Res.* **7,** 225–331.

Armstrong, W., and Boatman, D. J. (1967). Some field observations relating the growth of bog plants to conditions of soil aeration. *J. Ecol.* **55,** 101–110.

Armstrong, W., and Gaynard, T. J. (1976). The critical oxygen pressures for respiration in intact plants. *Physiol. Plant.* **37,** 200–206.

Armstrong, W., and Wright, E. J. (1975). The theoretical basis for the manipulation of flux data obtained by the cylindrical platinum electrode technique. *Physiol. Plant.* **35,** 21–26.

Armstrong, W., and Wright, E. J. (1976a). A polarographic assembly for multiple sampling of soil oxygen flux in the field. *J. Appl. Ecol.* **13,** 849–856.

Armstrong, W., and Wright, E. J. (1976b). An electrical analogue to simulate the oxygen relations of roots in anaerobic media. *Physiol. Plant.* **36,** 383–387.

Arnon, D. I. (1937). Ammonium and nitrate nitrogen nutrition of barley at different seasons in relation to hydrogen-ion concentrations, manganese, copper and oxygen supply. *Soil Sci.* **44,** 91–113.

Atkinson, D. E. (1969). Regulation of enzyme function. *Annu. Rev. Microbiol.* **23,** 47–68.

Atwell, B. J., Waters, I., and Greenway, H. (1982). The effect of oxygen and turbulence on elongation of coleoptiles of submergence-tolerant and -intolerant rice cultivars. *J. Exp. Bot.* **33,** 1030–1044.

Avadhani, P. N., Greenway, H., Lefroy, R., and Prior, L. (1978). Alcoholic fermentation and malate metabolism in rice germinating at low oxygen concentrations. *Aust. J. Plant Physiol.* **5,** 15–25.

Ball, N. E. (1969). Nastic responses. *In* "The Physiology of Plant Growth and Development" (M. B. Wilkins, ed.), pp. 275–300. McGraw-Hill, London.

Barber, D. A. (1961). Gas exchange between *Equisetum limosum* and its environment. *J. Exp. Bot.* **12,** 243–251.

Barber, D. A., Ebert, M., and Evans, N. T. S. (1962). The movement of ^{15}O through barley and rice plants. *J. Exp. Bot.* **13,** 397–403.

Barclay, A. M., and Crawford, R. M. M. (1982). Plant growth and survival under strict anaerobiosis. *J. Exp. Bot.* **33,** 541–549.

Batten, D. J., and Mullins, M. G. (1978). Ethylene and adventitious root formation in hypocotyl segments of etiolated mung bean [*Vigna radiata* (L.) Wilczek] seedlings. *Planta* **138,** 193–197.

Begg, J. E. (1980). Morphological adaptations of leaves to water stress. *In* "Adaptation of Plants to Water and High Temperature Stress" (N. C. Turner and P. J. Kramer, eds.), pp. 33–42. Wiley, New York.

Beletskya, E. K. (1977). Changes in metabolism of winter crops during their adaptation to flooding. *Sov. Plant Physiol. (Engl. Transl.)* **24,** 750–756.

Bendixen, L. E., and Peterson, M. L. (1962). Tropisms as a basis for tolerance of strawberry clover to flooding conditions. *Crop Sci.* **2,** 223–228.

Benjamin, L. R., and Greenway, H. (1979). Effects of a range of O_2 concentrations on porosity of barley roots and on their sugar and protein concentrations. *Ann. Bot. (London)* **43,** 383–391.

Bergman, H. F. (1920). Relation of aeration to the growth and activity of roots and its influence on the ecesis of plants in swamps. *Ann. Bot. (London)* **34,** 13–33.

Berry, L. J., and Norris, W. E. (1949). Studies on onion root respiration. I. Velocity of oxygen consumption in different segments of root at different temperatures as a function of partial pressure of oxygen. *Biochim. Biophys. Acta* **3,** 593–606.

Bertani, A., and Brambilla, I. (1982a). Effects of decreasing oxygen concentration on some aspects of protein and amino-acid metabolism in rice roots. *Z. Pflanzenphysiol.* **107,** 193–200.

Bertani, A., and Brambilla, I. (1982b). Effect of decreasing oxygen concentration on wheat roots. Growth and induction of anaerobic metabolism. *Z. Pflanzenphysiol.* **108,** 283–288.

Bertani, A., and Brambilla, I., and Menegas, F. (1980). Effect of anaerobiosis on rice seedlings: growth, metabolic rate, and fate of fermentation products. *J. Exp. Bot.* **31,** 325–331.

Bertani, A., Brambilla, I., and Menegas, F. (1981). Effect of anaerobiosis on carbohydrate content of rice roots. *Biochem. Physiol. Pflanz.* **176,** 835–840.

Bisseling, T., van Staveren, W., and van Kammen, A. (1980). The effect of waterlogging on the synthesis of the nitrogenase components in bacteroids of *Rhizobium leguminosarum* in root nodules of *Pisum sativum. Biochem. Biophys. Res. Commun.* **93,** 687–693.

Blackwell, P. S. (1983). Measurements of aeration in waterlogged soils; some improvements of techniques and their application to experiments using lysimeters. *J. Soil Sci.*, in press.

Blackwell, P. S., and Ayling, S. M. (1981). Changes in aeration following transient waterlogging of sandy loam and clay soils cropped with winter cereals. *Annu. Rep. Agric. Res. Counc. Letcombe Lab.* **1980,** 35–38.

Blackwell, P. S., and Wells, E. A. (1983). Limiting oxygen flux densities for oat root extension. *Plant Soil,* **73,** 129–139.

Bond, G. (1952). Some features of root growth in nodulated plants of *Myrica gale* L. *Ann. Bot. (London)* **16,** 467–475.

Bonner, W. D. (1973). Mitochondria and plant respiration. *In* "Phytochemistry" (L. P. Miller, ed.), pp. 221–261. Academic Press, New York.

Böttger, M. (1970). Hormonal regulation of petiolar abscission in *Coleus rehneltianus* explants. II. The possible role of abscisic acid in the abscission process. *Planta* **93,** 205–213.

Boyd, C. E. (1971). The limnological role of aquatic macrophytes and their relationship to reservoir management. *Am. Fish. Soc. Spec. Publ.* **8,** 153–166.

Bradford, K. J. (1982). Regulation of shoot responses to root stress by ethylene, abscisic acid, and cytokinin. *In* "Plant Growth Substances 1982" (P. F. Wareing, ed.), pp. 599–608. Academic Press, New York and London.

Bradford, K. J., and Dilley, D. R. (1978). Effects of root anaerobiosis on ethylene production, epinasty, and growth of tomato plants. *Plant Physiol.* **61,** 506–509.

Bradford, K. J., and Hsiao, T. C. (1982). Stomatal behaviour and water relations of waterlogged tomato plants. *Plant Physiol.* **70,** 1508–1513.

Bradford, K. J., and Yang, S. F. (1980a). Xylem transport of 1-aminocyclopropane-1-carboxylic acid, an ethylene precursor, in waterlogged tomato plants. *Plant Physiol.* **65,** 322–326.

Bradford, K. J., and Yang, S. F. (1980b). Stress-induced ethylene production in the ethylene-requiring tomato mutant 'Diageotropica'. *Plant Physiol.* **65,** 327–330.

Bragg, P. L., Henderson, F. K. G., and Ellis, F. B. (1984). Effect of mole-drainage in a clay soil. VI. Root and shoot growth of winter wheat, 1979–80. *J. Sci. Food Agric.*, in press.

Bristow, J. M. (1974). The structure and function of roots in aquatic vascular species. *In* "The Development and Function of Roots" (J. G. Torrey and D. T. Clarkson, eds.), pp. 221–236. Academic Press, New York and London.

Broadbent, F.E., and Clark, F. E. (1965). Denitrification. *In* "Soil Nitrogen" (W. V. Bartholomew and F. E. Clark, eds.), pp. 344–349. Am. Soc. Agron., Madison, Wisconsin.

Brouwer, R. (1954). The regulating influence of transpiration and suction tension on water and salt uptake of roots. *Acta Bot. Neerl.* **3,** 264–312.

Brouwer, R. (1977). The effect of soil waterlogging on various physiological processes in maize. *Phytotronic Newsl.* **15,** 75–80.

Bryant, A. E. (1934). Comparison of anatomical and histological differences between roots of barley grown in aerated and non-aerated culture solutions. *Plant Physiol.* **9,** 389–391.

Burg, S. P., and Thimann, K. V. (1959). The physiology of ethylene formation in apples. *Proc. Natl. Acad. Sci. USA* **45,** 335–344.

Burrows,W. J., and Carr, D. J. (1969). Effects of flooding the root system of sunflower plants on the cytokinin content in the xylem sap. *Physiol. Plant.* **22,** 1105–1112.

Butt, V. S. (1980). Direct oxidases and related enzymes. *In* "The Biochemistry of Plants" (D. D. Davies, ed.), Vol. 2, pp. 81–123. Academic Press, New York.

Campbell, R., and Drew, M. C. (1983). Electron microscopy of gas space (aerenchyma) formation in adventitious roots of *Zea mays* L. subjected to oxygen shortage. *Planta* **157,** 350–357.

Campbell, R. B., and Seaborn, G. T. (1972). Yield of flue-cured tobacco and levels of soil oxygen in lysimeters with different water-table depths. *Agron. J.* **64,** 730–733.

Cannell, R. Q., and Jackson, M. B. (1981). Alleviating aeration stresses. *In* "Modifying the Root Environment to Reduce Crop Stress" (G. F. Arkin and H. M. Taylor, eds.), pp. 141–192. Am. Soc. Agric. Eng., St. Joseph, Missouri.

Cannell, R. Q., Gales, K., and Suhail, B. A. (1977). Effects of waterlogging under field conditions on the growth of peas. *Annu. Rep. Agric. Res. Counc. Letcombe Lab.* **1976,** 67–69.

Cannell, R. Q., Gales, K., Snaydon, R. W., and Suhail, B. A. (1979). Effects of short-term waterlogging on the growth and yield of peas (*Pisum sativum*). *Ann. Appl. Biol.* **93,** 327–335.

Cannon, W. A. (1925). Physiological features of roots, with especial reference to the relation of roots to aeration of the soil. *Carnegie Inst. Washington Publ.* No. 368.

Carr, D. J., and Reid, D. M. (1968). The physiological significance of the synthesis of hormones in roots and their export to the shoot system. *In* "The Biochemistry and Physiology of Plant Growth Substances" (F. Wightman and G. Setterfield, eds.), pp. 1169–1185. Runge, Ottawa.

Chang, H. T., and Loomis, W. E. (1945). Effect of carbon dioxide on absorption of water and nutrients by roots. *Plant Physiol.* **20,** 221–232.

Cheeseman, J. M., and Hanson, J. B. (1979). Energy-linked potassium influx as related to cell potential in corn roots. *Plant Physiol.* **64,** 842–845.

Cheeseman, J. M., LaFayette, P. R., Gronewald, J. W., and Hanson, J. B. (1980). Effect of ATP-ase inhibitors on cell potential and K^+ influx in corn roots. *Plant Physiol.* **65,** 1139–1145.

Chirkova, T. V., Khoang, K. L., and Sinyutina, N. F. (1980). Ethanol action on lipid metabolism in roots of wheat and rice seedlings grown under different conditions of gas exchange. *Fiziol. Biokhim. Kult. Rast.* **12,** 24–28.

Chirkova, T. V., Khoang, K. L., and Blyudzin, Y. A. (1981). Effect of anaerobic conditions on fatty acid composition of wheat and rice root phospholipids. *Fiziol. Rast. (Moscow)* **28,** 358–366.

Clarkson, D. T., and Scattergood, C. B. (1982). Growth and phosphate transport in barley and tomato plants during the development of, and recovery from, phosphate stress. *J. Exp. Bot.* **33,** 865–875.

Clemens, J., Kirk, A.-M., and Mills, P. D. (1978). The resistance to waterlogging of three *Eucalyptus* species. Effect of waterlogging and an ethylene-releasing growth substance on *E. robusta, E. grandis* and *E. saligna. Oecologia* **34,** 125–131.

Coleman, W. K., Huxter, J. J., Reid, D. M., and Thorpe, T. A. (1980). Ethylene as an endogenous inhibitor of root regeneration in tomato leaf discs cultured *in vitro. Physiol. Plant.* **48,** 519–525.

Conway, V. M. (1937). Studies on the autecology of *Cladium mariscus* R. Br. III. The aeration of the subterranean parts of the plant. *New Phytol.* **36,** 64–96.

Cooke, G. W. (1976). A review of the effects of agriculture on the chemical composition and quality of surface and underground waters. *Tech. Bull. U.K. Min. Agric. Fish. Food* **32,** 5–57.

Cookson, C., and Osborne, D. J. (1978). The stimulation of cell extension by ethylene and auxin in aquatic plants. *Planta* **144,** 39–47.

Cookson, C., and Osborne, D. J. (1979). The effect of ethylene and auxin on cell wall extensibility of the semi-aquatic fern *Regnellidium diphyllum. Planta* **146,** 303–307.

Cossins, E. A. (1978). Ethanol metabolism in plants. *In* "Plant Life in Anaerobic Environments" (D. D. Hook and R. M. M. Crawford, eds.), pp. 169–202. Ann Arbor Sci. Publ., Ann Arbor, Michigan.

Coult, D. A., and Vallance, K. B. (1958). Observations on the gaseous exchange which takes place between *Menyanthes trifoliata* and its environment. *J. Exp. Bot.* **9,** 384–402.

Crawford, R. M. M. (1966). The control of anaerobic respiration as a determining factor in the distribution of the genus *Senecio. J. Ecol.* **54,** 403–413.

Crawford, R. M. M. (1967). Alcohol dehydrogenase activity in relation to flooding tolerance in roots. *J. Exp. Bot.* **18,** 458–464.

Crawford, R. M. M. (1978). Metabolic adaptations to anoxia. *In* "Plant Life in Anaerobic Environments" (D. D. Hook and R. M. M. Crawford, eds.), pp. 119–136. Ann Arbor Sci. Publ., Ann Arbor, Michigan.

Crawford, R. M. M. (1982). The anaerobic retreat as a survival strategy for aerobic plants and animals. *Trans. Bot. Soc. Edinburgh* **44,** 57–63.

Criswell, J. G., Ulysses, D. H., Quebedeaux, B., and Hardy, R. W. F. (1976). Adaptation of nitrogen fixation by intact soybean nodules to altered rhizosphere pO_2. *Plant Physiol.* **58,** 622–625.

Crocker, W., Zimmerman, P. W., and Hitchcock, A. E. (1932). Ethylene-induced epinasty of leaves and the relation of gravity to it. *Contrib. Boyce Thompson Inst.* **4,** 177–218.

Das, D. K., and Jat, R. L. (1972). Adaptability of maize to high soil water conditions. *Agron. J.* **64,** 849–850.

Das, D. K., and Jat, R. L. (1977). Influence of three soil-water regimes on root porosity and growth of four rice varieties. *Agron. J.* **69,** 197–200.

Davies, D. D. (1980). Anaerobic metabolism and the production of organic acids. *In* "The Biochemistry of Plants" (D. D. Davies, ed.), Vol. 2, pp. 581–611. Academic Press, New York.

Davies, D. D., Nascimento, K. H., and Patil, K. D. (1974). The distribution and properties of NADP malic enzyme in flowering plants. *Phytochemistry* **13,** 2417–2425.

Day, D. A., Arron, G. P., and Laties, G. G. (1980). Nature and control of respiratory pathways in plants: the interaction of cyanide-resistant respiration with the cyanide-sensitive pathway. *In* "The Biochemistry of Plants" (D. D. Davies, ed.), Vol. 2, pp. 197–241. Academic Press, New York.

de Wit, M. C. J. (1978). Morphology and function of roots and shoot growth of crop plants under oxygen deficiency. *In* "Plant Life in Anaerobic Environments" (D. D. Hook and R. M. M. Crawford, eds.), pp. 333–350. Ann Arbor Sci. Publ., Ann Arbor, Michigan.

Doubt, S. L. (1915). The responses of plants to illuminating gas. *Bot. Gaz. (Chicago)* **63,** 209–224.

Drew, M. C., and Lynch, J. M. (1980). Soil anaerobiosis, micro-organisms and root function. *Annu. Rev. Phytopathol.* **18,** 37–66.

Drew, M. C., and Sisworo, E. J. (1977). Early effects of flooding on nitrogen deficiency and leaf chlorosis in barley. *New Phytol.* **79,** 567–571.

Drew, M. C., and Sisworo, E. J. (1979). The development of waterlogging damage in young barley plants in relation to plant nutrient status and changes in soil properties. *New Phytol.* **82,** 301–314.

Drew, M. C., Jackson, M. B., and Giffard, S. C. (1979a). Ethylene-promoted adventitious rooting and development of cortical air spaces (aerenchyma) in roots may be adaptive responses to flooding in *Zea mays* L. *Planta* **147,** 83–88.

Drew, M. C., Sisworo, E. J., and Saker, L. R. (1979b). Alleviation of waterlogging damage to young barley plants by application of nitrate and a synthetic cytokinin, and comparison between the effects of waterlogging, nitrogen deficiency and root excision. *New Phytol.* **82,** 315–329.

Drew, M. C., Chamel, A., Garrec, J.-P., and Fourcy, A. (1980). Cortical air spaces (aerenchyma) in roots of corn subjected to oxygen stress. *Plant Physiol.* **65,** 506–511.

Drew, M. C., Jackson, M. B., Giffard, S. C., and Campbell, R. (1981). Inhibition by silver ions of gas space (aerenchyma) formation in adventitious roots of *Zea mays* L. subjected to exogenous ethylene or to oxygen deficiency. *Planta* **153,** 217–224.

Durell, W. D. (1941). The effect of aeration on growth of the tomato in nutrient solution. *Plant Physiol.* **16,** 327–341.

El-Beltagy, A. S., and Hall, M. A. (1974). Effect of water stress upon endogenous ethylene levels in *Vicia faba*. *New Phytol.* **73,** 47–60.

El-D., A. M. S., Salama, A., and Wareing, P. F. (1979). Effects of mineral nutrition on endogenous cytokinins in plants of sunflower (*Helianthus annuus* L.). *J. Exp. Bot.* **30,** 971–981.

Elkins, C. B., and Hoveland, C. S. (1977). Soil oxygen and temperature effect on tetany potential of three annual forage species. *Agron. J.* **69,** 626–628.

Ellmore, G. S. (1981). Root dimorphism in *Ludwigia peploides* (Onagraceae): structure and gas content of mature roots. *Am. J. Bot.* **68,** 557–568.

Emerson, F. W. (1921). Subterranean organs of bog plants. *Bot. Gaz. (Chicago)* **72,** 359–374.

Engelbrecht, L. (1972). Cytokinins in leaf-cuttings of *Phaseolus vulgaris* L. during their development. *Biochem. Physiol. Pflanz.* **163,** 335–343.

Erecinska, M., Stubbs, M., Miyata, Y., Ditre, C. M., and Wilson, D. F. (1977). Regulation of cellular metabolism by intracellular phosphate. *Biochim. Biophys. Acta* **462,** 20–35.

Fabijan, D., Yeung, E., Mukherjee, I., and Reid, D. M. (1981). Adventitious rooting in hypocotyls of sunflower (*Helianthus annuus*) seedlings. *Physiol. Plant.* **53,** 578–588.

Fletcher, R. A., and Osborne, D. J. (1966). Gibberellin as a regulator of protein and ribonucleic acid synthesis during senescence in leaf cells of *Taraxacum officinale*. *Can. J. Bot.* **44,** 739–745.

Francis, C. M., Devitt, A. C., and Steele, P. (1974). Influence of flooding on the alcohol dehydrogenase activity of roots of *Trifolium subterraneum* L. *Aust. J. Plant Physiol.* **1,** 9–13.

Fulton, J. M., and Erickson, A. E. (1964). Relation between soil aeration and ethyl alcohol accumulation in xylem exudate of tomatoes. *Soil Sci. Soc. Am. Proc.* **28,** 610–614.

Gahan, P. B. (1982). Cytochemical and ultrastructural changes in cell senescence and death. *In* "Growth Regulators in Plant Senescence" (M. B. Jackson, B. Grout, and I. A. Mackenzie, eds.), Monogr. No. 8, pp. 47–55. Br. Plant Growth Regul. Group, Wantage.

Garcia-Novo, F., and Crawford, R. M. M. (1973). Soil aeration, nitrate reduction and flooding tolerance in higher plants. *New Phytol.* **72,** 1031–1039.

Geisler, G. (1965). The morphogenetic effect of oxygen on roots. *Plant Physiol.* **40,** 85–88.

Gerdemann, J. W. (1968). Vesicular–arbuscular mycorrhiza and plant growth. *Annu. Rev. Phytopathol.* **6,** 397–418.

Gilbert, S. G., and Shive, J. W. (1942). The significance of oxygen in nutrient substrates for plants. I. The oxygen requirement. *Soil Sci.* **53,** 143–152.

Glinka, Z., and Reinhold, L. (1962). Rapid changes in permeability of cell membranes to water brought about by carbon dioxide and oxygen. *Plant Physiol.* **37,** 481–486.

Goodlass, G., and Smith, K. A. (1979). Effects of ethylene on root extension and nodulation of pea (*Pisum sativum* L.) and white clover (*Trifolium repens* L.). *Plant Soil* **51,** 387–395.

Green, M. S., and Etherington, J. R. (1977). Oxidation of ferrous iron by rice (*Oryza sativa* L.) roots: a mechanism for waterlogging tolerance? *J. Exp. Bot.* **28,** 679–690.

Greenwood, D. J. (1967a). Studies on the transport of oxygen through the stems and roots of vegetable seedlings. *New Phytol.* **66,** 337–347.

Greenwood, D. J. (1967b). Studies on oxygen transport through mustard seedlings (*Sinapis alba* L.). *New Phytol.* **66,** 597–606.

Greenwood, D. J., and Goodman, D. (1971). Studies on the supply of oxygen to the roots of mustard seedlings (*Sinapis alba* L.). *New Phytol.* **70,** 85–96.

Gregory, F. G., and Samantari, B. (1950). Factors concerned with the rooting responses of isolated leaves. *J. Exp. Bot.* **1,** 159–193.

Griffin, D. M. (1968). A theoretical study relating the concentration and diffusion of oxygen to the biology of organisms in soil. *New Phytol.* **67,** 561–577.

Grineva, G. M., and Nechiporenko, G. A. (1977). Distribution and conversion of sucrose U-C^{14} in corn plants under conditions of flooding. *Sov. Plant Physiol. (Engl. Transl.)* **24,** 32–37.

Grobbelaar, N., Clarke, B., and Hough, M. C. (1970). The inhibition of root nodulation by ethylene. *Agroplantae* **2,** 81–82.

Grobbelaar, N., Clarke, B., and Hough, M. C. (1971). The nodulation of nitrogen fixation of isolated roots of *Phaseolus vulgaris* L. III. The effect of carbon dioxide and ethylene. *Plant Soil* **special volume,** 215–223.

Guhman, H. (1924). Variations in the root system of the common everlast (*Gnaphalium polycephalum*). *Ohio J. Sci.* **24,** 199–208.
Haberkorn, H.-R., and Sievers, A. (1977). Response to gravity of roots growing in water. *Naturwissenschaften* **64,** 639–640.
Hamilton, W. A. (1979). Microbial energetics and metabolism. *In* "Microbial Ecology. A Conceptual Approach" (J. M. Lynch and N. J. Poole, eds.), pp. 22–44. Blackwell, Oxford.
Hansen, D. J., and Bendixen, L. E. (1974). Ethylene-induced tropism of *Trifolium fragiferum* L. stolons. *Plant Physiol.* **53,** 80–82.
Harberd, N. P., and Edwards, K. J. R. (1982). The effect of a mutation causing alcohol dehydrogenase deficiency on flooding tolerance in barley. *New Phytol.* **90,** 631–644.
Harley, J. L. (1972). "The Biology of Mycorrhiza," 2nd ed. Leonard Hill, London.
Harris, D. G., and van Bavel, C. H. M. (1957). Nutrient uptake and chemical composition of tobacco plants as affected by the composition of the root atmosphere. *Agron. J.* **49,** 176–181.
Harvey, E. M. (1913). The castor bean plant and laboratory air. *Bot. Gaz. (Chicago)* **56,** 439–442.
Healy, M. T., and Armstrong, W. (1972). The effectiveness of internal oxygen transport in a mesophyte (*Pisum sativum* L.). *Planta* **103,** 302–309.
Hetherington, A. M., Hunter, M. I. S., and Crawford, R. M. M. (1982). Contrasting effects of anoxia on rhizome lipids in *Iris* species. *Phytochemistry* **21,** 1275–1278.
Hiatt, A. J., and Lowe, R. H. (1967). Loss of organic acids, amino acids, K and Cl from barley roots treated anerobically and with metabolic inhibitors. *Plant Physiol.* **42,** 1731–1736.
Hiron, R. W. P., and Wright, S. T. C. (1973). The role of endogenous abscisic acid in the response of plants to stress. *J. Exp. Bot.* **24,** 769–781.
Hitchcock, C., and Nichols, B. W. (1971). "Plant Lipid Biochemistry." Academic Press, New York and London.
Hitchman, M. L. (1978). "Measurement of Dissolved Oxygen." Wiley, New York.
Hollis, J. P., and Rodriguez-Kabana, R. (1967). Fatty acids in Louisiana rice fields. *Phytopathology* **57,** 841–847.
Hollis, J. P., Allam, A. I., Pitts, G., Joshi, M. M., and Ibrahim, I. K. A. (1975). Sulfide diseases of rice on iron-excess soils. *Acta Phytopathol. Acad. Sci. Hung.* **10,** 329–341.
Hook, D. D., and Crawford, R. M. M. (eds.) (1978). "Plant Life in Anaerobic Environments." Ann Arbor Sci. Publ., Ann Arbor, Michigan.
Hook, D. D., and Scholtens, J. R. (1978). Adaptation and flood tolerance of tree species. *In* "Plant Life in Anaerobic Environments" (D. D. Hook and R. M. M. Crawford, eds.), pp. 299–331. Ann Arbor Sci. Publ., Ann Arbor, Michigan.
Hopkins, H. T. (1956). Absorption of ionic species of orthophosphate by barley roots: effects of 2,4-dinitrophenol and oxygen tension. *Plant Physiol.* **31,** 155–161.
Hopkins, H. T., Specht, A. W., and Hendricks, S. B. (1950). Growth and nutrient accumulation as controlled by oxygen supply to the plant roots. *Plant Physiol.* **25,** 193–209.
Horton, R. F., and Samarakoon, A. B. (1982). Petiole growth in the celery-leaved crowfoot (*Ranunculus sceleratus* L.): effects of auxin-transport inhibitors. *Aquatic Bot.* **13,** 97–104.
Horvath, I., Vigh, L., and Farkas, T. (1981). The manipulation of polar head group composition of phospholipids in the wheat Miranovskaja 808 affects frost tolerance. *Planta* **151,** 103–108.
Howeler, R. H. (1973). Iron-induced oranging disease of rice in relation to physico-chemical changes in a flooded oxisol. *Soil Sci. Soc. Am. Proc.* **37,** 898–903.
Huck, M. G. (1970). Variation in taproot elongation rate as influenced by composition of the soil air. *Agron. J.* **62,** 815–818.
Hunt, P. G., Campbell, R. B., Sojka, R. E., and Parsons, J. E. (1981). Flooding-induced soil and plant ethylene accumulation and water status response of field-grown tobacco. *Plant Soil* **59,** 427–439.
Ingram, L. O. (1976). Adaptation of membrane lipids to alcohols. *J. Bacteriol.* **125,** 670–678.

International Rice Research Institute (IRRI) (1979). Sudden wilting in Korea. *In* "IRRI Annual Report for 1978." IRRI, Los Baños, Philippines.

Ishizawa, K., and Esashi, Y. (1983). Cooperation of ethylene and auxin in the growth regulation of rice coleoptile segments. *J. Exp. Bot.* **34,** 74–82.

Jackson, M. B. (1971). "Ethylene and Plant Development with Special Reference to Abscission." D.Phil. Thesis, Univ. of Oxford.

Jackson, M. B. (1979a). Rapid injury to peas by soil waterlogging. *J. Sci. Food Agric.* **30,** 143–152.

Jackson, M. B. (1979b). Is the diageotropic tomato ethylene deficient? *Physiol. Plant.* **46,** 347–351.

Jackson, M. B. (1980). Aeration in the nutrient film technique of glasshouse crop production and the importance of oxygen, ethylene and carbon dioxide. *Acta Hortic.* **98,** 61–78.

Jackson, M. B. (1982). Ethylene as a growth promoting hormone under flooded conditions. *In* "Plant Growth Substances 1982" (P. F. Wareing, ed.), pp. 291–301. Academic Press, New York and London.

Jackson, M. B. (1983). Approaches to relieving aeration stress in waterlogged plants. *Pestic. Sci.* **14,** 25–32.

Jackson, M. B., and Barlow, P. W. (1981). Root geotropism and the role of growth regulators from the cap: a re-examination. *Plant Cell Environ.* **4,** 107–123.

Jackson, M. B., and Campbell, D. J. (1975a). Movement of ethylene from roots to shoots, a factor in the responses of tomato plants to waterlogged soil conditions. *New Phytol.* **74,** 397–406.

Jackson, M. B., and Campbell, D. J. (1975b). Ethylene and waterlogging effects in tomato. *Ann. Appl. Biol.* **81,** 102–105.

Jackson, M. B., and Campbell, D. J. (1976). Waterlogging and petiole epinasty in tomato: the role of ethylene and low oxygen. *New Phytol.* **76,** 21–29.

Jackson, M. B., and Campbell, D. J. (1979). Effects of benzyladenine and gibberellic acid on the responses of tomato plants to anaerobic root environments and to ethylene. *New Phytol.* **82,** 331–340.

Jackson, M. B., and Harney, P. M. (1970). Rooting cofactors, indoleacetic acid, and adventitious root initiation in mung bean cuttings (*Phaseolus aureus*). *Can. J. Bot.* **48,** 943–946.

Jackson, M. B., and Kowalewska, A. K. B. (1983). Positive and negative messages from roots induce foliar desiccation and stomatal closure in flooded pea plants. *J. Exp. Bot.* **34,** 493–506.

Jackson, M. B., and Osborne, D. J. (1970). Ethylene the natural regulator of leaf abscission. *Nature (London)* **225,** 1019–1022.

Jackson, M. B., and Palmer, J. H. (1981). Responses of sunflowers to soil flooding. *Plant Physiol.* **67,** 58 (Abstr.).

Jackson, M. B., Hartley, C. B., and Osborne, D. J. (1973). Timing abscission in *Phaseolus vulgaris* L. by controlling ethylene production and sensitivity to ethylene. *New Phytol.* **72,** 1251–1260.

Jackson, M. B., Gales, K., and Campbell, D. J. (1978). Effect of waterlogged soil conditions on the production of ethylene and on water relationships in tomato plants. *J. Exp. Bot.* **29,** 183–193.

Jackson, M. B., Drew, M. C., and Giffard, S. C. (1981). Effects of applying ethylene to the root system of *Zea mays* on growth and nutrient concentration, in relation to flooding tolerance. *Physiol. Plant.* **52,** 23–28.

Jackson, M. B., Herman, B., and Goodenough, A. (1982). An examination of the importance of ethanol in causing injury to flooded plants. *Plant Cell Environ.* **5,** 163–172.

Jackson, W. T. (1955). The role of adventitious roots in recovery of shoots following flooding of the original root systems. *Am. J. Bot.* **42,** 816–819.

Jackson, W. T. (1956). Flooding injury studied by approach graft and split root system techniques. *Am. J. Bot.* **43,** 496–502.

Jacoby, B., and Rudich, B. (1980). Proton–chloride symport in barley roots. *Ann. Bot. (London)* **46,** 493–498.

James, W. O. (1953). "Plant Respiration." Oxford Univ. Press (Clarendon), London.

Janes, B. E. (1974). The effect of variations in root environment on root growth and resistance to flow of water in intact plants. *In* "Mechanisms of Regulation of Plant Growth" (R. L. Bieleski, A. R. Ferguson, and M. M. Cresswell, eds.), Bull. 12, pp. 379–385. Royal Soc. of New Zealand, Wellington.

Jat, R. L., Dravid, M. S., Das, D. K., and Goswami, N. N. (1975). Effect of flooding and high soil water condition on root porosity and growth of maize. *J. Indian Soc. Soil Sci.* **23,** 291–297.

Jensen, C. R., Stolzy, L. H., and Letey, J. (1967). Tracer studies of oxygen diffusion through roots of barley, corn and rice. *Soil Sci.* **103,** 23–29.

Jeschke, W. D., and Jambor, W. (1981). Determination of unidirectional sodium fluxes in roots of intact sunflower seedlings. *J. Exp. Bot.* **32,** 1257–1272.

John, C. D. (1977). The structure of rice roots grown in aerobic and anaerobic environments. *Plant Soil* **47,** 269–274.

John, C. D., and Greenway, H. (1976). Alcoholic fermentation and activity of some enzymes in rice roots under anaerobiosis. *Aust. J. Plant Physiol.* **3,** 325–336.

John, C. D., Limpinuntana, V., and Greenway, H. (1974). Adaptation of rice to anaerobiosis. *Aust. J. Plant Physiol.* **1,** 513–520.

Jones, H. E. (1971). Comparative studies of plant growth and distribution in relation to waterlogging. II. An experimental study of the relationship between transpiration and the uptake of iron in *Erica cinerea* L. and *E. tetralix* L. *J. Ecol.* **59,** 167–178.

Jones, H. E., and Etherington, J. R. (1970). Comparative studies of plant growth and distribution in relation to waterlogging. I. The survival of *Erica cinerea* L. and *E. tetralix* L. and its apparent relationship to iron and manganese uptake in waterlogged soil. *J. Ecol.* **58,** 487–496.

Jones, R. L., and Phillips, I. D. J. (1966). Organs of gibberellin synthesis in light-grown sunflower plants. *Plant Physiol.* **41,** 1381–1386.

Jordan, W. D., Morgan, P. W., and Davenport, T. L. (1972). Water stress enhances ethylene-mediated leaf abscission. *Plant Physiol.* **50,** 756–758.

Juliano, J. B., and Aldana, M. J. (1937). Morphology of *Oryza sativa. Philipp. Agric.* **26,** 1–76.

Kang, B. G. (1979). Epinasty. *Encycl. Plant Physiol. New Ser.* **7,** 647–667.

Kar, S., Varade, S. B., Subramanyam, T. K., and Ghildyal, B. P. (1974). Nature and growth pattern of rice root system under submerged and unsaturated conditions. *Riso (Milan)* **23,** 173–179.

Karishnev, R. V. (1958). Resistance of summer wheat to excess water. *Sov. Plant Physiol. (Engl. Transl.)* **5,** 410–416.

Kawase, M. (1972). Effect of flooding on ethylene concentration in horticultural plants. *J. Am. Soc. Hortic. Sci.* **97,** 584–588.

Kawase, M. (1974). Role of ethylene in induction of flooding damage in sunflower. *Physiol. Plant.* **31,** 29–38.

Kawase, M. (1978). Aeration and waterlogging damages. *Hortic. Sci.* **13,** 370 (Abstr.).

Kawase, M. (1979). Role of cellulase in aerenchyma development in sunflower. *Am. J. Bot.* **66,** 183–190.

Kawase, M. (1981a). Anatomical and morphological adaptation of plants to waterlogging. *HortScience.* **16,** 30–34.

Kawase, M. (1981b). Effect of ethylene on aerenchyma development. *Am. J. Bot.* **68,** 651–658.

Kawase, M., and Whitmoyer, R. E. (1980). Aerenchyma development in waterlogged plants. *Am. J. Bot.* **67,** 18–22.

Keeley, J. E., and Franz, E. H. (1979). Alcoholic fermentation in swamp and upland populations of *Nyssa sylvatica:* temporal changes in adaptive strategy. *Am. Nat.* **113,** 587–591.

Keister, D. L., and Rao, V. R. (1977). Nitrogen fixation in *Rhizobium japonicum* × *planta. In* "Recent Developments in Nitrogen Fixation" (W. Newton, J. R. Postgate, and C. Rodriguez-Barrucco, eds.), pp. 419–430. Academic Press, New York and London.

Kende, H. (1964). Preservation of chlorophyll in leaf sections by substances obtained from root exudate. *Science (Washington, D.C.)* **145,** 1066–1067.

Kennedy, R. A., Barrett, S. C. H., Vander Zee, D., and Rumpho, M. E. (1980). Germination and seedling growth under anaerobic conditions in *Echinochloa crus-galli* (barnyard grass). *Plant Cell Environ.* **3,** 243–248.

Khan, A. G. (1974). The occurrence of mycorrhizas in halophytes, hydrophytes and xerophytes, and of *Endogone* spores in adjacent soils. *J. Gen. Microbiol.* **81,** 7–14.

King, R. W. (1976). Implications for plant growth of the transport of regulatory compounds in the phloem and xylem. *In* "Transport and Transfer Processes in Plants" (I. F. Wardlaw and J. B. Passioura, eds.), pp. 415–431. Academic Press, New York.

Kohl, J. G., Baierova, J., Radke, G., and Ramshorn, K. (1978). Regulative interaction between anaerobic metabolism and nitrogen assimilation as related to oxygen deficiency in maize roots. *In* "Plant Life in Anaerobic Environments" (D. D. Hook and R. M. M. Crawford, eds.), pp. 473–496. Ann Arbor Sci. Publ., Ann Arbor, Michigan.

Konings, H. (1982). Ethylene-promoted formation of aerenchyma in seedling roots of *Zea mays* L. under aerated and non-aerated conditions. *Physiol. Plant.* **54,** 119–124.

Konings, H., and Jackson, M. B. (1979). A relationship between rates of ethylene production by roots and the promoting or inhibiting effects of exogenous ethylene and water on root elongation. *Z. Pflanzenphysiol.* **92,** 385–397.

Kordan, H. A. (1976). Adventitious root initiation and growth in relation to oxygen supply in germinating rice seedlings. *New Phytol.* **76,** 81–86.

Kramer, D., Läuchli, A., Yeo, A. R., and Gullasch, J. (1977). Transfer cells in roots of *Phaseolus coccineus:* ultrastructure and possible function in exclusion of exclusion of sodium from the shoot. *Ann. Bot. (London)* **41,** 1031–1040.

Kramer, P. J. (1933). The intake of water through dead root systems and its relation to the problem of absorption by transpiring plants. *Am. J. Bot.* **20,** 481–492.

Kramer, P. J. (1940). Causes of decreased absorption of water by plants in poorly aerated media. *Am. J. Bot.* **27,** 216–220.

Kramer, P. J. (1951). Causes of injury to plants resulting from flooding of the soil. *Plant Physiol.* **26,** 722–736.

Kramer, P. J., and Jackson, W. T. (1954). Causes of injury to flooded tobacco plants. *Plant Physiol.* **29,** 241–245.

Krizek, D. T. (1981). Plant response to atmospheric stress caused by waterlogging. *In* "Breeding Plants for Less Favorable Environments" (M. N. Christiansen and C. Lewis, eds.), pp. 293–334. Wiley, New York.

Ku, H. S., Suge, H., Rappaport, L., and Pratt, H. K. (1970). Stimulation of rice coleoptile growth by ethylene. *Planta* **90,** 333–339.

Kulaeva, O. N. (1962). The effect of roots on leaf metabolism in relation to the action of kinetin on leaves. *Sov. Plant Physiol. (Engl. Transl.)* **9,** 182–189.

Kuo, C. G., and Chen, B. W. (1980). Physiological responses of tomato cultivars to flooding. *J. Am. Soc. Hortic. Sci.* **105,** 751–755.

Kursanov, A. L. (1963). Metabolism and the transport of organic substances in the phloem. *Adv. Bot. Res.* **1,** 209–279.

Kuznetsova, G. A., Kuznetsova, M. G., andGrineva, G. M. (1981). Characteristics of water exchange and anatomical and morphological structure of maize plants under flooding. *Fiziol. Rast. (Moscow)* **28,** 340–348 (English summary).

Laing, H. E. (1940). The composition of the internal atmosphere of *Nuphar advenum* and other water plants. *Am. J. Bot.* **27,** 861–868.

Lal, R., and Taylor, G. S. (1970). Drainage and nutrient effects in a field lysimeter study. II. Mineral uptake by corn. *Soil Sci. Soc. Am. Proc.* **34,** 245–248.

Lambers, H. (1976). Respiration and NADH-oxidation of the roots of flood-tolerant and flood-intolerant *Senecio* species as affected by anaerobiosis. *Physiol. Plant.* **37,** 117–122.
Lambers, H. (1980). The physiological significance of cyanide-resistant respiration in higher plants. *Plant Cell Environ.* **3,** 293–302.
Lambers, H., and Smakman, G. (1978). Respiration of the roots of flood-tolerant and flood-intolerant *Senecio* species: affinity for oxygen and resistance to cyanide. *Physiol. Plant.* **42,** 163–166.
Lambers, H., Steingrover, E., and Smakman, G. (1978). The significance of oxygen transport and of metabolic adaptation in flood-tolerance of *Senecio* species. *Physiol. Plant.* **43,** 277–281.
Larqué-Saavedra, A., Wilkins, H., and Wain, R. L. (1975). Promotion of cress root elongation in white light by 3,5-diiodo-4-hydroxybenzoic acid. *Planta* **126,** 269–272.
Laties, G. G. (1982). The cyanide-resistant, alternative path in higher plant respiration. *Annu. Rev. Plant Physiol.* **33,** 519–555.
Lee, R. B. (1978). Inorganic nitrogen metabolism in barley roots under poorly aerated conditions. *J. Exp. Bot.* **29,** 693–708.
Letey, J., Stolzy, L. H., Valoras, N., and Szuszkiewicz, T. E. (1962). Influence of oxygen diffusion rate on sunflower growth at various soil and air temperature. *Agron. J.* **54,** 538–540.
Letey, J., Stolzy, L. H., and Valoras, N. (1965). Relationships between oxygen diffusion rate and corn growth. *Agron. J.* **57,** 91–92.
Leyshon, A. J., and Sheard, R. W. (1974). Influence of short-term flooding on the growth and plant nutrient composition of barley. *Can. J. Soil Sci.* **54,** 463–473.
Livne, A., and Vaadia, Y. (1965). Stimulation of transpiration rate in barley leaves by kinetin and gibberellic acid. *Physiol. Plant* **18,** 658–664.
Lloyd, F. E. (1920). Environmental changes and their effect upon boll shedding in cotton (*Gossypium herbaceum*). *Ann. N.Y. Acad. Sci.* **29,** 1–11.
Lopez-Saez, J. F., Gonzalez-Bernaldez, F., Gonzalez-Fernadez, A., and Garcia-Ferrero, G. (1969). Effect of temperature and oxygen tension on root growth, cell cycle and cell elongation. *Protoplasma* **67,** 213–221.
Lüttge, V., and Pitman, M. G. (1976). Transport and energy. *Encycl. Plant Physiol. New Ser.* **2A,** 251–259.
MacDonald, I. R., and Gordon, D. C. (1978). An inhibitory effect of excess moisture on the early development of *Sinapis alba* L. seedlings. *Plant Cell Environ.* **1,** 313–316.
Malone, M., and Ridge, I. (1983). Ethylene-induced growth and proton excretion in the aquatic plant *Nymphoides peltata*. *Planta* **157,** 71–73.
Martin, M. H. (1968). Conditions affecting the distribution of *Mercurialis perennis* L. in certain Cambridgeshire woodlands. *J. Ecol.* **56,** 777–793.
Mason, T. G., and Phillis, E. (1936). Further studies of transport in the cotton plant. V. Oxygen supply and the activation of diffusion. *Ann. Bot. (London)* **50,** 455–499.
Massimino, D., Andre, M., Richaud, C., Daguenet, A., Masimino, J., and Vivali, J. (1981). The effect of a day at low irradiance of a maize crop. I. Root respiration and uptake of N, P and K. *Physiol. Plant.* **51,** 150–155.
McComb, A. J. (1965). The control of elongation in *Callitriche* shoots by environment and gibberellic acid. *Ann. Bot. (London)* **29,** 445–459.
McManmon, M., and Crawford, R. M. M. (1971). A metabolic theory of flooding tolerance. The significance of enzyme distribution and behaviour. *New Phytol.* **70,** 299–306.
McMichael, B. L., Jordon, W. R., and Powell, R. D. (1972). An effect of water stress on ethylene production by intact cotton petioles. *Plant Physiol.* **49,** 658–660.
McPherson, D. C. (1939). Cortical air spaces in the roots of *Zea mays* L. *New Phytol.* **38,** 190–202.
Mees, G. G., and Weatherley, P. E. (1957). The mechanism of water absorption by roots. II. The role of hydrostatic pressure gradients across the cortex. *Proc. R. Soc. London B* **141,** 381–391.

Meidner, H. (1967). The effect of kinetin on stomatal opening and the rate of intake of carbon dioxide in mature primary leaves of barley. *J. Exp. Bot.* **18,** 556–561.
Metraux, J.-P., deZacks, R., and Kende, H. (1982). The response of deep-water rice to submergence and ethylene. *Mich. State Univ.–DOE Plant Res. Lab. Annu. Rep.* **1981,** 80–81.
Minchin, F. R., and Pate, J. S. (1975). Effects of water, aeration and salt regime on nitrogen fixation in a nodulated legume—definition of an optimum root environment. *J. Exp. Bot.* **26,** 60–69.
Minchin, F. R., and Summerfield, R. J. (1976). Symbiotic nitrogen fixation and vegetative growth of cowpea [*Vigna unguiculata* (L.) Walp.] in waterlogged conditions. *Plant Soil* **45,** 113–127.
Mishra, B. N. (1973). On the physiological anatomy of *Ipomoea crassicaulis* (Benth.) Robinson. *Proc. Indian Sci. Congr. Assoc.* **60,** 389–390.
Mocquot, B., Pradet, A., and Litvak, S. (1977). DNA synthesis and anoxia in rice coleoptiles. *Plant Sci. Lett.* **9,** 365–371.
Mocquot, B., Prat, C., Mouches, C., and Pradet, A. (1981). Effect of anoxia on energy charge and protein synthesis in rice embryo. *Plant Physiol.* **68,** 636–640
Moldau, H. (1973). Effects of various water regimes on stomatal and mesophyll conductances of bean leaves. *Photosynthetica 7,* 1–7.
Morisset, C. (1973). Compartement des raeines isolées de *Lycopersicon esculentum* (Solanacées) cultivées '*in vitro*' et soumises à l'anoxie. *C.R. Hebd. Seances Acad. Sci. Ser. D* **276,** 311–314.
Morris, D. A., and Kadir, G. O. (1972). Pathways of auxin transport in the intact pea seedling (*Pisum sativum* L.). *Planta* **107,** 171–182.
Mullins, M. G. (1972). Auxin and ethylene in adventitious root formation in *Phaseolus aureus* (Roxb.). *In* "Plant Growth Substances 1970" (D. J. Carr, ed.), pp. 526–533. Springer-Verlag, Berlin.
Musgrave, A., and Walters, J. (1973). Ethylene-stimulated growth and auxin transport in *Ranunculus sceleratus* petioles. *New Phytol.* **72,** 783–789.
Musgrave, A., and Walters, J. (1974). Ethylene and buoyancy control rachis elongation of the semi-aquatic fern *Regnillidium diphyllum. Planta* **121,** 51–56.
Musgrave, A., Jackson, M. B., and Ling, E. (1972). *Callitriche* stem elongation is controlled by ethylene and gibberellin. *Nature (London) New Biol.* **238,** 93–96.
Needham, P. (1969). "Some Effects of Anaerobic Soil Conditions on Plant Growth and Metabolism." MSc. Thesis, Imperial Coll., Univ. of London.
Neilson-Jones, W. (1935). Organic soils and epinastic response. *Nature (London)* **136,** 554.
Nicholson, H. H., and Firth, D. H. (1958). The effect of ground water level on the yield of some common crops on a fen peat soil. *J. Agric. Sci.* **50,** 243–252.
Nuritdinov, N., and Vartapetian, B. B. (1980). Translocation of ^{14}C-sucrose in cotton plants under conditions of root anaerobiosis. *Fiziol. Rast. (Moscow)* **27,** 814–820.
Oliveira, L. (1977). Changes in the ultrastructure of mitochondria of roots of *Triticale* subjected to anaerobiosis. *Protoplasma* **91,** 267–280.
Olomu, M. O., and Racz, G. J. (1974). Effect of soil water and aeration on Fe and Mn utilization by flax. *Agron. J.* **68,** 523–526.
Olymbios, C. M., and Schwabe, W. W. (1977). Effects of aeration and soil compaction on growth of the carrot, *Daucus carota* L. *J. Hortic. Sci.* **52,** 485–500.
Onderdonk, J. J., and Ketcheson, J. W. (1973). Effect of soil temperature on direction of corn root growth. *Plant Soil* **39,** 177–186.
Opik, H. (1973). Effect of anaerobiosis on respiratory rate, cytochrome oxidase activity and mitochondrial structures in coleoptiles of rice (*Oryza sativa* L.). *J. Cell Sci.* **12,** 725–729.
Oppenheimer, H. R. (1960). Adaptation to drought: xerophytism. *In* "Plant–Water Relationships in Arid and Semi-Arid Conditions," pp. 105–138. UNESCO, Paris.
Osborne, D. J., Jackson, M. B., and Milborrow, B. V. (1972). Physiological properties of abscission accelerator from senescent leaves. *Nature (London) New Biol.* **240,** 98–101.

Palmer, J. H. (1975). Temperature sensitivity of the latent phase in ethylene-induced elongation. *Plant Physiol.* **55,** 580–581.

Pankhurst, C. E., and Sprent, J. I. (1975). Surface features of soybean root nodules. *Protoplasma* **85,** 85–98.

Parsons, L. R., and Kramer, P. J. (1974). Diurnal cycling in root resistance to water movement. *Physiol. Plant.* **30,** 19–23.

Phillips, I. D. J. (1964a). Root-shoot hormone relations. I. The importance of an aerated root system in the regulation of growth hormone levels in the shoot of *Helianthus annuus. Ann. Bot. (London)* **28,** 17–35.

Phillips, I. D. J. (1964b). Root-shoot hormone relations. II. Changes in endogenous auxin concentration produced by flooding of the root system in *Helianthus annuus. Ann. Bot. (London)* **28,** 37–45.

Phillips, J. W. (1947). Studies on fermentation in rice and barley. *Am. J. Bot.* **34,** 62–72.

Pitman, M. G. (1976). Ion uptake by plant roots. *Encycl. Plant Physiol. New Ser.* **2B,** 95–128.

Pradet, A., and Bomsel, J. L. (1978). Energy metabolism in plants under hypoxia and anoxia. *In* "Plant Life in Anaerobic Environments" (D. D. Hook and R. M. M. Crawford, eds.), pp. 89–118. Ann Arbor Sci. Publ., Ann Arbor, Michigan.

Pradet, A., and Raymond, P. (1983). Adenine nucleotide ratios and adenylate energy charge in energy metabolism. *Annu. Rev. Plant Physiol.*, in press.

Purvis, A. C., and Williamson, R. E. (1972). Effects of flooding and gaseous composition of the root environment on growth of corn. *Agron. J.* **64,** 674–678.

Quereshi, F. A., and Spanner, D. C. (1973). The effect of nitrogen on the movement of tracers down the stolon of *Saxifraga tormentosa,* with some observations on the influence of light. *Planta* **110,** 131–144.

Quinn, P. J., and Chapman, D. (1980). The dynamics of membrane structure. *CRC Crit. Rev. Biochem.* **8,** 1–117.

Quinn, P. J., and Williams. W. P. (1978). Plant lipids and their role in membrane function. *Prog. Biophys. Mol. Biol.* **34,** 109–173.

Railton, I. D., and Reid, D. M. (1973). Effects of benzyladenine on the growth of waterlogged tomato plants. *Planta* **111,** 261–266.

Ralston, E. J., and Imsande, J. (1982). Entry of oxygen and nitrogen into intact soybean nodules. *J. Exp. Bot.* **33,** 208–214.

Rao, K. P., and Rains, D. W. (1976). Nitrate absorption by barley. I. Kinetics and energetics. *Plant Physiol.* **57,** 55–58.

Raymond, P., and Pradet, A. (1980). Stabilization of adenine nucleotide ratios at various values by an oxygen limitation of respiration in germinating lettuce (*Lactuca sativa*) seeds. *Biochem. J.* **190,** 39–44.

Raymond, P., Bruzau, F., and Pradet, A. (1978). Étude du transport d'oxygène des parties aériennes aux racines à l'aide d'un parametre du métabolisme: la charge énergetique. *C.R. Hebd. Seances Acad. Sci. Ser. D* **286,** 1061–1063.

Read, D. J., Koucheki, H. K., and Hodgson, J. (1976). Vesicular–arbuscular mycorrhizae in natural vegetation systems. I. The occurrence of infection. *New Phytol.* **77,** 641–653.

Rebeille, F., Bligny, R., and Douce, R. (1980). Oxygen and temperature effects on the fatty acid composition of sycamore cells (*Acer pseudoplatanus* L.). *In* "Biogenesis and Function of Plant Lipids" (P. Mazliak, P. Benveniste, C. Costes, and R. Druce, eds.), pp. 203–206. Elsevier/North Holland, Amsterdam.

Reid, D. M., and Crozier, A. (1971). Effects of waterlogging on the gibberellin content and growth of tomato plants. *J. Exp. Bot.* **22,** 39–48.

Reid, D. M., Crozier, A., and Harvey, B. M. R. (1969). The effects of flooding on the export of gibberellins from the root to the shoot. *Planta* **89,** 376–379.

Richmond, A. E., and Lang, A. (1957). Effect of kinetin on protein content and survival of detached *Xanthium* leaves. *Science (Washington,D.C.)* **125,** 650–651.

Ridge, I., and Amarsinghe, I. (1981). Ethylene as a stress hormone: its role in submergence responses. *In* "Responses of Plants to Environmental Stress and Their Mediation by Plant Growth Substances," Abstr. pp. 13–14. Br. Plant Growth Regul. Group–Assoc. of Appl. Biol., Wantage.

Robards, A. W., Clarkson, D. T., and Sanderson, J. (1979). Structure and permeability of the epidermal/hypodermal layers of the sand sedge (*Carex arenaria,* L.). *Protoplasma* **101,** 331–347.

Rodionova, M. A., Kholodenko, N. Ya., and Grineva, G. M. (1977). Activity of adenylate kinase in corn plants under conditions of waterlogging. *Sov. Plant Physiol. (Engl. Transl.)* **24,** 652–656.

Rozema, J., and Blom, B. (1977). Effects of salinity and inundation on the growth of *Agrostis stolonifera* and *Juncus gerardii. J. Ecol.* **65,** 213–222.

Rumpho, M. E., and Kennedy, R. A. (1981). Anaerobic metabolism in germinating seeds of *Echinochloa crus-galli* (barnyard grass). Metabolite and enzyme studies. *Plant Physiol.* **68,** 165–168.

Russell, E. W. (1973). "Soil Conditions and Plant Growth," 10th ed. Longman, London.

Sachs, M. M., Freeling, M., and Okimoto, R. (1980). The anaerobic proteins of maize. *Cell* **20,** 761–767.

Saglio, P. H., and Pradet, A. (1980). Soluble sugars, respiration and energy charge during aging of excised maize root tips. *Plant Physiol.* **66,** 516–519.

Saglio, P. H., Raymond, P., and Pradet, A. (1980). Metabolic activity and energy charge of excised maize root tips under anoxia. Control by soluble sugars. *Plant Physiol.* **66,** 1053–1057.

Saif, S. R. (1981). The influence of soil aeration on the efficiency of vesicular–arbuscular mycorrhizae. I. Effect of soil oxygen on the growth and mineral uptake of *Eupatorium odoratum* L. inoculated with *Glomus macrocarpus. New Phytol.* **88,** 649–659.

Sartoris, G. B., and Belcher, B. A. (1949). The effect of flooding on flowering and survival of sugar cane. *Sugar* **44,** 36–39.

Selman, I. W., and Sandanam, S. (1972). Growth responses of tomato plants in non-aerated water culture to foliar sprays of gibberellic acid and benzyladenine. *Ann. Bot. (London)* **36,** 837–848.

Sheikh, K. H., and Rutter, A. J. (1969). The responses of *Molinia caerulea* and *Erica tetralix* to soil aeration and related factors. I. Root distribution in relation to soil porosity. *J. Ecol.* **57,** 713–726.

Sieben, W. H. (1964). [The effect of drainage condition of the soil on nitrogen supply and yield]. *Landbouwkd. Tijdschr.* **76,** 784–802.

Singh, R., and Ghildyal, B. P. (1980). Soil submergence effects on nutrient uptake, growth and yield of five corn cultivars. *Agron. J.* **72,** 737–741.

Smith, A. M., and ap Rees, T. (1979a). Effects of anaerobiosis on carbohydrate oxidation by roots of *Pisum sativum. Phytochemistry* **18,** 1453–1458.

Smith, A. M., and ap Rees, T. (1979b). Pathways of carbohydrate fermentation in the roots of marsh plants. *Planta* **146,** 327–334.

Smith, K. A., and Jackson, M. B. (1974). Ethylene, waterlogging and plant growth [review article]. *Annu. Rep. Agric. Res. Counc. Letcombe Lab.* **1973,** 60–75.

Smith, K. A., and Robertson, P. D. (1971). Effect of ethylene on root extension of cereals. *Nature (London)* **234,** 148–149.

Smith, K. A., and Russell, R. S. (1969). Occurrence of ethylene, and its significance, in anaerobic soil. *Nature (London)* **222,** 769–771.

Smucker, A. J. M., and Adler, F. (1980). Accumulation and loss of toxic assimilates by plant roots. *Abstr. Am. Soc. Agron. Mtg., 72nd,* p. 93.

Sojka, R. E., and Stolzy, L. H. (1980). Soil-oxygen effects on stomatal response. *Soil Sci.* **130,** 350–358.

Sojka, R. E., Stolzy, L. H., and Kaufman, M. R. (1975). Wheat growth related to rhizosphere temperature and oxygen levels. *Agron. J.* **67,** 591–596.

Solomos, T. (1977). Cyanide-resistant respiration in higher plants. *Annu. Rev. Plant Physiol.* **28,** 279–297.

Søndergaard, M., and Laegaard, S. (1977). Vesicular–arbuscular mycorrhiza in some aquatic vascular plants. *Nature (London)* **268,** 232–233.

Sprent, J. I. (1969). Prolonged reduction of acetylene by detached soybean nodules. *Planta* **88,** 372–375.

Sprent, J. I. (1971). Effects of water stress on nitrogen fixation in root nodules. *Plant Soil* **special volume,** 225–228.

Sprent, J. I., and Gallacher, A. (1976). Anaerobiosis in soybean root nodules under water stress. *Soil Biol. Biochem.* **8,** 317–320.

Stanley, C. D., Kaspar, T. C., and Taylor, H. M. (1980). Soybean top and root response to temporary water-tables imposed at three different stages of growth. *Agron. J.* **72,** 341–346.

Stelzer, R., and Läuchli, A. (1977). Salt- and flooding tolerance of *Puccinellia peisonis.* I. The effect of NaCl- and KCl-salinity on growth at varied oxygen supply to the root. *Z. Pflanzenphysiol.* **83,** 35–42 (in German, with English summary.)

Stelzer, R., and Läuchli, A. (1980). Salt- and flooding tolerance of *Puccinellia peisonis.* IV. Root respiration and the role of aerenchyma in providing atmospheric oxygen to the roots. *Z. Pflanzenphysiol.* **97,** 171–178.

Stoltz, L. P., and Hess, C. E. (1966). The effects of girdling upon root initiation, carbohydrates and amino acids. *Proc. Am. Soc. Hortic. Sci.* **89,** 734–743.

Strickland, R. W. (1968). The effect of drainage on physiological disorders of rice grown on the sub-coastal plains of the Adelaide River, N. T. *Aust. J. Exp. Agric. Anim. Husb.* **8,** 212–222.

Stumpf, P. K. (1980). Biosynthesis of saturated and unsaturated fatty acids. *In* "The Biochemistry of Plants" (P. K. Stumpf, ed.), Vol. 4, pp. 177–204. Academic Press, New York.

Sugano, T., Oshino, N., and Chance, B. (1974). Mitochondrial function under hypoxic conditions. The steady states of cytochrome *c* reduction and of energy metabolism. *Biochim. Biophys. Acta* **347,** 340–358.

Szlovak, S. (1975). A study of flooding effect on maize transpiration at two nutrient levels. *Acta Bot. Acad. Sci. Hung.* **21,** 167–174.

Teal, J. M., and Kanwisher, J. W. (1966). Gas transport in the marsh grass, *Spartina alterniflora. J. Exp. Bot.* **17,** 355–361.

Thimann, K. V. (1980). The senescence of leaves. *In* "Senescence in Plants" (K. V. Thimann, ed.), pp. 85–115. CRC Press, Boca Raton, Florida.

Thomas, D. S., and Rose, A. H. (1979). Inhibitory effect of ethanol on growth and solute accumulation by *Saccharomyces cerevisiae* as affected by plasma-membrane lipid composition. *Arch. Microbiol.* **122,** 49–55.

Thomas, D. S., Hossack, J. A., and Rose, A. H. (1978). Plasma-membrane lipid composition and ethanol tolerance in *Saccaromyces cerevisiae. Arch. Microbiol.* **117,** 239–245.

Thomas, H., and Stoddart, J. L. (1980). Leaf senecence. *Annu. Rev. Plant Physiol.* **31,** 83–111.

Tjepkema, J. (1977). The role of oxygen diffusion from the shoots and nodule roots in nitrogen fixation by root nodules of *Myrica gale. Can. J. Bot.* **56,** 1365–1371.

Torrey, J. G., and Callaham, D. (1978). Determinate development of nodule roots in actinomycete-induced root nodules of *Myrica gale. Can. J. Bot.* **56,** 1357–1364.

Trought, M. C. T., and Drew, M. C. (1980a). The development of waterlogging damage in wheat seedlings (*Triticum aestivum* L.). I. Shoot and root growth in relation to changes in the concentrations of dissolved gases and solutes in the soil solution. *Plant Soil* **54,** 77–94.

Trought, M. C. T., and Drew, M. C. (1980b). The development of waterlogging damage in young wheat plants in anaerobic solution cultures. *J. Exp. Bot.* **31,** 1573–1585.

Trought, M. C. T., and Drew, M. C. (1980c). The development of waterlogging damage in wheat seedlings (*Triticum aestivum* L.). II. Accumulation and redistribution of nutrients by the shoot. *Plant Soil* **56,** 187–199.

Trought, M. C. T., and Drew, M. C. (1981). Alleviation of injury to young wheat plants in anaerobic solution culture in relation to the supply of nitrate and other inorganic nutrients. *J. Exp. Bot.* **32,** 509–522.

Turkova, N. S. (1944). Growth reactions in plants under excessive watering. *Dokl. Akad. Nauk SSSR Ser. Biol.* **42,** 87–90.

Turner, F. T., Chen, C. C., and McCauley, G. N. (1981). Morphological development of rice seedlings in water at controlled oxygen levels. *Agron. J.* **73,** 566–570.

Turner, J. F., and Turner, D. H. (1980). The regulation of glycolysis and the pentose phosphate pathway. *In* "The Biochemistry of Plants" (D. D. Davies, ed.), Vol. 2, pp. 279–316. Academic Press, New York.

Vámos, R. (1957). Chemical examination of the water of flooded rice fields. *Nature (London)* **180,** 1484–1485.

Vámos, R., and Köves, E. (1972). Role of light in the prevention of the poisoning action of hydrogen sulphide in the rice plant. *J. Appl. Ecol.* **9,** 519–525.

VanderZee, D., and Kennedy, R. A. (1981). Germination and seedling growth in *Echinochloa crus-galli* var. *oryzicola* under anoxic conditions: structural aspects. *Am. J. Bot.* **68,** 1269–1277.

Van Hoorn, J. W. (1958). Results of a ground water level experimental field with arable crops on clay soil. *Neth. J. Agric. Sci.* **6,** 1–10.

Van't Woudt, B. D., and Hagan, R. M. (1957). Crop responses to excessively high soil moisture levels. *In* "Drainage of Agricultural Lands" (J. N. Luthin, ed.), pp. 514–611. Am. Soc. Agric. Eng., Madison, Wisconsin.

Varade, S. B., Stolzy, L. H., and Letey, J. (1970). Influence of temperature, light intensity, and aeration on growth and root porosity of wheat, *Triticum aestivum*. *Agron. J.* **62,** 505–507.

Vartapetian, B. B. (1982). Pasteur effect visualized by electron microscopy. *Naturwissenschaften* **69,** 99.

Vartapetian, B. B., Andreeva, I. N., Kozlova, G. I., and Agapova, L. P. (1977). Mitochondrial ultrastructure in roots of mesophyte and hydrophyte at anoxia and after glucose feeding. *Protoplasma* **91,** 243–256.

Vartapetian, B. B., Mazliak, P., and Lance, C. (1978a). Lipid biosynthesis in rice coleoptiles grown in the presence or in the absence of oxygen. *Plant Sci. Lett.* **13,** 321–328.

Vartapetian, B. B., Andreeva, I. N., and Nuritdinov, N. (1978b). Plant cells under oxygen stress. *In* "Plant Life in Anaerobic Environments" (D. D. Hook and R. M. M. Crawford, eds.), pp. 13–88. Ann Arbor Sci. Publ., Ann Arbor, Michigan.

Vlamis, J., and Davis, A. R. (1944). Effects of oxygen tension on certain physiological responses of rice, barley, and tomato. *Plant Physiol.* **19,** 33–51.

Waggoner, P. E. (1966). Decreasing transpiration and the effect upon growth. *In* "Plant Environment and Efficient Water Use" (W. H. Pierre, D. Kirkham, J. Pesek, and R. Shaw, eds.), pp. 49–68. Am. Soc. Agron. and Soil Sci. Soc. Am., Madison, Wisconsin.

Wample, R. L., and Reid, D. M. (1975). Effect of aeration on the flood-induced formation of adventitious roots and other changes in sunflower (*Helianthus annuus* L.). *Planta* **127,** 263–270.

Wample, R. L., and Reid, D. M. (1978). Control of adventitious root production and hypocotyl hypertrophy of sunflower (*Helianthus annuus*) in response to flooding. *Physiol. Plant.* **44,** 351–358.

Wample, R. L., and Reid, D. M. (1979). The role of endogenous auxins and ethylene in the formation of adventitious roots and hypocotyl hypertrophy in flooded sunflower plants (*Helianthus annuus* L.). *Physiol. Plant.* **45,** 219–226.

Wang, T. L., and Woolhouse, H. W. (1982). Hormonal aspects of senescence in plant development. *In* "Growth Regulators in Plant Senescence" (M. B. Jackson, B. Grout, and I. A. Mackenzie, eds.), Monogr. No. 8, pp. 5–25. Br. Plant Growth Regul. Group, Wantage.

Wardle, K. (1982). Stomatal responses and foliar senescence. *In* "Growth Regulators in Plant Senecence" (M. B. Jackson, B. Grout, and I. A. Mackenzie, eds.), Monogr. No. 8, pp. 85–92. Br. Plant Growth Regul. Group, Wantage.

Watt, T. A., and Haggar, R. J. (1980). The effect of height of water-table on the growth of *Holcus lanatus* with reference to *Lolium perenne*. *J. Appl. Ecol.* **17,** 423–430.

Webb, T. (1982). "Some Aspects of Root Growth, Respiration and Internal Aeration in Wetland and Non-Wetland species." Ph.D. Thesis, Univ. of Hull, England.

Webb, T., and Armstrong, W. (1983). The effects of anoxia and carbohydrates on the growth and viability of rice, pea and pumpkin roots. *J. Exp. Bot.*, **34,** 579–603.

Webster, B. D. (1973). Ultrastructural studies of abscission in *Phaseolus:* ethylene effects on cell walls. *Am. J. Bot.* **60,** 436–447.

Webster, P. L. (1980). 'Stress' protein synthesis in pea root meristem cells? *Plant Sci. Lett.* **20,** 141–145.

Webster, P. L., and Van't Hof, J. (1969). Dependence on energy and aerobic metabolism of initiation of DNA synthesis and mitosis by G_1 and G_2 cells. *Exp. Cell Res.* **55,** 88–94.

Webster, P. W. D., and Eavis, B. W. (1972). Effects of flooding on sugarcane growth. I. Stage of growth and duration of flooding. *Proc. Int. Sugar Cane Technol. Congr. 14th,* pp. 708–714.

Wenkert, W., Fausey, N. R., and Watters, H. D. (1981). Flooding responses in *Zea mays* L. *Plant Soil* **62,** 351–366.

Wesseling, J. (1974). Crop growth and wet soils. *In* "Drainage for Agriculture" (J. van Schilfgaarde, ed.), pp. 7–37. Am. Soc. Agron., Madison, Wisconsin.

West, D. W., and Taylor, J. A. (1980). The response of *Phaseolus vulgaris* L. to root-zone anaerobiosis, waterlogging and high sodium chloride. *Ann. Bot. (London)* **46,** 51–60.

Wheeler, A. W. (1971). Auxins and cytokinins exuded during formation of roots by detached primary leaves and stems of dwarf French bean (*Phaseolus vulgaris* L.). *Planta* **98,** 128–135.

Whitlow, T. H., and Harris, R. W. (1979). Flood tolerance in plants: a state-of-the-art review. *U.S. Army Office Chief Eng. Tech. Rep.* No. E79, pp. 1–257.

Whitney, J. B., Jr. (1942). "Effects of the Composition of the Soil Atmosphere on the Absorption of Water by Plants." Ph.D. Thesis, Ohio State Univ.

Wiedenroth, E.-M. (1981). The distribution of ^{14}C-photoassimilates in wheat seedlings under root anaerobiosis and DNP application to the roots. *In* "Structure and Function of Plant Roots" (R. Brouwer and O. Gasparikova, eds.), pp. 389–393. Nijhoff/Junk, The Hague.

Wiedenroth, E.-M., and Poskuta, J. (1981). The influence of oxygen deficiency in roots on CO_2 exchange rates of shoots and distribution of ^{14}C-photoassimilates of wheat seedlings. *Z. Pflanzenphysiol.* **103,** 459–467.

Wignarajah, K., and Greenway, H. (1976). Effects of anaerobiosis on activities of alcohol dehydrogenase and pyruvate decarboxylase in roots of *Zea mays*. *New Phytol.* **77,** 575–584.

Wignarajah, K., Greenway, H., and John, C. D. (1976). Effect of waterlogging on growth and activity of alcohol dehydrogenase in barley and rice. *New Phytol.* **77,** 585–592.

Wilkins, H., Alejar, A. A., and Wilkins, S. M. (1978). Some effects of halogenated hydroxybenzoic acids on seedling growth. *In* "Opportunities for Chemical Plant Growth Regulation," Monogr. No. 21, pp. 83–94. Br. Crop Protection Council, Croydon.

Willey, C. R. (1970). Effect of short periods of anaerobic and near-anaerobic conditions on water uptake by tobacco roots. *Agron. J.* **62,** 224–229.

Williamson, R. E. (1968). Influence of gas mixtures on cell division and root elongation of broad bean, *Vicia faba* L. *Agron. J.* **60,** 317–321.

Williamson, R. E. (1970). Effect of soil gas composition and flooding on growth of *Nicotiana tabacum* L. *Agron. J.* **62,** 80–83.

Wright, M., and Osborne, D. J. (1974). Abscission in *Phaseolus vulgaris*. The positional differentiation and ethylene induced expansion growth of specialised cells. *Planta* **120,** 163–170.

Wright, S. T. C. (1972). Physiological and biochemical responses to wilting and other stress conditions. *In* "Crop Processes in Controlled Environments" (A. R. Rees, K. E. Cockshull, D. W. Hand, and R. G. Hurd, eds.), pp. 349–361. Academic Press, London and New York.

Xu, X. d., and Lou, C. h. (1980). Presence of roots—a prerequisite for normal function of stomata on sweet potato leaves. *Bull. Beijing Agric. Univ.* **first issue,** 44–45 (In Chinese with English abstr.).

Yeo, A. R., Kramer, D., Läuchli, A., and Gullasch, J. (1977). Ion distribution in salt-stressed mature *Zea mays* roots in relation to ultrastructure and retention of sodium. *J. Exp. Bot.* **28,** 17–29.

Yu, P. T., Stolzy, L. H., and Letey, J. (1969). Survival of plants under prolonged flooded conditions. *Agron. J.* **61,** 844–847.

Zablotowicz, R. M., and Focht, D. D. (1979). Denitrification and anaerobic, nitrate-dependent acetylene reduction in cowpea rhizobium. *J. Gen. Microbiol.* **111,** 445–448.

Zablotowicz, R. M., Eskew, D. L., and Focht, D. D. (1978). Denitrification in *Rhizobium. Can. J. Microbiol.* **24,** 757–760.

Zimmerman, P. W. (1930). Oxygen requirements for root growth of cuttings in water. *Am. J. Bot.* **17,** 842–861.

Zimmerman, P. W., and Hitchcock, A. E. (1933). Initiation and stimulation of adventitious roots caused by unsaturated hydrocarbon gases. *Contrib. Boyce Thompson Inst.* **5,** 351–369.

Zobel, R. W. (1973). Some physiological characteristics of the ethylene-requiring tomato mutant diageotropica. *Plant Physiol.* **52,** 385–389.

CHAPTER 4

Responses of Woody Plants to Flooding

T. T. KOZLOWSKI
Department of Forestry
University of Wisconsin
Madison, Wisconsin

I. INTRODUCTION

Flooding of soils that support forest, orchard, and shade trees is a common occurrence. Many trees grow in swamps and on floodplains that are inundated during much of the year, and others grow where they are temporarily flooded. The poor soil aeration that accompanies flooding of soil is associated with physiological changes in woody plants that variously influence their growth and species composition in forest stands. Poor soil aeration is not limited to flooded

Flooding and Plant Growth

ISBN 0-12-424120-4

soil but is a problem with many trees and shrubs growing in fine-textured soils (Kozlowski, 1982). With few exceptions growth of trees is adversely affected by flooding of soil for a few weeks or more during the growing season. However, if trees are flooded during the dormant season, and if the flood water recedes before the growing seasons starts, growth of trees may not be adversely affected and may even be stimulated (Broadfoot, 1967).

II. SPECIES DISTRIBUTION, COMPOSITION, AND SUCCESSION IN RESPONSE TO FLOODING

Very few species of woody plants are adapted to grow on permanently flooded sites. Prominent among these are *Taxodium distichum, Nyssa aquatica,* species of *Salix,* and the mangroves. *Taxodium distichum* is dominant in certain swamp forests in the southeastern United States where soil water is plentiful and fairly permanent (Fowells, 1965). *Nyssa aquatica* occurs most commonly in low, wet flats or sloughs on the floodplain of alluvial streams. Its best development occurs in the Coastal Plain bottomlands of the southeastern United States and in the *Taxodium* swamps of western Louisiana and southeastern Texas.

Species composition of wetlands is influenced by such factors as flooding frequency and duration, a variety of soil factors, and flood tolerance of seedlings. Because a complex gradient of ecological factors is oriented more or less perpendicularly to a river channel, species composition may be expected to vary with this gradient. As distance from a river channel increases, the soil generally becomes less friable, and depth to gley, soil drainage, and soil aeration decrease. Frye and Quinn (1979) found that depth to the gley horizon was more closely correlated with distribution of vegetation along the Raritan River in New Jersey than was any other factor studied (mineral nutrients, soil texture, cation-exchange capacity, and pH). The position of the gley layer appeared to exert its effects by restricting root growth to the well-drained surface soil. This was reflected in greater species diversity and biomass along the natural levee and a progressive decrease with distance from the river channel. Buchholz (1981) found that minor drainages intersecting floodplains significantly influenced flooding and drainage patterns, thereby affecting species composition.

The remainder of this section gives a few regional examples of species composition and successional trends on various types of wetlands.

A. Southern United States

Bottomlands of the southern United States support the following major forest types:

1. *Taxodium–Nyssa* Type. *Taxodium distichum* and *Nyssa aquatica* are the major species in first bottoms and terraces in deep sloughs and swamps in the

lower Mississippi River Valley. In nonalluvial swamps of the Coastal Plain, *Nyssa sylvatica* var. *biflora* is associated with *Taxodium,* but all three species sometimes grow in pure stands. Either drainage or continuous sedimentation may eventually alter these sites so that *Taxodium* and *Nyssa* are replaced by such species as *Fraxinus pennsylvanica, Ulmus americana, Acer rubrum, Celtis laevigata, Carya aquatica, Quercus lyrata, Q. nuttallii, Q. laurifolia,* and *Diospyros virginiana* (Putnam, 1951; Putnam *et al.*, 1960; Smith and Linnartz, 1980).

2. *Populus–Salix* Type. Pure stands of *Populus deltoides* or *Salix nigra* invade recent alluvial deposits. *Populus* occurs on higher, better-drained and coarser sediments; *Salix* on low-lying, wet, and fine-textured soil. Both types are temporary. *Populus* is replaced by *Acer saccharinum, Ulmus americana, Celtis laevigata, Platanus occidentalis, Carya illinoensis, Fraxinus pennsylvanica,* and *Acer negundo. Liquidambar styraciflua* and *Quercus rubra* are major species in later successional stages.

In low areas *Salix nigra* stands are succeeded by *Taxodium distichum, Fraxinus pennsylvanica, Ulmus americana, Acer rubrum,* and *Celtis laevigata.* These species are eventually followed by *Quercus lyrata, Carya aquatica,* and *Diospyros virginiana.* With increased sedimentation these sites may be invaded by *Q. nuttallii* and *Q. nigra.*

When sloughs, swamps, and oxbows fill with sediment, the pioneer invader is *Salix nigra.* Where flooded conditions prevail for most of the year, *Salix* is replaced by *Taxodium* and *Nyssa.* Under less continuous flooding, *Salix* is replaced by *Quercus lyrata, Carya aquatica, Fraxinus pennsylvanica, Diospyros virginiana,* and *Q. nuttallii* (Putnam *et al.*, 1960).

3. Mixed Bottomland Hardwoods. This type occurs on stream bottoms and terraces throughout the southeastern United States. Several successional stages may be found, depending on the pattern of alluvium deposition. The species in each successional stage vary with the frequency of flooding.

On lower river fronts with fine-textured soil, *Salix nigra* is a pioneer and is succeeded by *Taxodium distichum* and *Fraxinus pennsylvanica.* Much later these species are replaced by *Quercus lyrata, Carya aquatica,* and *Diospyros virginiana.* If sedimentation continues and the site changes to a low flat, *Quercus nuttallii, Q. phellos,* and *Q. nigra* may also invade. The best and highest sites in first bottoms support *Q. falcata* var. *pagodaefolia* and *Liquidambar styraciflua.*

B. Central United States

The central-region forests are complex, and their composition differs widely with variations in soils and drainage. In bottomlands and floodplains the principal components are *Ulmus americana, Fraxinus pennsylvanica, Populus de-*

ltoides, and *Acer saccharinum,* growing singly or together. Important associates include *Celtis occidentalis, Platanus occidentalis, Salix nigra,* and *A. negundo.* Individual stands may vary from pioneer to climax stages of succession. However, heavy cutting of timber and clearing of land for agriculture have eliminated most of the climax stands (Merritt, 1980).

In southern Illinois, species composition of floodplain forests varies with duration of flooding. The heavy, poorly drained soils that are flooded for several months each year are dominated by *Acer rubrum, Fraxinus pennsylvanica, Liquidambar styraciflua,* and *Ulmus americana.* Species on moderately heavy and poorly drained soils with intermediate flooding include *Asimina triloba, Liquidambar styraciflua, Quercus michauxii, Q. muehlenbergii, Ulmus americana, Q. pagodaefolia,* and *Q. shumardii.* Species characteristic of well-drained soils with infrequent flooding are *Acer saccharum, Ulmus rubra,* and *Tilia americana* (Robertson *et al.,* 1978).

C. Lake States

The entire Lake States region of the United States, which includes large parts of the states of Minnesota, Wisconsin, and Michigan, is dotted with swamps and bogs. The hydric soils of the region may be organic, consisting of peats and mucks, or they may be gley loams and clays. Poorly drained mineral soils occur in parts of northeastern Minnesota and as lacustrine clays along the south shore of Lake Superior in northeastern Wisconsin. These soils support stands of *Picea mariana, Abies balsamea,* and *Thuja occidentalis;* the better-drained, adjacent areas support *Pinus resinosa, P. strobus,* and a variety of species of northern hardwoods (Hansen, 1980).

An important wetland forest type in the Lake States is the swamp conifer type in which *Picea mariana* dominates. In rich swamp forests the associates of *Picea mariana* include *Thuja occidentalis, Fraxinus nigra,* and *Larix laricina.* A shrub layer of *Alnus rugosa* is present except where *Thuja* is dense. This type merges with mixed conifer–deciduous forests on mineral soils along the margins and with poor swamp forest toward the interior. The transition from a rich to a poor swamp forest is usually preceded by a change in dominance from *Thuja* to *Larix* and with elimination of *Fraxinus, Abies,* and *Betula papyrifera.* There is also a change in shrubs from *Alnus* to *B. pumila* (Heinselman, 1970).

The ecological processes of Lake States wetlands may involve a hydrarch succession of lake filling, with *Picea mariana* forest representing a late successional stage and sometimes succeeded by *Thuja occidentalis* or swamp hardwoods. Where water is in excess and drainage poor, paludification occurs in which invasion of former upland areas by sedges, sphagnums, and bog forests takes place. *Picea mariana* and *Larix laricina* may also be involved in the succession. On acid swamps the *P. mariana* type is quite stable except when it is logged or burned (Hansen, 1980).

D. Mangroves

About three-fourths of the world's coastlines between 25°N and 25°S support ~20 woody species of mangroves. These shrubs to small trees occur in the tidal zone in water with a salt concentration of about 3.5%. Mangrove communities extend from the coast into the estuaries and up rivers a short distance beyond tidal influences (Daubenmire, 1978).

The mangroves are distributed among eight families and 12 genera. The most important genera are *Rhizophora, Avicennia, Laguncularia, Conocarpus, Bruguiera, Ceriops, Sonneratia, Xylocarpus, Aegiceros,* and *Lemnitzera.* The different species of mangroves tend to grow in zones depending on variations in the environmental gradient from the sea to the land. Each species has a distinct range of tolerance of depth of water and duration of tidal flooding. Salt resistance appears to be important in competition among mangrove species, and hence in zonation. *Avicennia* appears to be a weak competitor but has high salt tolerance, so small plants of this species are found far from the sea. In comparison, *Sonneratia,* a strong competitor, has low salt tolerance and is confined to outer fringes. In east Africa, from the sea inward, mangroves occur in sequential zones of *Sonneratia, Rhizophora, Ceriops,* and *Avicennia* (Walter, 1973). In Florida four zones are common, with *Rhizophora mangle* next to the open water, followed by *Avicennia nitida, Laguncularia racemosa,* and *Conocarpus erectus* (Daubenmire, 1978). The mangroves survive flooding by seawater through various morphological and physiological adaptations as discussed by Wainwright (Chapter 9 in this volume).

III. FACTORS INFLUENCING RESPONSES TO FLOODING

Responses to flooding may vary widely, with differences according to species and genetic constitution, age, properties of the floodwater, and duration of flooding. These are discussed separately.

A. Species and Genetic Constitution

The flood tolerance of different species, ecotypes, and plants of different provenances varies greatly (Kozlowski, 1982). Hall and Smith (1955) found that some species of trees survived continuous flooding for two growing seasons and others were killed by less than 4 weeks of flooding. Tolerance ratings have been based on different criteria, including growth responses of trees, amount of injury sustained, or survival. Angiosperms as a group are more flood tolerant than gymnosperms. However, a few gymnosperms are very flood tolerant.

1. *Forest and Shade Trees*

a. Angiosperms. *Nyssa aquatica* and *Salix* spp. grow in wet soils and are much more tolerant of inadequate soil aeration and flooding than species such as *Cornus florida, Liriodendron tulipifera,* and *Liquidambar styraciflua,* which grow in well-drained soil (Kramer and Kozlowski, 1979).

Flood tolerance of 14 bottomland species varied in the following order from most to least tolerant: *Acer saccharinum, Cephalanthus occidentalis, Acer negundo, Salix nigra, Populus deltoides, Fraxinus pennsylvanica, Ulmus americana, Quercus palustris, Platanus occidentalis,* and *Q. falcata* var. *pagodaefolia* (Hosner, 1960). Growth of seedlings of the wet-site species (*Nyssa aquatica, Q. palustris, F. pennsylvanica,* and *Platanus occidentalis*) varied widely. *Nyssa* and *Fraxinus* grew best in continuously saturated soils; *Quercus* and *Platanus* grew best in soil returned to the moisture equivalent daily (Dickson *et al.*, 1965).

Flood-tolerant species in Illinois include *Acer saccharinum, Populus deltoides, Platanus occidentalis, Salix nigra, Quercus macrocarpa, Gleditsia triacanthos, Acer negundo, Q. bicolor, Q. palustris,* and *Diospyros virginiana.* Somewhat tolerant species are *Cercis canadensis, Juglans nigra, Q. imbricaria, Carya ovata, Celtis occidentalis, Ulmus americana,* and *Fraxinus pennsylvanica.* Species least tolerant to flooding are *Q. velutina, Prunus serotina,* and *Sassafras albidum* (Bell and Johnson, 1974).

Considerable variation has been shown in flood tolerance of closely related species. For example, *Nyssa aquatica* tolerates flooding better than *N. sylvatica* (Hall and Smith, 1955). *Betula papyrifera* seedlings were more adversely affected by 5 weeks of flooding, and were slower to recover after the soil was drained, than seedlings of *B. nigra* (Table I). Height growth during flooding was negligible in *B. papyrifera* but substantial in *B. nigra.* Leaf epinasty and abscission were observed only in *B. papyrifera.* Flooding also induced lenticel hypertrophy and formation of adventitous roots in *B. nigra,* but not in *B. papyrifera* (Norby and Kozlowski, 1983).

Eucalyptus species varied in the following order from most to least flood tolerant: *E. grandis, E. robusta,* and *E. saligna* (Clemens *et al.*, 1978). *Eucalyptus camaldulensis* seedlings adapted better than those of *E. globulus* to flooding (Sena Gomes and Kozlowski, 1980c). Flood tolerance of *Eucalyptus* also varied among ecotypes and provenances (Karschon and Zohar, 1972, 1975; Ladiges and Kelso, 1977).

TABLE I

Effect of Flooding for 5 Weeks on Growth of *Betula papyrifera* and *Betula nigra* Seedlings[a,b]

Plant part	*Betula papyrifera* Unflooded	*Betula papyrifera* Flooded	*Betula nigra* Unflooded	*Betula nigra* Flooded
Roots				
dry wt (mg)	1356	237	1662	404
$\overline{RGR}$	0.345	−0.009	0.435	0.150
Stems				
dry wt (mg)	422	139	1025	780
$\overline{RGR}$	0.321	0.096	0.392	0.336
Leaves				
Dry wt (mg)	1685	608	1119	813
$\overline{RGR}$	0.318	0.111	0.232	0.164
Root/shoot				
dry wt/dry wt	0.654	0.323	0.783	0.258
$\overline{RGR}/\overline{RGR}$	1.088	−0.192	1.489	0.647
Leaf area				
cm^2	258	87	180	142

[a]From Norby and Kozlowski (1983).
[b]$\overline{RGR}$, Average relative growth rate.

b. Gymnosperms. As mentioned, a few species of gymnosperms are very flood tolerant. *Taxodium distichum* often grows vigorously in standing water (Krinard and Johnson, 1976). *Sequoia sempervirens* grows well on flooded alluvial flats in California (Stone and Vasey, 1968). *Pinus echinata, P. taeda,* and *P. rigida* are also somewhat flood tolerant (Hunt, 1951). *Pinus contorta* is much more tolerant than *Picea sitchensis* to flooding (Coutts and Philipson, 1978a,b).

Tolerance to flooding for up to 48 days varied in the following order: *Abies balsamea* > *Picea mariana* > *P. glauca* > *Pinus strobus* > *P. resinosa*. However, all of these species were killed by flooding for more than 48 days (Ahlgren and Hansen, 1957). Solution-culture experiments indicated the following order of species tolerance to poor soil aeration: *Picea mariana* > *Pinus banksiana* > *Picea glauca* (Zinkan *et al.*, 1974). *Thuja plicata* and *Pinus contorta* seedlings were highly tolerant of summer flooding for 4–8 weeks, *Picea sitchensis* and *Tsuga heterophylla* were intermediately tolerant, and *Pseudotsuga menziesii* was very intolerant (Minore, 1968).

Only a few studies have been conducted on flood tolerance of tropical gymnosperms. However, *Pinus pinaster* in Australia survived flooding better than *P.*

radiata (Poutsma and Simpfendorfer, 1963). *Pinus halepensis* seedlings adapted very poorly to flooding (Sena Gomes and Kozlowski, 1980d).

2. *Fruit Trees*

Several investigators have found wide variations in tolerance of fruit trees to waterlogging. In France, apples and pears were most tolerant to soil inundation, plums intermediate, and peaches and cherries least tolerant (Remy and Bidabe, 1962). Whereas pear, mango, and guava trees were very tolerant, citrus, loquat and papaw were very sensitive (Jawanda, 1961).

Rowe and Beardsell (1973) ranked fruit trees in the following order of decreasing tolerance to waterlogging: extremely tolerant (quince), very tolerant (pear), tolerant (apple), intermediately tolerant (citrus and plum), intermediately sensitive (cherry), sensitive (apricot, peach, and almond), and very sensitive (olive).

Closely related trees often show considerable variation in sensitivity to flooding. For example, of several *Prunus* species, *P. japonica* was most tolerant to soil inundation and *P. armeniaca* the least tolerant. Among intermediate-response species, *P. salicina* and *P. cerasifera* were tolerant and *P. persica, P. mume, P. tomentosa, P. pauciflora* (*pseudocerasus*), and *P. subhirtella* less tolerant (Mizutani *et al.,* 1979).

The aforementioned flood-tolerance ratings should be viewed with caution, however, because of wide variations in flood tolerance among rootstocks on which fruit trees are grown. In California, for example, apricots, almonds, and peaches on peach rootstocks were very sensitive to flooding. On apricot rootstocks, apricots were less sensitive. Apricots, plums, and prunes on myrobalan rootstock, walnut on black walnut rootstock, and fig. grape, quince, and olive tolerated spring floods well but were injured by flooding later in the season (Van't Woudt and Hagan, 1957). Red Delicious apple scions on MM.111 and seedling rootstocks were more sensitive to June flooding than those on M.27 or MM.106 stocks. Trees on M.26 rootstocks were least affected. Flooding during the dormant season had little effect on growth of scions, irrespective of rootstock (Rom and Brown, 1979). Growth of peach scions grafted on *Prunus japonica, P. salicina,* and *P. cerasifera* was reduced by waterlogging, but the trees survived. *Prunus japonica* rootstocks exhibited the greatest flood tolerance. Peach scions on *P. persica, P. armeniaca,* and *P. mume* stopped growing under waterlogged conditions, and most of the trees died within a week or two after the excess water drained away (Mizutani *et al.,* 1979).

Rowe and Beardsell (1973) reviewed the literature on sensitivity of rootstocks of fruit trees to flooding of soil and compiled the following rating for apple rootstocks: fairly resistant (M.1, M.3, M.6, M.7, M.13, M.14, M.15, M.16, Crab C, Jonathan), moderately sensitive (M.4, M.9, M.26), very sensitive (M.2, MM.104, MM.109), and extremely sensitive (M.779, M.789, M.793, Northern

TABLE II

Sensitivity of *Prunus* spp. Rootstocks to Waterlogging[a]

Sensitivity to waterlogging	Rootstock	Species
Resistant	Damas de Toulouse	*P. domestica*
	Damas GF1869	*P. domestica*
	GF8-1	*P. cerasifera* cv. Marianna
	S2544-2	
Moderately resistant	GF31 hybrid	*P. cerasifera* × *P. salicina*
	Myrob B	*P. cerasifera*
	P936	*P. cerasifera*
	P938	*P. cerasifera*
	P855	*P. cerasifera*
	P34	*P. cerasifera*
	St. Julian A	*P. domestica*
	St. Julian GF355-2	*P. domestica*
	Brompton	*P. domestica*
	Ciruelo 43	*P. domestica*
Moderately sensitive	S37	
	S2540	*P. salicina*
	S2541	
	S300	
Sensitive	S2514	
	S2508	
	S763	
	S2538	
Extremely sensitive	Apricot	*P. armeniaca*
	St. Lucie 39	*P. mahaleb*
	Cherry	*P. avium*
	Peach	*P. persica*
	GF305	*P. davidiana*

[a]From Rowe and Beardsell (1973).

Spy). Sensitivity of citrus rootstocks to waterlogging varied as follows: tolerant (Rough lemon and Trifoliata), intermediate (Cleopatra, Mandarin, Sweet orange); and sensitive (Sour orange and Sweet lime). Ranking of tolerance to waterlogging of *Prunus* rootstocks is given in Table II.

B. Age of Trees

Older trees generally tolerate flooding much better than seedlings or saplings of the same species. When *Populus* ×*canadensis* plantations of varying age (1–4, 5–6, and 8–15 years old) were flooded for 150 days, all trees were injured but injury was least in the oldest trees (Popescu and Necsulescu, 1967). Because old trees usually have their crowns above water, they are understandably subjected to less severe conditions than are young seedlings. Flowing water not only softens the soil but often washes it away. As flooding continues, seedlings may be pushed over, buried in mud, or uprooted and floated away.

Several studies show that certain species rated as very flood tolerant may be very intolerant in the seedling stage. For example, seedlings of *Taxodium distichum,* a flood-tolerant species, died after 2 weeks of complete submergence (Demaree, 1932). *Fraxinus pennsylvanica* is widely rated as flood tolerant; it commonly occurs on land that is subject to periodic flooding and grows vigorously even when flooded for much of the growing season (Broadfoot and Williston, 1973). Nevertheless, 30 days of flooding in stagnant water greatly reduced growth of leaves, stems, and roots of *F. pennsylvanica. Quercus shumardii, Liquidambar styraciflua, Celtis occidentalis,* and *Q. falcata* var. *pagodaefolia* died after 20 days of complete submersion (Hosner, 1960), but some or all seedlings of these species survived when flooded to the root collar (Hosner and Boyce, 1962).

C. Properties of Floodwater

Trees are injured much more by standing water than by flowing water. For example, height growth and root growth of *Pinus echinata* and *P. rigida* var. *serotina* seedlings were reduced more by flooding with standing water than with moving water (Hunt, 1951). Two-year-old *Nyssa sylvatica* var. *biflora* and *N. aquatica* seedlings were continuously flooded to a depth of 20 cm above the soil surface with moving water, to a 20-cm depth above the soil surface with stagnant water, or to the soil surface with moving water. Height growth and dry-weight increment of both species were lowest in the stagnant water, which also had the lowest O_2 and highest CO_2 contents (Harms, 1973). In another experiment height growth of *N. sylvatica* var. *biflora* and *N. aquatica* seedlings in moving water was about twice that of seedlings in stagnant water, and dry-weight increment of seedlings in moving water was two to five times greater (Hook *et al.*, 1970b). That water itself is not injurious is shown by vigorous growth of most plants with roots immersed in aerated nutrient solutions.

D. Duration of Flooding

Species composition as well as growth and survival of trees are very responsive to differences in duration of flooding during the growing season. Floodplains are composed of different levels that vary widely in the seasonal duration of flooding, with successively higher levels flooded for shorter periods. In a group of floodplain sites flooded annually, the duration of flooding ranged from 6 to 40% of the year (Bedinger, 1981).

On floodplains of the southeastern United States, swamps, sloughs, oxbows, and other areas of deep water support species such as *Nyssa aquatica, Taxodium distichum,* and *Planera aquatica.* The next most poorly drained sites support *Quercus lyrata* and *Carya aquatica.* Under even shorter duration of flooding, common species include *Q. laurifolia, Acer rubrum, Ulmus americana, Fraxinus pennsylvanica, Celtis laevigata,* and *C. occidentalis.* Ridges are dominated by *Liquidambar styraciflua* if low, or by *N. sylvatica, Carya* spp., and *Q. alba* if somewhat higher and with a shorter wet period (Clark and Benforado, 1981).

Of six species of bottomland trees studied in laboratory experiments, only *Salix nigra* seedlings survived 32 days of flooding. When completely submerged for 16 days, however, many *Fraxinus pennsylvanica,* some *Liquidambar styraciflua,* and a few *Acer negundo* seedlings survived. *Populus deltoides* and *Acer saccharinum* seedlings survived only when flooded for less than 16 days. The rate of recovery of the surviving trees also varied. In general, *S. nigra* and *F. pennsylvanica* recovered faster from flooding than did the other four species (Hosner, 1958).

A study of 39 species of woody plants near a reservoir margin showed that all species were killed when the root crowns were flooded for more than half the growing season, but trees flooded for shorter periods showed various degrees of flood tolerance (Hall and Smith, 1955). According to Whitlow and Harris (1979), colonization of reservoir shorelines is generally unlikely when flood duration exceeds 40% of the growing season. However, many swamps are flooded for more than 40% of the growing season and support various woody plants. As mentioned, Hall *et al.* (1946) emphasized that some flood-tolerant trees could survive continuous flooding for as much as two growing seasons.

IV. THE NATURE OF RESPONSES TO FLOODING

Flooding during the growing season affects trees at all stages of development. Responses include inhibition of seed germination, shoot, cambial, and root growth, arrested reproductive growth, morphological changes, and often, death of trees. Growth reduction, which is mediated through altered physiological processes, often is not apparent for a long time after the initiation of flooding. Sometimes the reduced growth and injury occur so long after flooding that other causes are sought (Kozlowski, 1982).

A. Seed Germination and Seedling Establishment

Seedling establishment and regeneration of forest stands may be stimulated or arrested by flooding, depending on the species and duration of flooding. Seeds of bottomland species such as *Nyssa aquatica, N. sylvatica* var. *biflora,* and *Taxodium distichum* remain viable when submerged in water for long periods. Al-

TABLE III

Effect of Flooding on Growth of *Betula papyrifera* Seedlings[a,b]

Days	Treatment	Seedlings		Roots		Stems		Leaves		Branches		Root–shoot ratio
		Grams	$\overline{RGR}$	Grams	RGR	Grams	$\overline{RGR}$	Grams	$\overline{RGR}$	Grams	$\overline{RGR}$	
0		$3.4 \pm 0.6_a$	—	$1.4 \pm 0.3_a$	—	$0.7 \pm 0.1_a$	—	$1.3 \pm 0.2_a$	—	—	—	$0.69 \pm 0.07_a$
20	Unflooded	7.0 ± 0.8	0.72	2.2 ± 0.3	0.45	1.3 ± 0.2	0.61	2.7 ± 0.3	0.73	$0.8 \pm 0.3_b$	—	$0.46 \pm 0.07_b$
20	Flooded	$3.4 \pm 0.5_a$	0	$1.2 \pm 0.2_a$	−0.15	$0.8 \pm 0.1_a$	0.13	$1.3 \pm 0.3_a$	0	$0.1 \pm 0.1_a$	—	$0.59 \pm 0.09_{ad}$
40	Unflooded	13.5 ± 2.0	0.65	4.8 ± 0.9	0.78	2.9 ± 0.5	0.80	$4.3 \pm 0.9_b$	0.46	$1.5 \pm 0.9_{bc}$	0.62	$0.54 \pm 0.08_{bcd}$
40	Flooded	$3.8 \pm 0.5_a$	0.11	$1.2 \pm 0.2_a$	0	$0.8 \pm 0.1_a$	0	$1.6 \pm 0.3_a$	0.20	$0.2 \pm 0.1_a$	0.69	$0.47 \pm 0.07_{bc}$
60	Unflooded	18.4 ± 2.9	0.31	8.0 ± 1.6	0.51	3.6 ± 0.4	0.22	$5.1 \pm 1.2_b$	0.17	$1.7 \pm 0.6_c$	0.13	$0.79 \pm 0.15_a$
60	Flooded	$3.6 \pm 0.5_a$	−0.05	$1.2 \pm 0.2_a$	0	$0.8 \pm 0.1_a$	0	$1.5 \pm 0.3_a$	−0.06	$0.1 \pm 0.1_a$	−0.69	$0.49 \pm 0.10_b$

[a] Data are means and standard error of dry weights of various plant parts or relative growth rates (RGR's) for 0–20, 20–40, and 40–60 days. The ratio of the dry weights of roots to shoots is also given. From Tang and Kozlowski (1982c).

[b] Values followed by the same subscript letter do not differ significantly (.01 level) in the same rank, except for $\overline{RGR}$.

though seeds of these species usually do not germinate under water, they do so readily when the floodwaters recede and the surface soil is exposed. Seeds of *Populus deltoides, Salix nigra,* and *Platanus occidentalis* often germinate under water, making it possible for these species to become established on sandbars and new sediments when the floodwaters recede. In contrast, seeds of *Fraxinus pennsylvanica, Acer negundo, Quercus shumardii,* and *Q. falcata* var. *pagodaefolia* lose their viability when submerged for short periods (Hosner, 1957, 1962). In the southern bottomlands of the United States, flooding early in the growing season favors regeneration of many light-seeded species because competing vegetation is killed and seeds germinate readily in the moist-silt seedbed (Hosner, 1962).

Flood-tolerance characteristics of seedlings are very important in determining species occurrence on wet sites. Seedlings of *Taxodium distichum* withstand long periods of flooding. *Acer saccharinum* and *Cephalanthus occidentalis* seedlings can withstand longer periods of flooding then those of *Celtis laevigata* or *Quercus falcata* var. *pagodaefolia* (Hosner, 1958, 1960). Seedlings of *Fraxinus pennsylvanica, Nyssa aquatica,* and *Salix* spp. are much more tolerant to soil inundation than those of *Ulmus americana, Liquidambar styraciflua, Celtis laevigata,* and *Liriodendron tulipifera.*

Flood tolerance of seedlings often varies widely in closely related species. On open, unflooded sites, rapid growth of *Betula papyrifera* seedlings is associated with its capacity for free growth characterized by initiation and expansion of new leaves during much of the growing season (Clausen and Kozlowski, 1965; Kozlowski and Clausen, 1966). Flooding, however, inhibits formation and growth of new leaves and induces abscission of old leaves. As a result of extensive leaf abscission in *B. papyrifera* seedlings, fewer leaves were present after a flooding period than before flooding was initiated. The drastic reduction in leaf area was associated with a marked decrease in height, cambial, and root growth as well as death of many seedlings (Table III). In comparison, growth of seedlings of *B. nigra,* a wetter-site species, was reduced much less by flooding. In addition, *B. nigra* seedlings adapted to flooding by (1) production of hypertrophied lenticels, which may assist in exchange of dissolved gases in floodwater and in release of toxic compounds produced by flooded seedlings, and (2) production of adventitious roots, which may increase absorption of water. Flooded *B. papyrifera* seedlings did not develop these adaptations (Norby and Kozlowski, 1983).

B. Shoot Growth

Flooding adversely influences shoot growth by inhibiting internode elongation, leaf initiation, and leaf expansion and by inducing leaf senescence, injury, and abscission.

Fig. 1. Effect of flooding for 60 days on growth of *Betula papyrifera* seedlings (left) and flooding for 45 days on growth of *Pinus banksiana* seedlings (right). Unflooded seedling of each species is on the left.

1. *Height Growth*

As seen in Fig. 1, waterlogging of soil retards height growth of many species. Flooding had a stunting effect on young seedlings of *Alnus rugosa, Platanus occidentalis, Betula nigra, Ulmus americana, U. alata,* and *Acer rubrum.* When soils were drained after flooding, *B. nigra* and *Acer rubrum* seedlings recovered very rapidly, *P. occidentalis* rapidly, and *U. alata* at a moderate rate (McDermott, 1954). Flooding for 39 days reduced height growth of *U. americana* seedlings, with the rate of growth of unflooded plants three to five times that of flooded plants. Most height growth of flooded seedlings occurred during the first 15 days, whereas height growth of unflooded seedlings continued throughout the experimental period. Leaf expansion followed a similar pattern, but leaves of flooded plants continued to expand after height growth ceased (Newsome *et al.*, 1982). Flooding with standing water also inhibited height growth of seedlings of *Quercus macrocarpa* (Tang and Kozlowski, 1982a), *Eucalyptus camaldulensis* and *E. globulus* (Sena Gomes and Kozlowski, 1980c), and *P. occidentalis* (Tang and Kozlowski, 1982b).

A few examples of reduced height growth of gymnosperm species following flooding include *Picea glauca, P. mariana, Pinus resinosa, P. strobus, P. banksiana,* and *Abies balsamea* (Ahlgren and Hansen, 1957), *Pinus elliotti* vars. *elliottii* and *densa* (McMinn and McNab, 1971), and *Pinus echinata, P. taeda,* and *P. serotina* (Hunt, 1951), and *Pinus halepensis* (Sena Gomes and Kozlowski, 1980d).

Height growth of some flood-tolerant species is increased by flooding, provided that the soil is not inundated with stagnant water. For example, height growth of *Nyssa aquatica* seedlings was greater when the soil was continuously saturated than when the soil was watered to moisture equivalent daily (Dickson *et al.*, 1965). In comparison, height growth of *N. aquatica* seedlings growing in moving water was about twice that of seedlings in stagnant water (Hook *et al.*, 1970b).

2. *Leaf Growth*

Flooding of soil reduces leaf area by inhibiting leaf initiation and expansion as well as by inducing leaf abscission. For example, waterlogging of soil inhibited expansion of leaves of *Betula papyrifera* seedlings and arrested formation of new leaves (Table III). After 40 days the number of leaves on unflooded *Platanus occidentalis* seedlings increased by 115%, and on flooded seedlings by only 21% (Tang and Kozlowski, 1982b). Flooding also inhibited leaf initiation and expansion in *Ulmus americana* (Newsome *et al.*, 1982). Inundation of soil also greatly reduced formation of secondary needles in *Pinus halepensis* (Sena Gomes and Kozlowski, 1980d).

3. *Leaf Abscission*

Leaves of many flooded plants often become chlorotic, senesce early, and abscise. For example, premature leaf shedding was induced by flooding of peach and pecan trees (Marth and Gardner, 1939; Alben, 1958). Flooding also induced extensive leaf abscission in *Betula papyrifera* seedlings. As a result, fewer leaves were present on flooded plants after 60 days of flooding than before flooding was initiated. Whereas the number of leaves on unflooded plants had approximately doubled, on flooded plants it decreased by more than half. The shed leaves represented a wide range of sizes from small leaves at the stem base to large leaves in a mid- to upper-stem location. The remaining attached leaves in the upper stem were generally small new leaves (Tang and Kozlowski, 1982c). Norby and Kozlowski (1983) also noted that flooding induced leaf shedding in *B. papyrifera* seedlings but not in the more flood-tolerant *B. nigra*.

Yelenosky (1964) reported considerable variation among species in time of leaf abscission in response to flooding. *Liriodendron tulipifera* seedlings lost all their leaves within 2 weeks of flooding, *Quercus alba* and *Acer saccharinum* within 3 weeks, *Gleditsia triacanthos* within 4 weeks, and *Ulmus americana* after 8 weeks or more.

Ethylene appears to play an important role in causing leaf abscission in flooded plants. Three lines of evidence support this:

1. The amount of ethylene in plants is increased by flooding. This results from accelerated production of ethylene in the plants (Kawase, 1972a,b, 1981; Tang and Kozlowski, 1982a,b) and translocation into plants of soil-borne ethylene, production of which is increased by microorganisms (Smith and Restall, 1971).
2. Exposure of unflooded plants to ethylene induces leaf epinasty, senescence, and abscission (Osborne, 1973).
3. Application to plants of ethrel (2-chloroethylphosphonic acid), an ethylene-releasing chemical, causes leaf abscission in deciduous trees (Cummins and Fiorino, 1969; Hartmann *et al.*, 1970; Sterrett *et al.*, 1973, 1974a,b; Cooper and Henry, 1973)

C. Cambial Growth

The effects of flooding on cambial growth are complex and vary from inhibition to acceleration. Measurements of diameter changes of stems as estimates of cambial growth of flooded trees are complicated by stem swelling and hypertrophy. Nevertheless, in flood-tolerant species cambial growth often is reduced appreciably by flooding. In flooded *Betula papyrifera* seedlings, a reduction in leaf area was followed by decreased cambial growth (Tang and Kozlowski, 1982c). Cambial growth is reduced by flooding older trees (*Fraxinus excelsior;*

Kassas, 1951). In bottomland species the cambial growth response varies appreciably with the extent and duration of flooding. Green (1947) reported that when several species of trees were flooded for only short periods, there was little effect on cambial growth. However, impoundment of long duration caused an increase in cambial growth for ~2 years, followed by a rapid decrease.

Several examples of increased cambial growth of flood-tolerant trees following temporary soil inundation are available. Flooding increased diameter growth of *Nyssa aquatica* in Alabama (Silker, 1948) and South Carolina (Klawitter, 1964). On a year-to-year basis diameter growth of 90-year-old *N. sylvatica* var. *biflora* trees was positively related to increasingly higher water-table levels (Langdon *et al.*, 1978). In a Mississippi River Valley stand that had been flooded continuously for 4 years with less than 30 cm of water, diameter growth of all broadleaved species was increased during the first year. During the second growing season, however, all species of *Ulmus, Celtis laevigata, Gleditsia triacanthos,* and *Diospyros virginiana* died. Most trees of *Quercus phellos, Q. nigra, Q. lyrata, Q. nuttallii, Fraxinus pennsylvanica,* and *Liquidambar styraciflua* were dead after 3 years (Broadfoot and Williston, 1973).

D. Root Growth

The growth of existing roots, formation of new ones, and root viability are very sensitive to availability of O_2. As early as 1925 Cannon concluded that roots needed 8–10% O_2 for fast growth but that roots of many species could make some growth with 2% O_2 or less. At O_2 concentrations between O.1 and 3%, the roots of apple trees survived but did not grow. Some root growth occurred when O_2 concentrations were between 5 and 10%, but higher concentrations were needed for rapid growth. Initiation of new roots required about 12% O_2 (Boynton *et al.*, 1938), and formation of new roots was inhibited when the O_2 concentration dropped below 15%. When the O_2 concentration decreased below 10% so few new roots formed that growth of shoots was subsequently reduced (Boynton, 1940). Roots of apple, prune, and peach trees grew best when the nutrient solutions in which they were immersed were aerated with gas containing 20% O_2 (Boynton and Compton, 1943).

Woody roots are much more tolerant than nonwoody roots to flooding. All nonwoody (primary) roots of rooted cuttings of *Pinus contorta* and *Picea sitchensis* were killed in waterlogged soil, whereas all the woody roots of *Pinus contorta* and most of those of *Picea sitchensis* survived (Coutts, 1982).

In poorly aerated soil, root growth often is restricted to the soil surface. This results in poorly anchored trees that are subject to wind throw and to drought injury because their roots occupy too small a volume of soil to supply water to the shoots during a drought. In England, *Picea sitchensis* trees planted in wet soils develop a superficial rooting habit. The roots grow deeper during the summer but die rapidly, and the water table later rises and O_2 deficiency occurs below 10–20 cm (Fraser and Gardiner, 1967). On upland peaty soils in England, anaerobiosis occurred in all except the top 10–15 cm of the soil in winter (Armstrong *et al.*, 1976). In fruit orchards a rise in the water table often leads to root mortality (Head, 1973).

The decrease in root–shoot ratio by flooding, reflecting greater reduction in growth of roots than of shoots, often predisposes trees to drought injury when the floodwater and water table recede. Flooding injury to trees is common around irrigation reservoirs that inundate large tracts of forest land for periods of days to several weeks (Whitlow and Harris, 1979).

Soil inundation often results in decay of roots, primarily as a result of invasion of *Phytophthora* fungi. Flooding almost completely stopped root growth and caused decay of the original roots of *Pinus halepensis* seedlings. After 40 days of flooding, the dry weight of the root system was lower than before flooding was initiated (Sena Gomes and Kozlowski, 1980d). In saturated soil, decay of *Citrus* roots was caused by *Phytophthora* fungi, with the amount of decay increasing as the length of saturation time increased. However, when the soil was not saturated, root decay resulted from lack of O_2 and not as a result of the presence of *Phytophthora* (Stolzy *et al.*, 1965).

The activity of root-rotting fungi in wet soils is associated with their capacity to tolerate low O_2 contents (Mitchell and Zentmyer, 1971), exudation of fungal substrates under anaerobic conditions (Rittenhouse and Hale, 1971), and weakening of the host tree. Zentmyer (1961) demonstrated that roots of avocado and other hosts of *Phytophthora cinnamomi* exuded a chemical that attracted motile zoospores. Exudation of the chemical, and consequent attraction of the spores, was greater with rapid- than with slow-growing roots.

1. *Root Regeneration*

Many species intolerant to flooding lose part of the original root system by decay and do not regenerate new roots. In comparison, flood-tolerant species often regenerate new roots on the original root system, on the submerged part of the stem, or on both. Flood-induced stem roots often emerge through hypertrophied lenticels (Fig. 2). Roots that form in the soil originate at the points to which the original roots die back when flooded (Coutts and Philipson, 1978a,b). Often the new roots are more succulent and permeable than the original roots (Hook *et al.*, 1971).

In flooded *Eucalyptus camaldulensis* seedlings the succulent adventitious roots had very few branches and their white color contrasted sharply with the dark brown color of the original roots (Sena Gomes and Kozlowski, 1980b). Submerged portions of stems of *Ulmus americana* seedlings developed two forms of adventitious roots: (1) a dense, freely branching form and (2) a loose,

Fig. 2. Hypertrophied lenticels and adventitious roots on submerged *Fraxinus pennsylvanica* stem. Arrow indicates the height to which the stem was flooded. From Sena Gomes and Kozlowski (1980a).

unbranched form. The branched roots usually entered the soil close to the stem, whereas most of the unbranched roots floated on the water for a short distance before penetrating the soil. Both types were present on most seedlings, but when only one type was present, it was the densely branched form. Production of adventitious roots was asymmetrical around the stem (Newsome *et al.*, 1982).

Flood-induced roots sometimes are negatively geotropic. Within 48 hr after flooding of *Melaleuca quinquenervia* seedlings was initiated, numerous hairlike roots as well as a few thick roots developed on the original root system. Some adventitious roots also formed on submerged portions of the stem. The hairlike roots eventually formed a dense mat of upwardly growing roots with tips protruding above the water level (Fig. 3). Both formation of roots and loss of root polarity have been attributed to accelerated ethylene production by flooded plants (Abeles, 1973; Kawase, 1981).

As emphasized by Gill (1970), there has been some disagreement about the adaptive significance of adventitious roots in flooded plants. However, the weight of evidence indicates that such roots supplement absorption in the somewhat aerobic zone, whereas the original root system does not function normally (Fig. 4). Adventitious rooting is characteristic of certain species only and gener-

Fig. 3. Numerous, upwardly growing adventitous roots on *Melaleuca quinquenervia* seedling flooded for 37 days. Unflooded seedling is on the right. From Sena Gomes and Kozlowski (1980b).

ally is associated with species that appear to tolerate flooding best. For example, flooding tolerance of *Eucalyptus grandis, E. robusta,* and *E. saligna* is related to capacity of these species to form adventitious roots (Clemens *et al.*, 1978). Extensive production of new roots from the upper portion of the tap root, lateral roots, and submerged portions of the stem of *Melaleuca quinquenervia* seedlings is consistent with the aggressiveness and vigorous growth of this species on wet sites (Sena Gomes and Kozlowski, 1980b). However, Gill (1975) did not attribute the high flood tolerance of *Alnus glutinosa* primarily to the absorbing capacity of its many adventitious roots. Gill concluded that absorption of water and mineral nutrients as well as synthesis of hormones by adventitious roots were important adaptations to flooding.

Adaptation to flooding by production of adventitious roots appears to be greater in *Eucalyptus camaldulensis* than in *E. globulus* seedlings. This was demonstrated by less reduction of growth in *E. camaldulensis* by flooding, a leveling off of the difference in growth of flooded and unflooded seedlings after a critical period (e.g., 20 days, by which time adventitious roots were abundant), and production of adventitious roots from submerged portions of stems as well as from the original root system. In *E. globulus* seedlings, adventitious roots did not form on submerged parts of the stem (Sena Gomes and Kozlowski, 1980c). In *Fraxinus pennsylvanica* seedlings the formation of adventitious roots on submerged stems appeared to be an important adaptation for flood tolerance. Absorption of water was almost twice as high in flooded plants with adventitious

Fig. 4. Effect of flooding on root growth of *Platanus occidentalis* seedling. Unflooded seedling is on the left. Note the presence of adventitious roots on the previously submerged portion of the stem. From Tang and Kozlowski (1982b).

roots as in flooded plants that had not developed such roots. After ~2 weeks of stomatal closure following flooding, the stomata began to reopen at about the time adventitious roots began to form. As more adventitious roots formed, the stomata opened further (Sena Gomes and Kozlowski, 1980a). In flooded *Nyssa sylvatica* var. *biflora,* adventitious roots occurred in the upper floodwater where the O_2 content was higher and the amounts of toxic compounds lower than in the soil. Under flooding conditions the adventitious roots oxidized the rhizosphere whereas unflooded roots did not. The combined adaptations of increased anaerobic respiration, oxidation of the rhizosphere, and high tolerance of CO_2 by the adventitious roots appeared to account for flood tolerance of this species (Hook *et al.*, 1970a, 1971; Hook and Scholtens, 1978).

2. *Mycorrhizae*

Mycorrhizae are strongly aerobic (Marks and Foster, 1973); hence the number and type of mycorrhizal associations are influenced by flooding. In peat bogs, mycorrhizae occurred only near the surface (Mikola, 1973), and submerged layers of peat and abandoned beaver flowages were devoid of mycorrhiza-forming fungi (Wilde, 1954). Waterlogging of soil caused deterioration of mycorrhizae of *Pinus radiata* and *Pseudotsuga menziesii* seedlings and also reduced phosphorus uptake and succinate dehydrogenase activity of mycorrhizae (Gadgil, 1972).

Temporary flooding in the spring did not alter the kinds of ecto- and endomycorrhizal fungi associated with *Liquidambar styraciflua, Quercus nuttallii, Q. phellos,* and *Q. lyrata* trees. However, flooding greatly reduced the numbers of fungi present and prevented formation of new populations. After the floodwaters receded, the fungal populations increased and were back to normal by the end of the growing season (Filer, 1975). Formation of mycorrhizae is also inhibited in clay soils because roots do not extend to deep layers. In pot-grown plants, mycorrhizal frequency is higher in the well-aerated roots at the pot periphery than in the inner roots.

E. Biomass and Dry-Weight Changes

There are many examples of reduction of dry-weight increase of upland species by flooding and only a few will be cited. The dry weight of *Ulmus americana* seedlings flooded for 39 days was only about half that of unflooded seedlings. The dry weight of leaves of flooded plants was also about half that of leaves of unflooded plants. This difference largely reflected variations in leaf number, but average dry weight per leaf also differed between flooded and unflooded plants. The most dramatic effect of flooding was a drastic reduction in dry-weight increment of roots, even though new adventitious roots comprised a third to a half of the dry weight of the root systems of flooded plants (Newsome *et al.*, 1982).

Betula papyrifera adapted very poorly to flooding for up to 60 days. In unflooded seedlings the dry weights of leaves increased approximately five times, of stems four times, and of roots five times (Table III). In flooded seedlings the dry-weight increment of attached leaves, and of stems, was negligible and dry weight of roots decreased, reflecting death of some roots without compensatory root growth. Relative growth rates were appreciable in unflooded seedlings but negligible in flooded seedlings (Tang and Kozlowski, 1982c).

Flooding for 30 days greatly reduced dry-weight increment of *Quercus macrocarpa* seedlings (Table IV). Whereas the dry weight of roots of unflooded plants more than doubled, in flooded plants it changed negligibly. Inhibition of

TABLE IV

Effect of Flooding for 30 Days on Growth of *Quercus macrocarpa* Seedlings[a]

Plant part	Before flooding	30 days after flooding		Change (%)	
		Unflooded	Flooded	Unflooded	Flooded
Dry weight (g)					
Seedlings	4.2 ± 0.5	8.2 ± 1.7	4.8 ± 1.3	95.2	14.3
Leaves	1.4 ± 0.2	2.0 ± 1.0	1.9 ± 0.6	42.9	35.7
Stems	0.5 ± 0.1	1.0 ± 0.3	0.7 ± 0.3	100.0	40.0
Roots	2.3 ± 0.3	5.2 ± 1.0	2.2 ± 0.6	126.1	−4.3
Root–shoot ratio	1.2 ± 0.2	1.8 ± 0.5	0.7 ± 0.3	50.0	−41.7
Seedling height (cm)	10.7 ± 1.6	12.5 ± 3.7	10.8 ± 2.8	16.8	0.9
Number of leaves per plant	6.6 ± 1.1	7.3 ± 1.9	7.0 ± 1.1	10.6	6.1

[a]From Tang and Kozlowski (1982a).

dry-weight increment of stems of flooded plants was smaller than in roots. Dry weights of leaves of flooded and unflooded plants were similar (Tang and Kozlowski, 1982a).

Inundation of soil for 40 days reduced dry-weight increment of *Eucalyptus camaldulensis* and *E. globulus* seedlings (Sena Gomes and Kozlowski, 1980c). Dry-weight increase was reduced proportionally more in shoots than in roots, primarily because of the addition of many adventitious roots (Fig. 5). In comparison, flooding reduced root growth much more than it reduced shoot growth of *Pinus halepensis,* which did not produce new adventitious roots (Sena Gomes and Kozlowski, 1980d). The weight of roots after 40 days of flooding was less than at the time flooding was initiated. Hence flooding not only inhibited root growth but also caused decay of a part of the original root system. Growth of young *Picea glauca, Picea mariana,* and *Pinus banksiana* plants was reduced and mortality increased when the O_2 content of the nutrient solution in which roots were immersed was decreased (Zinkan *et al.*, 1974). Oxygen between 2.0 and 3.3 ppm limited seedling growth; O_2 above that range was less critical.

Dry-weight increment of seedlings of bottomland species is also reduced when the soil is inundated with standing water. For example, flooding for 30 days significantly reduced dry-weight increment of 8- and 10-week-old *Fraxinus pennsylvanica* seedlings, with growth inhibited more in the younger plants (Sena Gomes and Kozlowski, 1980a). In plants of both ages, dry-weight increment was reduced most in roots (Table V). Flooding of *Melaleuca quinquenervia* seedlings for up to 30 days did not significantly influence dry-weight increment of seedlings, stems, or roots. However, flooding for 60–90 days reduced dry-weight increment of whole plants as well as leaves and stems. Dry-weight increment of

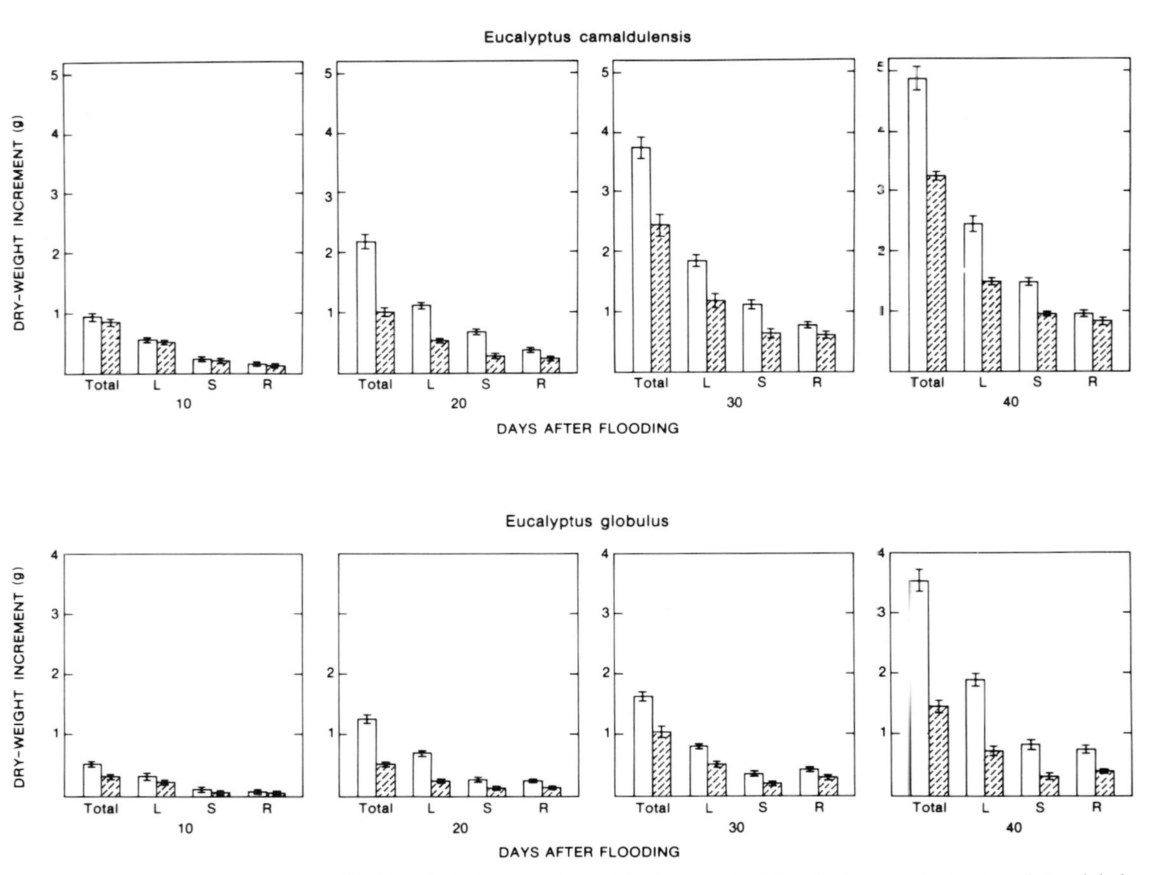

Fig. 5. Effect of flooding for 10, 20, 30 and 40 days on dry-weight increment of *Eucalyptus camaldulensis* and *E. globulus* seedlings. Total, whole plants; L, leaves; S, stem; R, roots. Bars (□, unflooded; ▨, flooded) indicate standard errors of the mean. From Sena Gomes and Kozlowski (1980c).

TABLE V

Effect of Flooding for 30 Days on Growth of *Fraxinus pennsylvanica* Seedlings of Two Age Classes[a]

Age (weeks)	Treatment	Dry-weight increment (g)								Root–shoot ratio
		Leaves	Percent of control	Stems	Percent of control	Roots	Percent of control	Total	Percent of control	
8	Control	2.47 ± 0.14		2.84 ± 0.04		2.19 ± 0.20		7.51 ± 0.28		0.41
	Flooded	0.98 ± 0.05[b]	39.6	0.94 ± 0.09[b]	33.0	0.36 ± 0.03[b]	16.4	2.29 ± 0.16[b]	30.4	0.19
10	Control	3.95 ± 0.31		3.31 ± 0.29		3.76 ± 0.29		11.03 ± 2.43		0.53
	Flooded	2.39 ± 0.34[b]	60.5	2.60 ± 0.28[c]	78.5	1.21 ± 0.15[b]	32.1	6.22 ± 0.74[b]	56.3	0.26

[a]From Sena Gomes and Kozlowski (1980a).
[b]Values significantly different from control ($p = .01$).
[c]Not significant.

TABLE VI

Effect of Flooding for 15, 30, 60, or 90 Days on Dry-Weight Increment (g) of *Melaleuca quinquenervia* Seedlings[a]

	Days after flooding							
	15		30		60		90	
Plant part	Unflooded	Flooded	Unflooded	Flooded	Unflooded	Flooded	Unflooded	Flooded
Leaves	0.70 ±0.05	0.57[b] ±0.05	2.68 ±0.17	2.40[b] ±0.16	10.07 ±1.00	6.36[c] ±0.30	15.83 ±1.05	8.74[c] ±0.48
Stems	0.19 ±0.02	0.18[b] ±0.02	1.27 ±0.08	1.13[b] ±0.07	5.55 ±0.25	3.98[c] ±0.25	10.68 ±0.56	6.84[c] ±0.32
Roots	0.17 ±0.01	0.14[b] ±0.01	0.65 ±0.05	0.75[b] ±0.06	2.79 ±0.20	2.73[b] ±0.17	6.99 ±0.58	6.52[b] ±0.87
Whole plants	1.06 ±0.08	0.89[b] ±0.08	4.60 ±0.28	4.28[b] ±0.28	18.40 ±1.29	13.07[c] ±0.59	33.50 ±1.84	22.11[c] ±1.49

[a]From Sena Gomes and Kozlowski (1980b).
[b]Not significant.
[c]Values significantly different at the 1% level.

the whole root system was not significantly affected by flooding, because the reduced growth and mortality of the original roots were compensated by growth of copious adventitious roots (Table VI; Fig. 3).

F. Morphological Changes

Although species vary widely in morphological changes following flooding, typical responses in some species include stem hypertrophy and development of aerenchyma tissue, hypertrophy of lenticels on stems, initiation of adventitious roots on submerged portions of stems (Section IVD), regeneration of new roots from the original roots, and formation of "knee" roots and pneumatophores.

Many species of trees respond to flooding by producing hypertrophied lenticels (Fig. 2), which may facilitate exchange of dissolved gases in the flood water. Although the hypertrophied lenticels of flooded plants collapse when exposed to the air, they apparently are more pervious to gas exchange than nonflooded lenticels because of larger intercellular spaces and many breaks in the closing layers (Hook *et al.*, 1970a; Hook and Scholtens, 1978). Lenticels apparently are connected by continuity of intercellular spaces in cortical and phloem tissues. This is confirmed by the observation that air can be continuously pulled through submerged lenticels by application of a slight vacuum to the flooded part of the stem if some of the lenticels are exposed to the atmosphere (Gill, 1970). Armstrong (1968) showed that O_2 readily entered stems of woody plants through lenticels located a few centimeters above the water line. Occlusion of such lenticels with lanolin resulted in decreased O_2 uptake. In some species (e.g., *Salix alba*) the flood-induced lenticels not only assist in aeration of the stem and roots but also serve as openings through which toxic compounds associated with anaerobiosis (e.g., ethanol, acetaldehyde, and ethylene) are released. In comparison, lenticels of *Populus petrowskiana* did not release toxic compounds (Chirkova and Gutman, 1972).

The structure of lenticels varies appreciably in unflooded and flooded woody plants. For example, stem lenticels of unflooded *Nyssa sylvatica* var. *biflora* seedlings are round, only slightly hypertrophied, and have few closing layers. The degree of hypertrophy and number of closing layers increase in intermittently flooded plants, and closing layers are absent in continuously flooded plants.

As early as 1920, Hahn *et al.* reported that hypertrophied lenticels formed on roots and stems of a variety of gymnosperms growing in wet soils. Species affected included *Pinus ponderosa, P. coulteri, P. rigida, P. resinosa, P. banksiana, P. virginiana, P. sylvestris, P. caribaea, P. strobus, P. monticola, P. excelsa, Picea canadensis, P. rubens, P. mariana, P. pungens, Abies balsamea, Tsuga canadensis, Larix laricina, Taxus cuspidata, T. brevifolia,* and *Araucaria bidwellii.* Within 5–10 days after flooding was initiated, hypertrophied lenticels formed on submerged portions of stems of seedlings of *Salix nigra* and *Populus deltoides* (Pereira and Kozlowski, 1977), *Fraxinus pennsylvanica* (Sena Gomes and Kozlowski, 1980a), *Eucalyptus camaldulensis* and *E. globulus* (Sena Gomes and Kozlowski, 1980c), *Ulmus americana* (Newsome *et al.*, 1982), *Quercus macrocarpa* (Tang and Kozlowski, 1982a), and *Platanus occidentalis* (Tang and Kozlowski, 1982b).

Lenticel hypertrophy apparently is induced by ethylene that forms in response to flooding (Kozlowski, 1982). The formation of such lenticels involves dissolution of cell walls as well as hypertrophy and proliferation of cells and rounding of protoplasts. Often, all living cells between the phellogen and cambium are affected. According to Kawase (1981), a chain of events occurs in which ethylene production is accelerated in flooded plants, followed by increased cellulase activity. Cellulase may play a role in softening cell walls, leading to stem hypertrophy and development of aerenchyma tissue. Cellulase activity alone probably does not cause cell enlargement, however, because all cortical cells are not affected similarly following application of cellulase to stem sections. Kawase (1979) proposed that competition for water exists between adjacent cortical cells that have been growing, or are ready to grow, after their cell walls are softened by cellulase. The stronger cells survive and grow; the weaker cells become dehydrated and plasmolyzed and finally are killed.

A number of species of trees produce various forms of "knee" roots or pneumatophores when subjected to periodic flooding. The tendency to produce such roots usually is greatly suppressed on upland sites. For example, *Anthocleista nobilis* and *Xylopia staudtii* form pneumatophores in flooded areas but not on well-drained sites (Gill, 1970). Examples of woody plants that produce knee roots or pneumatophores in wet areas include species of *Taxodium, Avicennia, Sonneratia, Mitragyna, Phoenix, Pterocarpus, Amoora, Carapa, Heritiera, Ploiorum, Cratoxylon, Tristania, Symphonia,* and *Terminalia* (Kozlowski, 1971). Pneumatophores often bear lenticels that serve as aerating channels for submerged roots. The arched stilt roots of mangroves also contain lenticels and function as aerating organs. In contrast, the conical, vertical knees of *Taxodium distichum* develop because of active, localized cambial activity on the upper surfaces of roots, which are better aerated than the lower surfaces. They do not appear to act as aerating organs that supply O_2 to submerged roots (Kramer and Kozlowski, 1979).

G. Mortality

There are many examples of failure of trees to survive prolonged flooding during the growing season. Mortality varies greatly with duration of flooding, species, and age of trees. Often, trees eventually killed by flooding grow very rapidly for some time when they are first flooded.

Bottomland species survive flooding better than upland species. For example, in an impounded flood-control lake in Oklahoma several upland species such as *Quercus rubra, Q. prinus, Q. stellata,* and *Robinia pseudoacacia* exhibited higher mortality rates than the bottomland species *Populus deltoides, Fraxinus pennsylvanica,* and *Acer negundo* (Harris, 1975).

Continuous flooding for 4 years in the Mississippi River Valley near Vicksburg, Mississippi greatly decreased survival of several species of forest trees (Fig. 6). *Quercus falcata* var. *pagodaefolia* died at the end of the first year of flooding; *Ulmus* spp., *Celtis laevigata, Gleditsia triacanthos,* and *Diospyros virginiana* died during the second growing season. *Quercus phellos* and *Q. nigra* survived a little longer. Some declined in vigor during the second year but few died until the third year; some grew rapidly for 3 years but died in the fourth year. *Quercus lyrata* and *Q. nuttallii, Fraxinus pennsylvanica,* and *Liquidambar styraciflua* were the most tolerant species, but most of these were also dead after 3 years (Broadfoot and Williston, 1973). Cuttings of *Populus deltoides* and seedlings of *L. styraciflua, Nyssa aquatica, Platanus occidentalis,* and *F. pennsylvanica* were planted in a slack-water clay in Mississippi in two consecutive years and were inundated soon after the leaves emerged. During each year survival after flooding was consistently high for *Nyssa, Fraxinus,* and *Platanus,* low for *Populus,* and intermediate for *Liquidambar* (Baker, 1977).

Fig. 6. Trees killed by flooding following impoundment as a result of dam construction. U.S. Forest Service photograph.

Loucks and Keen (1973) studied responses of seedlings of 10 species of forest trees to 1–4 weeks of flooding. All species survived 2 weeks of flooding. Whereas all *Fraxinus pennsylvanica, Taxodium distichum,* and *Acer saccharinum* seedlings survived 3 weeks of flooding, survival of the other species ranged from 44 to 67%. Survival after 4 weeks of flooding varied from 0 to 100%. It was highest in *F. pennsylvanica, T. distichum,* and *A. saccharinum;* intermediate in *Carya illinoensis, Populus deltoides, Gleditsia triacanthos,* and *Quercus macrocarpa;* and lowest in *Acer negundo, Ulmus pumila,* and *Juglans nigra.*

Kennedy and Krinard (1974) summarized the effects on hardwood forests and plantations of a major flood of the Mississippi River during the spring and summer of 1973. Trees less than 1 year old died rapidly, whereas trees 1 year old or older survived where flooding occurred in the first 2 months of the growing season. However, *Liriodendron tulipifera* was an exception, and trees 1–15 years old were killed; *Populus deltoides* plantings established in the winter of 1972–1973 were virtually destroyed. Taxa such as *Liquidambar styraciflua, Quercus nuttallii, Q. nigra,* and *Q. falcata* var. *pagodaefolia* survived 60–65 days of flooding, even though they had gone through only one growing season and were only ⅓–⅔ m high. *Fraxinus pennsylvanica, Populus deltoides,* and *Platanus occidentalis* trees were also in good condition.

Dellinger *et al.* (1976) investigated the effects of continuous spring and summer flooding of bottomland hardwoods in southern Illinois. Major species included *Quercus palustris, Ulmus americana, Acer saccharinum, Q. bicolor,* and *Populus deltoides.* Substantial mortality occurred in nearly all species subjected to continuous flooding for 83 days after the leaves emerged (129 days after the onset of flooding in mid-March).

Brink (1954) studied the effects of extensive flooding in the Fraser River Valley of British Columbia on growth of trees in forests than had not been previously flooded for decades. Much of the flooded lowland forest was a subclimax mixed conifer–hardwood community. Of the gymnosperms, *Pseudotsuga menziesii* succumbed most readily to flooding; *Tsuga heterophylla* also was very susceptible; *Thuja plicata* became chlorotic but did not exhibit further injury; *Picea sitchensis* survived flooding well. Wide variations in mortality of flooded broad-leaved trees also were evident. *Prunus emarginata* rarely survived a flood of a few days' duration. Some *Acer macrophyllum* trees died, especially if they were growing rapidly when flooded. *Alnus rubra,* which died in large numbers, was one of the most susceptible trees to flooding; *A. sinuata* tolerated flooding better; *Cornus stolonifera* and *C. nuttallii* readily survived flooding.

Brink (1954) classified flood tolerance of cultivated species in the following categories:

1. Very high mortality and extensive injury: *Ilex aquifolium, Corylus avellana, Syringa vulgaris, Sorbus aucuparia, Prunus laurocerasus*

2. Lower mortality than species in category 1 but often killed or severely injured: *Crataegus oxycantha, Buxus sempervirens, Rubus procerus*
3. Low mortality but with some leaf chlorosis: *Rubus laciniatus*, pears (*Prunus* spp.), *Rosa* spp., *Vitis* spp., *Rhododendron* spp., *Robinia pseudoacacia*
4. No mortality and usually no obvious injury: *Malus* spp., *Juglans* spp., *Acer negundo, Castanea* spp.

REFERENCES

Abeles, F. B. (1973). "Ethylene in Plant Biology." Academic Press, New York.

Ahlgren, C. E., and Hansen, H. L. (1957). Some effects of temporary flooding on coniferous trees. *J. For.* **55,** 647–650.

Alben, A. O. (1958). Waterlogging of subsoil associated with scorching and defoliation of Stuart pecan trees. *Proc. Am. Soc. Hortic. Sci.* **72,** 219–223.

Armstrong, W. (1968). Oxygen diffusion from the roots of woody species. *Physiol. Plant.* **21,** 539.

Armstrong, W., Booth, T. C., Priestley, P., and Read, D. J. (1976). The relationship between soil aeration, stability, and growth of Sitka spruce (*Picea sitchensis* Bong. Carr.) on upland peaty gleys. *J. Appl. Ecol.* **13,** 585–591.

Baker, J. B. (1977). Tolerance of planted hardwoods to spring flooding. *South. J. Appl. For.* **1,** 23–25.

Bedinger, M. S. (1981). Hydrology of bottomland hardwood forests of the Mississippi embayment. *In* "Wetlands of Bottomland Hardwood Forests" (J. R. Clark and J. Benforado, eds.), pp. 161–176. Elsevier, New York.

Bell, D. T., and Johnson, F. L. (1974). Flood-caused tree mortality around Illinois reservoirs. *Trans. Ill. Acad. Sci.* **67,** 28–37.

Boynton, D. (1940). Soil atmosphere and the production of new rootlets by apple tree root systems. *Proc. Am. Soc. Hortic. Sci.* **37,** 19–26.

Boynton, D., and Compton, O. C. (1943). Effect of oxygen pressure in aerated nutrient solutions on production of new roots and growth of roots and top by fruit trees. *Proc. Am. Soc. Hortic. Sci.* **42,** 53–58.

Boynton, D., DeVilliers, J. I., and Reuther, W. (1938). Are there different cortical oxygen levels for the different phases of root activity? *Science (Washington, D.C.)* **88,** 569–570.

Brink, V. C. (1954). Survival of plants under flood in the lower Fraser River Valley, B. C. *Ecology* **35,** 94–95.

Broadfoot, W. M. (1967). Shallow-water impoundment increases soil moisture and growth of hardwoods. *Soil Sci. Soc. Am. Proc.* **31,** 562–564.

Broadfoot, W. M., and Williston, H. L. (1973). Flooding effects on southern forests. *J. For.* **71,** 584–587.

Buchholz, K. (1981). Effects of minor drainages on woody species distributions in a successional floodplain forest. *Can. J. For. Res.* **11,** 671–676.

Cannon, W. A. (1925). Physiological features of roots with special reference to the relation of roots to aeration of the soil. *Carnegie Inst. Washington Publ.* No. 368.

Chirkova, T. V., and Gutman, T. S. (1972). Physiological role of branch lenticels in willow and poplar under conditions of root anaerobiosis. *Sov. Plant Physiol. (Eng. Transl.)* **19,** 289–295.

Clark, J. R., and Benforado, J. (eds.) (1981). "Wetlands of Bottomland Hardwood Forests." Elsevier, New York.

Clausen, J. J., and Kozlowski, T. T. (1965). Heterophyllous shoots in *Betula papyrifera*. *Nature (London)* **205,** 1030–1031.

Clemens, J., Kirk, A. M., and Mills, P. D. (1978). The resistance to water-logging of three *Eucalyptus* species, effect of flooding and on ethylene-releasing growth substances on *E. robusta, E. grandis,* and *E. saligna. Oecologia* **34,** 125–131.

Cooper, W. C., and Henry, W. H. (1973). Chemical control of fruit abscission. *In* "Shedding of Plant Parts" (T. T. Kozlowski, ed.), pp. 475–524. Academic Press, New York.

Coutts, M. P. (1982). The tolerance of tree roots to waterlogging. V. Growth of woody roots of Sitka spruce and Lodgepole pine in waterlogged soil. *New Phytol.* **90,** 467–476.

Coutts, M. P., and Philipson, J. J. (1978a). The tolerance of tree roots to water-logging. I. Survival of Sitka spruce and Lodgepole pine. New Phytol. 80, 63–69.

Coutts, M. P., and Philipson, J. J. (1978b). The tolerance of tree roots to water-logging. II. Adaptation of Sitka spruce and Lodgepole pine to waterlogged soil. *New Phytol.* **80,** 71–77.

Cummins, J. N., and Fiorino, P. (1969). Preharvest defoliation of apple nursery stock using Ethrel. *HortScience* **4,** 339–341.

Daubenmire, R. (1978). "Plant Geography." Academic Press, New York.

Dellinger, G. P., Brink, E. L., and Allmon, A. D. (1976). Tree mortality caused by flooding at two midwestern reservoirs. *Proc. Annu. Conf. Southeast. Fish Game Commissioners 3rd,* pp. 645–648.

Demaree, D. (1932). Submerging experiments with *Taxodium. Ecology* **13,** 258–262.

Dickson. R. E., Hosner, J. F., and Hosley, N. W. (1965). The effects of four water regimes upon the growth of four bottomland tree species. *For Sci.* **11,** 299–305.

Filer, T. H. (1975). Mycorrhizae and soil microflora in a green tree reservoir. *For. Sci.* **24,** 36–39.

Fowells, H. A. (1965). Silvics of forest trees of the United States. *U.S. Dept. Agric. For. Serv. Agric. Handb.* No. 271.

Fraser, A. I., and Gardiner, J. B. H. (1967). Rooting and stability in Sitka spruce. *U.K. Bull. For. Comm.* No. 40.

Frye, R. J., and Quinn, J. A. (1979). Forest development in relation to topography and soils in a flood-plain of the Raritan River, New Jersey. *Bull. Torrey Bot. Club* **106,** 334–345.

Gadgil, P. D. (1972). Effect of waterlogging on mycorrhizas of radiata pine and Douglas-fir. *N.Z. J. For. Sci.* **2,** 222–226.

Gill, C. J. (1970). The flooding tolerance of woody species—a review. *For. Abstr.* **31,** 671–688.

Gill, C. J. (1975). The ecological significance of adventitious rooting as a response to flooding in woody species, with special reference to *Alnus glutinosa* L. Gaertn. *Flora (Jena)* **164,** 85–97.

Green, W. E. (1947). Effect of water impoundment on tree mortality and growth. *J. For.* **45,** 118–120.

Hahn, G. G., Hartley, C., and Rhoads, A. S. (1920). Hypertrophied lenticels on the roots of conifers and their relation to moisture and aeration. *J. Agric. Res.* **20,** 253–265.

Hall, T. F., and Smith, G. E. (1955). Effects of flooding on woody plants. West Sandy dewatering project, Kentucky Reservoir. *J. For.* **53,** 281–285.

Hall, T. F., Penfound, W. T., and Hess, A. D. (1946). Water level relationships of plants in the Tennessee Valley with particular reference to malaria control. *J. Tenn. Acad. Sci.* **21,** 18–59.

Hansen, H. L. (1980). The Lake States region. *In* "Regional Silviculture of the United States" (J. W. Barrett, ed.), pp. 67–105. Wiley, New York.

Harms, W. R. (1973). Some effects of soil type and water regime on growth of tupelo seedlings. *Ecology* **54,** 188–193.

Harris, M. D. (1975). Effects of initial flooding on forest vegetation at two Oklahoma lakes. *J. Soil Water Conserv.* **30,** 294–295.

Hartmann, H. T., Tombesi, A., and Whisler, A. (1970). Promotion of ethylene evolution and fruit abscission in the olive by 2-chloroethylphosphonic acid and cycloheximide. *J. Am. Soc. Hortic. Sci.* **95,** 635–639.

Head, G. C. (1973). Shedding of roots. *In* "Shedding of Plant Parts" (T. T. Kozlowski, ed.), pp. 237–293. Academic Press, New York.

Heinselman, M. L. (1970). Landscape evolution, peatland types and the environment in the Agassiz Peatlands Natural Area, Minnesota. *Ecol. Monogr.* **46,** 59–84.

Hook, D. D., and Scholtens, J. R. (1978). Adaptations and flood tolerance of tree species. *In* ''Plant Life in Anaerobic Environments'' (D. D. Hook and R. M. M. Crawford, eds.), pp. 299–331. Ann Arbor Sci. Publ., Ann Arbor, Michigan.

Hook, D. D., Brown, C. L., and Kormanik, P. P. (1970a). Lenticels and water root development of swamp tupelo under various flooding conditions. *Bot. Gaz. (Chicago)* **131,** 217–224.

Hook, D. D., Langdon, O. G., Stubbs, J., and Brown, C. L. (1970b). Effects of water regimes on the survival, growth, and morphology of tupelo seedlings. *For. Sci.* **16,** 304–311.

Hook, D. D., Brown, C. L., and Kormanik, P. P. (1971). Inductive flood tolerance in swamp tupelo [*Nyssa sylvatica* var. *biflora* (Walt.) Sarg.]. *J. Exp. Bot.* **22,** 78–89.

Hosner, J. F. (1957). Effects of water upon the seed germination of bottomland trees. *For. Sci.* **3,** 67–71.

Hosner, J. F. (1958). The effects of complete inundation upon seedlings of six bottomland tree species. *Ecology* **39,** 371–373.

Hosner, J. F. (1960). Relative tolerance to complete inundation of fourteen bottomland tree species. *For. Sci.* **6,** 246–251.

Hosner, J. F. (1962). The southern bottomland region. *In* ''Regional Silviculture'' (J. W. Barrett, ed.), pp. 296–333. Academic Press, New York.

Hosner, J. F., and Boyce, S. G. (1962). Relative tolerance to water saturated soil of various bottomland hardwoods. *For. Sci.* **8,** 180–186.

Hunt, F. M. (1951). Effect of flooded soil on growth on pine seedlings. *Plant Physiol.* **26,** 363–368.

Jawanda, J. S. (1961). The effect of waterlogging on fruit trees: Punjab. *Hortic. J.* **1,** 150–152.

Karschon, R., and Zohar, Y. (1972). Effects of flooding on ecotypes of *Eucalyptus viminalis*. *Isr. Div. For. Leaflet* No. 45. Bet Dagan, Israel.

Karschon, R., and Zohar, Y. (1975). Effects of flooding and of irrigation water salinity on *Eucalyptus camaldulensis* Dehn. From three seed sources. *Isr. Div. For. Agric. Res. Organ. Leaflet* No. 54. Ilanot, Israel.

Kassas, M. (1951). Studies in the ecology of Chippenham Fen. II. Recent history of the Fen from evidence of historical records, vegetation analysis, and tree ring analysis. *J. Ecol.* **39,** 19–32.

Kawase, M. (1972a). Effect of flooding on ethylene concentration in horticultural plants. *J. Am. Soc. Hortic. Sci.* **9,** 548–588.

Kawase, M. (1972b). Submersion increases ethylene and stimulates rooting in cuttings. *Proc. Int. Plant Prop. Soc.* **22,** 360–366.

Kawase, M. (1979). Cellulase activity in waterlogged herbaceous horticultural crops. *HortScience* **16,** 30–34.

Kawase, M. (1981). Anatomical and morphological adaptations of plants to water-logging. *HortScience* **16,** 8–12.

Kennedy, H. E., Jr., and Krinard, R. M. (1974). 1973 Mississippi river floods impact on natural hardwood forests and plantations. *U.S. For. Serv. Res. Note* No. SO–177.

Klawitter, R. A. (1964). Water tupelos like it wet. *South. Lumberman* **209,** 108–109.

Kozlowski, T. T. (1971). ''Growth and Development of Trees,'' Vol. 2. Academic Press, New York.

Kozlowski, T. T. (1982). Water supply and tree growth. II. Flooding. *For. Abstr.* **43,** 145–161.

Kozlowski, T. T., and Clausen, J. J. (1966). Shoot growth characteristics of heterophyllous woody plants. *Can. J. Bot.* **44,** 827–843.

Kramer, P. J., and Kozlowski, T. T. (1979). ''Physiology of Woody Plants.'' Academic Press, New York.

Krinard, R. M., and Johnson, R. L. (1976). 21-Year growth and development of bald cypress planted on a flood-prone site. *U.S. For. Serv. South. For. Exp. Stn. Res. Note* No. SO–217.

Ladiges, P. Y., and Kelso, A. (1977). The comparative effects of waterlogging on two populations of *Eucalyptus viminalis* Labill. and one population of *E. ovata* Labill. *Aust. J. Bot.* **25,** 159–169.

Langdon, O. G., DeBell, D. S., and Hook, D. D. (1978). Diameter growth of swamp tupelo: seasonal pattern and relations to water table level. *Proc. North Am. For. Biol. Workshop 5th,* pp. 326–333.

Loucks, W. L., and Keen, R. A. (1973). Submersion tolerance of selected seedling trees. *J. For.* **71,** 496–497.

Marks, G. C., and Foster, R. C. (1973). Structure, morphogenesis, and ultrastructure of ectomycorrhizae. *In* "Ectomycorrhizae" (G. C. Marks and T. T. Kozlowski, eds.), pp. 1–41. Academic Press, New York.

Marth, P. C., and Gardner, T. E. (1939). Evaluation of a variety of peach seedling stocks with respect to "wet feet" tolerance. *Proc. Am. Soc. Hortic. Sci.* **37,** 335–337.

McDermott, R. E. (1954). Effects of saturated soil on seedling growth of some bottomland hardwood species. *Ecology* **35,** 36–41.

McMinn, J. W., and McNab, W. H. (1971). Early growth and development of slash pine under drought and flooding. *U.S. Dept. Agric. Serv. Res. Paper* No. SE–89.

Merritt, C. (1980). The Central region. *In* "Regional Silviculture of the United States" (J. W. Barrett, ed.), pp. 107–143. Wiley, New York.

Mikola, P. (1973). Applications of mycorrhizal symbiosis in forestry practice. *In* "Ectomycorrhizae" (G. C. Marks and T. T. Kozlowski, eds.), pp. 383–411. Academic Press, New York.

Minore, D. (1968). Effects of artificial flooding on seedling survival and growth of six northwestern tree species. *U.S. Dept. Agric. For. Serv. Res. Note* No. PNW–92.

Mitchell, D. J., and Zentmyer, G. A. (1971). Effects of oxygen and carbon dioxide tensions on growth of several species of *Phytophthora. Phytopathology* **61,** 787–791.

Mizutani, F., Yamada, M., Sugiura, A., and Tomana, T. (1979). Differential water tolerance among *Prunus* species and the effect of waterlogging on the growth of peach scions on various root stocks. *Engeigaku Kenkyu Shuroku* (*Stud. Inst. Hortic. Kyoto Univ.*) **9,** 28–35.

Newsome, R. D., Kozlowski, T. T., and Tang, Z. C. (1982). Responses of *Ulmus americana* seedlings to flooding of soil. *Can. J. Bot.,* **60,** 1688–1695.

Norby, R. J., and Kozlowski, T. T. (1983). Flooding and SO_2 stress interaction in *Betula papyrifera* and *B. nigra* seedlings. *For. Sci.,* **29**.

Osborne, D. J. (1973). Internal factors regulating abscission. *In* "Shedding of Plant Parts" (T. T. Kozlowski, ed.), pp. 125–147. Academic Press, New York.

Pereira, J. S., and Kozlowski, T. T. (1977). Variations among woody angiosperms in response to flooding. *Physiol. Plant.* **41,** 184–192.

Popescu, I., and Necsulescu, H. (1967). The harmful effect of prolonged inundation on plantations of black poplars in the Braila Marshes. *Rev. Padurilor* **82,** 20–23.

Poutsma, T., and Simpfendorfer, K. J. (1963). Soil moisture conditions and pine failure at Waarre, near Port Campbell, Victoria. *Aust. J. Agric. Res.* **13,** 426–433.

Putnam, J. A. (1951). Management of bottomland hardwoods. *U.S. Dept. Agric. For. Serv. Occ. Paper* No. 116.

Putnam, J. A., Furnival, G. M., and McKnight, J. S. (1960). Management and inventory of southern hardwoods. *U.S. Dept. Agric., Agric. Handb.* No. 181.

Remy, P., and Bidabe, B. (1962). Root asphyxia and collar rot in pome fruit trees. The influence of the rootstock. *Proc. Congr. Pomol. 92nd,* pp. 17–28.

Rittenhouse, R. L., and Hale, M. G. (1971). Loss of organic compounds from roots. II. Effect of oxygen and carbon dioxide tension on release of sugars from peanut roots under axenic conditions. *Plant Soil* **35,** 311–321.

Robertson, P. A., Weaver, G. T., and Cavanaugh, J. A. (1978). Vegetation and tree species patterns near the northern terminus of the southern floodplain forest. *Ecol. Monogr.* **48,** 249–267.

Rom, C., and Brown, S. A. (1979). Water tolerance of apples on clonal rootstocks and peaches on seedling rootstocks. *Compact Fruit Tree* **12,** 30–33.

Rowe, R. N., and Beardsell, D. V. (1973). Waterlogging of fruit trees. *Hortic. Abstr.* **43,** 533–548.

Sena Gomes, A. R., and Kozlowski, T. T. (1980a). Growth responses and adaptations of *Fraxinus pennsylvanica* seedlings to flooding. *Plant Physiol.* **66,** 267–271.

Sena Gomes, A. R., and Kozlowski, T. T. (1980b). Responses of *Melaleuca quinquenervia* seedlings to flooding. *Physiol. Plant.* **49,** 373–377.

Sena Gomes, A. R., and Kozlowski, T. T. (1980c). Effects of flooding on growth of *Eucalyptus camaldulensis* and *E. globulus* seedlings. *Oecologia* **46,** 139–142.

Sena Gomes, A. R., and Kozlowski, T. T. (1980d). Responses of *Pinus halepensis* seedlings to flooding. *Can. J. For. Res.* **10,** 308–311.

Silker, T. H. (1948). Planting of water-tolerant trees along margins of fluctuating-level reservoirs. *Iowa State Coll. J. Sci.* **22,** 431–448.

Smith, D. W., and Linnartz, N. E. (1980). The southern hardwood region. *In* "Regional Silviculture of the United States" (J. W. Barrett, ed.), pp. 145–230. Wiley, New York.

Smith, K. A., and Restall, S. W. F. (1971). The occurrence of ethylene in anaerobic soil. *J. Soil Sci.* **22,** 430–443.

Sterrett, J. P., Leather, G. R., and Tozer, W. E. (1973). Defoliation response of woody seedlings to endothall/ethephon. *HortScience* **8,** 387–388.

Sterrett, J. P., Leather, G. R., and Tozer, W. E. (1974a). An explanation for the synergistic interaction of endothall and ethephon on foliar abscission. *J. Am. Soc. Hortic. Sci.* **99,** 395–397.

Sterrett, J. P., Leather, G. R., Tozer, W. E., Foster, W. D., and Webb, D. T. (1974b). Foliar abscission of woody plants with combinations of endothall and ethephon. *Weed Sci.* **22,** 608–614.

Stolzy, L. H., Letey, J., Klotz, L. J., and DeWolfe, T. A. (1965). Soil aeration and root-rotting fungi as factors in decay of citrus feeder roots. *Soil Sci.* **99,** 403–406.

Stone, E. C., and Vasey, R. B. (1968). Preservation of coast redwood on alluvial flats. *Science (Washington, D.C.)* 159, 157–161.

Tang, Z. C., and Kozlowski, T. T. (1982a). Some physiological and morphological responses of *Quercus macrocarpa* seedlings to flooding. *Can. J. For. Res.* **10,** 308–311.

Tang, Z. C., and Kozlowski, T. T. (1982b). Physiological, morphological, and growth responses of *Platanus occidentalis* seedlings to flooding. *Plant Soil* **66,** 243–255.

Tang, Z. C., and Kozlowski, T. T. (1982c). Some physiological and growth responses of *Betula papyrifera* seedlings to flooding. *Physiol. Plant.* **55,** 415–420.

Van't Woudt, B. D., and Hagan, R. M. (1957). Crop responses at excessively high soil moisture levels. *In* "Drainage of Agricultural Lands" (J. N. Luthin, ed.), pp. 514–578. Academic Press, New York.

Walter, H. (1973). "Vegetation of the Earth." Springer-Verlag, New York.

Whitlow, T. H., and Harris, R. W. (1979). Flood tolerance in plants: a state of the art review. *U.S. Army Corps Eng. Tech. Rep.* No. E–72–2. U.S.A.C.E. Waterways Exp. Stn. Environ. Lab., Vicksburg, Mississippi.

Wilde, S. A. (1954). Mycorrhizal fungi: their distribution and effect on tree growth. *Soil Sci.* **78,** 23–31.

Yelenosky, G. (1964). Tolerance of trees to deficiencies of soil aeration. *Proc. Int. Shade Tree Conf. 40th,* pp. 127–147.

Zentmyer, G. A. (1961). Chemotaxis of zoospores for root exudates. *Science (Washington, D.C.)* **133,** 1595–1596.

Zinkan, C. G., Jeglum, J. K., and Harvey, D. E. (1974). Oxygen in water culture influences growth and nutrient uptake of jack pine, black spruce, and white spruce seedlings. *Can. J. Plant Sci.* **54,** 553–558.

CHAPTER 5

Effect of Flooding on Water, Carbohydrate, and Mineral Relations

T. T. KOZLOWSKI
Department of Forestry
University of Wisconsin
Madison, Wisconsin

S. G. PALLARDY
School of Forestry
University of Missouri
Columbia, Missouri

I. INTRODUCTION

The effects of environmental stresses on growth of plants are mediated through changes in physiological processes (Kramer and Kozlowski, 1979). For example, inundation of soil usually sets in motion a sequential and complicated series of metabolic disturbances in plants that eventually lead to alterations in growth. Furthermore, morphological changes or death of flooded plants are preceded by abnormal physiological events. This chapter presents an overview of the effects

ISBN 0-12-424120-4

of flooding of soil on some important aspects of water, carbohydrate, and mineral relations of plants. The effects of soil inundation on changes in plant growth hormones, and the influence of such changes on plant growth, are discussed by Reid and Bradford (Chapter 6 in this volume).

II. WATER RELATIONS

Soil inundation is followed by changes in stomatal aperture, transpiration, and absorption of water. Sometimes flooding also influences turgidity of plant tissues.

A. Stomatal Aperture

Closing of stomata is one of the earliest plant responses to soil inundation (Kozlowski, 1982). However, the specific pattern of stomatal closure varies with species, age of leaves, and, in amphistomatous species, is complicated with the presence of stomata on both the adaxial and abaxial leaf surfaces.

Flooding of soil caused a sharp decrease in stomatal conductance of *Phaseolus vulgaris* (Moldau, 1973). In *Quercus macrocarpa* and *Platanus occidentalis* seedlings, flooding was followed by stomatal closure within a day or two and stomata remained closed for at least 30 days, the duration of the experiment (Tang and Kozlowski, 1982a, 1982b; Fig. 1). Flooding also closed stomata of *Betula papyrifera* seedlings for at least 14 days. Observations were discontinued after 14 days because leaves of flooded plants were yellow and stomata had lost their regulatory function (Tang and Kozlowski, 1982b). Diurnal changes occurred in stomatal aperture in both unflooded and flooded plants, with the daily amplitude much greater in flooded plants (Pereira and Kozlowski, 1977; Tang and Kozlowski, 1982c).

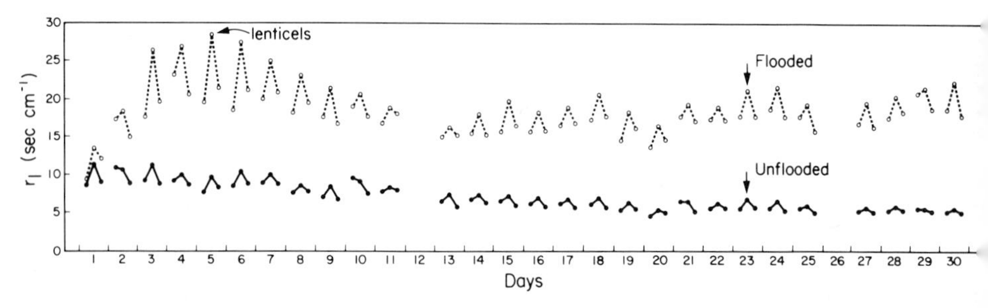

Fig. 1. Effect of flooding for 30 days on leaf diffusive resistance r_l of *Quercus macrocarpa* seedlings. An increase in r_l indicates stomatal closure; a decrease indicates stomatal opening. From Tang and Kozlowski (1982a).

In seedlings of the flood-tolerant species *Fraxinus pennsylvanica,* stomata began to close within a day or two after flooding was initiated, and they continued to close progressively for the next 7 days (Fig. 2). For the following 7 days average daily stomatal aperture remained about the same. Large diurnal variations were superimposed on this general pattern. Stomata of flooded seedlings began to reopen after ~15 days of flooding. Stomatal reopening was closely correlated with production and growth of adventitious roots on submerged portions of stems. Stomata began to reopen when such roots first appeared, and they continued to open over a 2-week period during which more adventitious roots formed, elongated, and branched, thereby progressively increasing the absorbing surface for water and minerals. Leaf conductance was only slightly lower in plants that had been flooded for 30 days, compared with that of unflooded plants (Kozlowski and Pallardy, 1979). Induction of stomatal reopening after a critical period of flooding has also been reported for *Populus deltoides* (Regehr *et al.*, 1975) and *Melaleuca quinquenervia* (Sena Gomes and Kozlowski, 1980b).

Stomatal responses of *Ulmus americana* seedlings to flooding varied with leaf age. Flooding caused stomatal closure in leaves that were fully expanded by the time the plants were first flooded. However, flooding did not induce stomatal closure of leaves that expanded during the flooding period. In both unflooded and flooded plants the stomata were more open in leaves that completed expansion during the flooding period (Newsome *et al.*, 1982).

In amphistomatous species the responses of adaxial and abaxial stomata to flooding often vary greatly. In *Populus deltoides* and *Salix nigra,* flooding

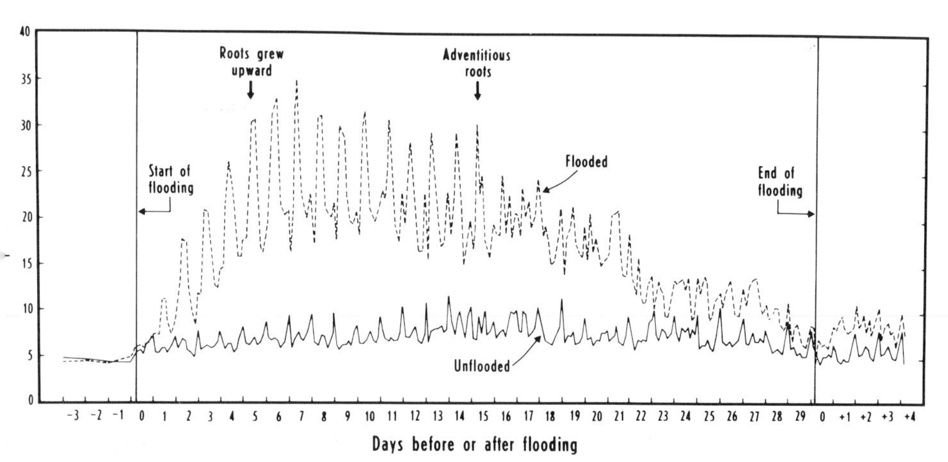

Fig. 2. Effect of flooding for 30 days on leaf diffusive resistance r_l of *Fraxinus pennsylvanica* seedlings. An increase in r_l indicates stomatal closure; a decrease indicates stomatal opening. From Sena Gomes and Kozlowski (1980a).

rapidly induced stomatal closure on the adaxial leaf surface. It also caused stomata of the abaxial epidermis of *P. deltoides* to close, but not the abaxial stomata of *S. nigra* (Pereira and Kozlowski, 1977; Fig. 3). Flooding induced stomatal closure on both the adaxial and abaxial leaf surfaces of the flood-tolerant species *Melaleuca quinquenervia*. However, the differences in mean stomatal aperture of flooded and unflooded plants were small (Sena Gomes and Kozlowski, 1980b).

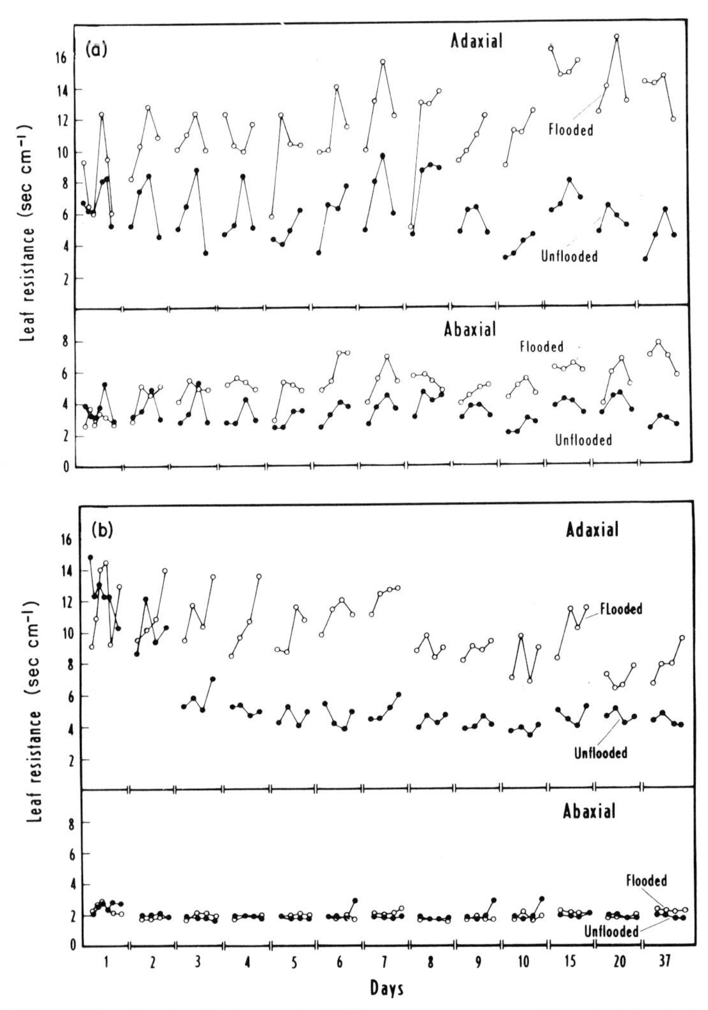

Fig. 3. Effect of flooding for 37 days on leaf diffusion resistance r_1 of the adaxial and abaxial leaf surfaces of (a) *Populus deltoides* and (b) *Salix nigra* seedlings. An increase in r_1 indicates stomatal closure; a decrease indicates stomatal opening. From Pereira and Kozlowski (1977).

B. Causes of Stomatal Closure in Flooded Plants

The mechanism of stomatal closure is not fully understood, but both nutritional and hormonal influences appear to be involved. Closure of stomata following flooding may be associated with migration of potassium ions from guard cells, resulting in osmotic changes that lead to loss of guard-cell turgor. During stomatal opening a net influx of potassium ions into guard cells has been shown for many species of plants (Allaway and Milthorpe, 1976). Stomata of *Acer saccharinum* seedlings were more open than those supplied with the same mineral nutrients, but without potassium (Noland and Kozlowski, 1979). Some investigators have shown a linear dependency of stomatal opening on potassium content of guard cells (Humble and Raschke, 1971; Raschke, 1975).

Several studies of accumulation of mineral nutrients have shown reduction in potassium concentration in leaves when soil oxygen is limiting (Hammond *et al.*, 1955; Harris and van Bavel, 1957). For example, potassium content of *Citrus sinensis* seedlings was decreased when the plants were grown under low soil O_2 levels (Labanauskas *et al.*, 1971). Absorption of potassium by *Zea mays* roots was greatly reduced by low O_2 or high CO_2 concentration. The effect of added CO_2 varied with the partial pressure of O_2 (Hammond *et al.*, 1955). Oxygen deficiency decreased uptake of potassium and caused stomatal closure of *Nicotiana tabacum* (Harris and van Bavel, 1957). With high potassium nutrition, stomatal aperture and photosynthesis of *Medicago sativa* increased (Cooper *et al.*, 1967). Specific data are lacking, however, on how long roots would have to be deprived of soil O_2 before changes in potassium flux would cause stomata to close (Sojka and Stolzy, 1980). More research is needed on this question.

Stomatal closure in flooded plants may also be associated with production of abscisic acid (ABA) in leaves. This conclusion is based on three lines of evidence:

1. Flooding induces rapid increases in ABA content of leaves. For example, flooding of *Phaseolus* seedlings induced a rapid rise in leaf ABA (Hiron and Wright, 1973). The ABA contents of leaves of *Juglans nigra, J. hindsii, J. regia,* and *J. hindsii* × *J. regia* also increased severalfold after flooding. Maximum ABA contents were reached after the first 12–18 hours of flooding (Shaybany and Martin, 1977).
2. There is good correlation between buildup of ABA and stomatal closure (Mansfield and Davies, 1981).
3. When ABA is applied exogenously to leaves the stomata close (Jones and Mansfield, 1970; Kriedemann *et al.*, 1972).

When Davies and Kozlowski (1975a,b) applied a 10^{-1} *M* solution of ABA to leaves of *Acer saccharum, Fraxinus americana,* and *Citrus mitis* seedlings, the rate of transpiration decreased as a result of stomatal closure (Table I). Stomatal response times suggest that *Fraxinus* stomata are less sensitive to applied ABA than are stomata of *Citrus* or *Acer*.

TABLE I

Effect of Applied Abscisic Acid (ABA) on Transpiration and Stomatal Resistance of Three Species of Woody Angiosperms[a]

Observations	*Citrus mitis*	*Acer saccharum*	*Fraxinus americana*
Lag between time of ABA treatment and onset of transpiration reduction (min)	1.6 ± 0.7	7.0 ± 1.1	30.4 ± 6.1
Time between ABA treatment and minimal transpiration rates (min)	11.6 ± 1.2	41.0 ± 4.9	114.0 ± 6.8
Transpiration rate			
Control plants			
Original (mg dm^{-2} hr^{-1})	697 ± 43	598 ± 36	637 ± 40
Final (mg dm^{-2} hr^{-1})	672 ± 37	593 ± 48	622 ± 37
ABA-treated plants			
Original (mg dm^{-2} hr^{-1})	718 ± 28	644 ± 23	627 ± 48
Final (mg dm^{-2} hr^{-1})	295 ± 42	254 ± 35	252 ± 43
Final transpiration (% of original rate)	41.8	39.6	41.6
Transpiration resistance, ABA-treated plants			
Original (sec cm^{-1})	4.3 ± 1.3	4.2 ± 1.3	3.9 ± 1.1
Final (sec cm^{-1})	13.0 ± 2.3	14.6 ± 1.4	14.2 ± 1.9

[a]Data are means and standard errors. From Davies and Kozlowski (1975a).

The regulation of stomata by ABA may involve interference with ionic transport between guard cells and adjacent cells, or with metabolic events. The result of the action of ABA is inhibition of uptake of potassium ions and disappearance of guard-cell starch under conditions normally conducive to stomatal opening (Mansfield and Jones, 1971). The effects of ABA on flooded plants are discussed in more detail by Reid and Bradford (Chapter 6 in this volume).

There is some evidence that ethylene may be directly or indirectly involved in inducing stomatal closure in flooded plants of certain species. For example, Pallas and Kays (1982) reported that ethylene at hormonally significant levels induced stomatal closure, and hence lower CO_2 uptake in *Arachis hypogaea*. Conductance of abaxial stomata was affected more than that of adaxial stomata. Pallas and Kays (1982) suggested that ethylene might affect stomatal aperture by (1) inducing changes in guard-cell membranes, thereby altering ion and/or water efflux, or (2) altering the rate of carbon fixation, leading to elevated CO_2 concentrations and subsequent stomatal closure. In contrast to the marked effect of ethylene on *Arachis,* no effect was apparent on stomata of *Phaseolus vulgaris, P. coccinea, Pisum sativum, Solanum tuberosum, Mimosa pudica,* or *Trifolium repens*. Pallaghy and Raschke (1972) noted that stomata of *Zea mays* and *Pisum sativum* did not respond to ethylene concentrations between 1 and 10^5 μl $liter^{-1}$.

They concluded that the mechanism of CO_2 action on stomata of these species differs from that competitively inhibiting morphogenetic response to ethylene. Much more research is needed on the mechanism of stomatal closure in flooded plants.

1. Stomatal Closure Associated with Leaf Dehydration

The changes in guard cells that lead to stomatal closure in flooded plants may occur with or without appreciable dehydration of leaves. Stomatal closure in some flooded plants has been attributed to leaf desiccation caused by decreased absorption of water as a result of lowered permeability of roots. When CO_2 was bubbled through a nutrient solution for 10 min of each hour, absorption of water by plants growing in the solution was reduced by 14–50%. When air was subsequently bubbled through the solution, absorption of water increased appreciably (Chang and Loomis, 1945).

The sudden rapid wilting (''flopping'') of tobacco and tomato leaves following a downpour of rain has been attributed to a temporary decrease in permeability of roots. Prolonged flooding caused injury and death of roots because of deficient aeration, but this was complicated by the activity of microorganisms that destroyed the roots (Kramer and Jackson, 1954). Yelenosky (1964) reported both a reduction in transpiration and dehydration of leaves of flooded *Liriodendron tulipifera* seedlings, further emphasizing that flooding induces leaf water deficits by decreasing permeability of roots.

The permeability of roots of flooded plants to water is decreased by both high CO_2 or low O_2 concentrations, leading to lowering of absorption of water. The direct inhibitory effect of high CO_2 concentration is very rapid; O_2 deficiency is generally slower acting (Slatyer, 1967).

Exposure of *Helianthus annuus* and *Lycopersicon esculentum* root systems to a high CO_2 concentration reduced transpiration by 34–52%. Bubbling O_2-free nitrogen gas through the water surrounding the roots reduced transpiration of *Lycopersicon* by only 8% (Kramer, 1940). Saturating the soil with CO_2 reduced transpiration of *Clethra alnifolia, Ilex glabra, Myrica cerifera,* and *Quercus alba* by 50–70% (Caughey, 1945). Kramer (1940) concluded that the rapid reduction in water absorption caused by a high concentration of CO_2 in the soil water resulted largely from decreased passive absorption caused by physical changes in the protoplasm and protoplasmic membranes. Hence the resistance to water movement across the root cortex was increased, leading to shoot desiccation and stomatal closure.

2. Stomatal Closure without Leaf Dehydration

In contrast to the studies just mentioned, several investigators showed that stomatal closure in certain flooded plants was not the result of leaf dehydration (e.g., reduced leaf ψ). For example, in *Eucalyptus camaldulensis, E. globulus,* and *Ulmus americana* seedlings, the bulk leaf ψ was significantly higher in flooded than in unflooded plants. In *Populus deltoides* and *Salix nigra,* leaf ψ did not differ significantly between flooded and unflooded plants (Pereira and Kozlowski, 1977). Leaf ψ of *Quercus macrocarpa* seedlings was much higher in flooded than in unflooded plants, yet stomata of flooded plants were more closed (Fig. 4). These observations are consistent with those of Regehr *et al.* (1975), who reported that *Populus deltoides* plants remained turgid throughout 28 days of flooding. They concluded that decreased absorption of water resulting from increased root resistance was offset by simultaneous increase in stomatal closure, thereby maintaining cell turgor.

A variety of experiments show that reduction in soil O_2 can cause stomatal closure even with no change in bulk leaf ψ. Stolzy *et al.* (1961) provided indirect evidence that stomatal closure was induced as the soil O_2 diffusion rate (ODR) declined, with the reduction independent of soil water status. A range of soil O_2 levels was achieved by passing various partial pressures of O_2 through sealed root containers in which well-watered *Lycopersicon esculentum* plants were growing. Plants in soils with ODR's of 16–24 × 10^{-8} g cm^{-2} min^{-1} were not injured by peroxyacetyl nitrate (PAN) or ozone. Plants growing in soils with ODR's of 34–90 × 10^{-8} g cm^{-2} min^{-1} were moderately to severely damaged, apparently because their more open stomata resulted in greater pollutant uptake. In another experiment, Stolzy *et al.* (1964) manipulated soil O_2 by passing nitrogen-gas mixtures over the soil-grown roots. Low soil ODR for 8.5 hr decreased the rate of photosynthesis by half, again indicating stomatal closure.

More direct evidence of the effect of soil O_2 on stomatal aperture was obtained by Sojka *et al.* (1975). Well-watered *Triticum aestivum* plants were grown in equilibrium with 0, 4, and 21% O_2 mixtures passed over the soil surface, creating a range of soil ODR's. Stomata closed at low O_2 levels. Subsequent studies

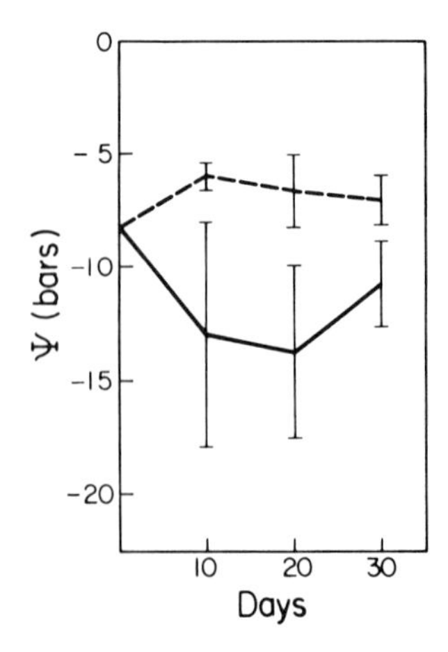

Fig. 4. Effect of flooding for 30 days on leaf ψ of *Quercus macrocarpa* seedings. ——, Unflooded; - - -, flooded. From Tang and Kozlowski (1982a).

demonstrated that low ODR in soil induced stomatal closure in *Helianthus, Gossypium,* and *Simmondsia* when soil moisture was readily available (Sojka and Stolzy, 1981). Stomatal closure was more readily induced by low ODR as soil temperatures increased. At high soil temperatures the respiration rate of roots (i.e., O_2 requirement) increases. Hence high soil temperature induces an O_2 deficiency, resulting in greater stomatal closure.

Coutts (1981) showed that more than one specific mechanism may be involved in controlling stomatal aperture of flooded plants. Coutts found that changes in water relations of flooded *Picea sitchensis* seedlings differ appreciably between dormant and actively growing plants. Dormant seedlings exhibited a gradual reduction in transpiration and increased shoot water deficit over the flooding period. After the soil was drained the leaf ψ increased to approximately that of unflooded plants, but transpiration did not increase until the roots began to grow. Water relations of flooded, actively growing plants were more complex. There was a very rapid reduction in transpiration after initiation of flooding, accompanied by a short period of plant water deficit, followed by a period of increasing transpiration but without a plant water deficit. Finally, there was another reduction in transpiration, and plant water deficits increased as the seedlings died. Coutts (1981) explained the variations in water relations of dormant and active *Picea sitchensis* seedlings as follows. In dormant seedlings the gradual buildup of leaf water defects suggested a progressive reduction in absorption of water, presumably because of reduced permeability of root membranes. In living roots changes in permeability can be reversed by varying the O_2 supply (Willey, 1970). Hence restoration of leaf ψ in dormant flooded seedlings would be expected because many roots remained alive. The decrease in ψ (to -1.5 MPa) that developed in dormant seedlings during waterlogging would not be expected to decrease transpiration as much as it did because a ψ of -2.0 MPa or less is required for substantial stomatal closure in *Picea sitchensis* (Beadle *et al.*, 1979). This indicates that a motivating stimulus, in addition to leaf water deficit, is involved in inducing stomatal closure.

In waterlogged, actively growing seedlings the sudden brief increase in shoot water deficit and reduced transpiration is consistent with a pattern of reduction in water absorption, probably because of lowered root permeability. The subsequent increase in transpiration may have been associated with hormonal changes. The final increase in water deficit in active, flooded *Picea sitchensis* seedlings was attributed to blockage in the xylem at the root or stem base.

Coutts's experiments indicate that flooding of soil can alter shoot water relations by effects on absorption of water, water loss from leaves, and the translocation pathway. Coutts showed that in the same plant stomatal control mechanisms may vary with duration of flooding, condition of the root system, and growth stage (e.g., dormant versus actively growing plants). As emphasized by Coutts (1981), variations in water relations of flooded plants that have been attributed to species characteristics sometimes merely reflect variations in the condition of the root system at the time of flooding. Hence preoccupation with a single mechanism of control of stomatal aperture in flooded plants probably is no longer justified.

C. Transpiration

Flooding of soil decreases the rate of transpiration. For example, transpiration of young apple trees was reduced within a few days after the soil was inundated (Childers and White, 1942). Transpiration of *Juniperus virginiana, Quercus alba, Q. lyrata,* and *Cornus florida* was also rapidly reduced by flooding. In contrast, transpirational water loss remained high in the flood-tolerant species *Taxodium distichum*. In some experiments there was a sudden increase in transpiration just after flooding, followed by a subsequent reduction (Fig. 5). The temporary increase suggested that soil moisture was somewhat limiting to water absorption at the time the plants were flooded (Parker, 1950). Kramer (1951) found that flooding reduced transpiration most in *Nicotiana tabacum,* an intermediate amount in *Lycopersicon esculentum,* and least in *Helianthus annuus*. It appeared that rapid development of adventitious roots in *Helianthus* largely

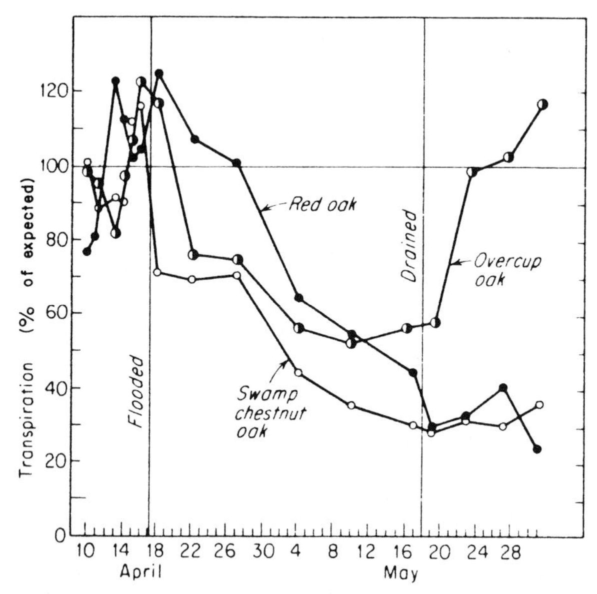

Fig. 5. Effect of flooding for 30 days on transpiration of *Quercus rubra* (red oak), *Q. lyrata* (overcup oak), and *Q. michauxii* (swamp chestnut oak) seedlings. The seedlings lost most of their leaves as a result of flooding, but *Q. lyrata* was leafing out again when the soil was drained and its rate of transpiration recovered rapidly. From Parker (1950).

compensated for loss of the original root system, whereas in *Nicotiana* the original roots died before enough adventitious roots developed to effectively replace them.

III. CARBOHYDRATE RELATIONS

Flooding of soil is followed by changes in rates of photosynthesis and translocation of photosynthetic products. These are discussed separately.

A. Photosynthesis

Plants that are readily injured by flooding also show a rapid reduction in the rate of photosynthesis. Photosynthesis of *Pseudotsuga menziesii* seedlings began to decrease within 4–5 hr after flooding (Zaerr, 1983). The rate of photosynthesis of four citrus rootstocks began to decline progressively within 1 day after flooding and was reduced by 80 to more than 90% within 10 days, depending on the rootstock. Stomatal closure appeared to be mainly responsible for the rapid decrease of photosynthesis (Phung and Knipling, 1976). Photosynthesis of *Malus* trees was also reduced within a few days by flooding (Childers and White, 1942), and a few days of root submersion substantially reduced photosynthesis of *Carya illinoensis* (Loustalot, 1945). The rate of photosynthesis of flooded trees growing in sand was reduced to about 10% of the control rate and was negligible in flooded trees growing in a heavier soil. When the waterlogged soil was drained, photosynthesis increased but did not return to normal for several days. Regehr *et al.* (1975) noted that the decrease in the rate of photosynthesis following flooding of *Populus deltoides* seedlings closely followed reduction in transpiration rates (Fig. 6). Complete inundation of the root system for 28 days reduced photosynthesis by about half. Photosynthesis decreased rapidly from a

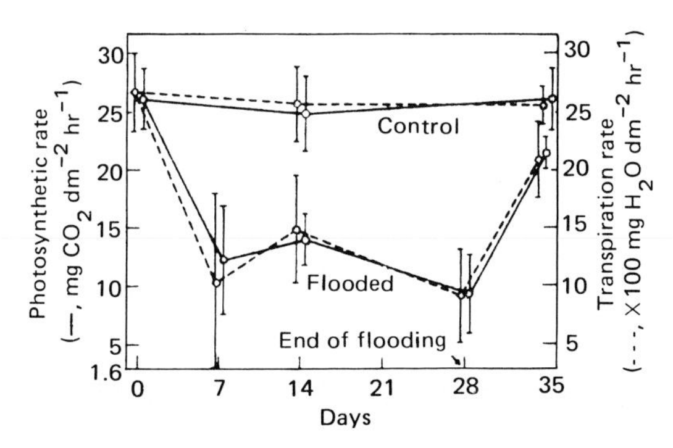

Fig. 6. Effect of flooding for 28 days on the rate of photosynthesis and transpiration of *Populus deltoides* seedlings. From Regehr *et al.* (1975).

maximum at leaf ψ between -0.3 and -0.8 MPa to near zero at -1.1 MPa. The rate of photosynthesis increased to the preflooding rate within a week after the end of flooding.

Flooding also reduced photosynthesis in the relatively flood-tolerant species *Acer saccharinum*. Absorption of CO_2 by leaves was measured for seedlings that were completely inundated and then removed from the water or under continuous soil saturation. Control seedlings were maintained with soil at field capacity of moisture. Under continuously saturated soil conditions, the rate of photosynthesis dropped within a few days to approximately 75% of that of control seedlings. After 10 days of complete inundation, the rate of photosynthesis was reduced to less than 25% of that of control plants (Sipp and Bell, 1974).

Some genetic differentiation in photosynthetic response to flooding was shown by McGee *et al.* (1981). They compared effects of soil inundation on *Populus deltoides* plants from sand-dune, strip-mine, and floodplain habitats. After 11 days of flooding, photosynthesis of the floodplain population decreased to 70% of the original rates, and to 50% by day 16. Photosynthesis of strip-mine plants declined sharply to 38% of the original rate after 11 days, and to 32% after 16 days. In contrast, the rate of photosynthesis of dune plants was unchanged after 11 days of flooding but decreased precipitously by day 16. This response of dune plants was associated with stomatal opening early during flooding, followed by delayed stomatal closure.

B. Causes of Reduction in Photosynthesis

Photosynthesis of flooded plants appears to be reduced first primarily because of stomatal closure and later by a reduction in photosynthetic capacity as well. Hence, in the longer term reduction in photosynthesis is partly traceable to changes in carboxylation enzymes, reduced chlorophyll content of leaves, early leaf senescence and abscission, and reduced leaf area (the latter a result of inhibition of leaf formation and expansion) (Kozlowski, 1982).

Stomatal resistance to CO_2 uptake by leaves often provides a major limitation for photosynthesis. The importance of stomatal control is shown by high correlations between stomatal aperture and photosynthesis (Holmgren *et al.*, 1965; Kriedemann, 1971). Childers and White (1942) found a parallel reduction in photosynthesis and transpiration of flooded apple trees.

Flooded plants often become chlorotic (Rodriguez and Gausman, 1973). For example, soil inundation reduced chlorophyll content of barley (*Hordeum vulgare*) plants (Mikhailova, 1977). The importance of chlorophyll content to photosynthesis is emphasized by decline in both chlorophyll and photosynthesis in mineral-deficient plants. A high correlation has often been found between the chlorophyll content of leaves and CO_2 uptake (Keller and Koch, 1962, 1964).

Bradford (1982) found that stomatal closure was not the only factor limiting photosynthesis of flooded *Lycopersicon esculentum* plants. Bradford concluded that a combination of ABA accumulation and decreasing cytokinin supply may induce stomatal closure. The decrease in cytokinins also resulted in lowered ribulose bisphosphate regeneration capacity. The end result was a reduction in photosynthesis, without reduction in bulk leaf ψ.

C. Translocation of Photosynthetic Products

Flooding of soil not only depresses the rate of photosynthesis but also reduces the rate of translocation of photosynthetic products from sources to various sinks. For example, flooding decreased flow of ^{14}C-labeled assimilates from leaves to roots of *Glycine max* (Zvareva and Bartkov, 1976). Several early investigators (work reviewed by Curtis and Clark, 1950) showed that a deficiency of O_2 impeded phloem transport of carbohydrates, but the rate of translocation often recovered after several days. Using a more sensitive method, Sij and Swanson (1973) demonstrated that anaerobic conditions almost immediately decreased the rate of movement of ^{14}C-labeled assimilates through petioles of *Cucurbita melo-pepo* var. *torticollis* plants. Translocation declined to 35–45% of the pretreatment rate within 30–40 min. The decrease was temporary, and the rate recovered to the pretreatment rate in 60–90 min. Such recovery was followed by a second and more permanent decline in the translocation rate.

Rapid inhibition by anoxia of carbohydrate translocation in *Saxifraga sarmentosa* stolons (Qureshi and Spanner, 1973) as well as many early examples of inhibition of translocation suggest that anaerobiosis induces some poorly understood conditions that cause blockage of translocation. The mechanism does not appear to involve a deficiency of ATP or metabolic energy. Coulsen *et al.* (1972) did not show a stoichiometric relationship between the rate of sucrose translocation and respiration or ATP turnover. It has been suggested that accumulation of toxic products of anaerobic respiration is involved, but more research is needed on the mechanism (Geiger and Savonick, 1975).

IV. MINERAL RELATIONS

The ways in which flooding influences plant mineral nutrition are complex, being determined by several concomitant flooding effects on the soil, initial soil conditions, and nutrient-absorption mechanisms as well as other physiological processes and responses of the particular plant species under study. Several generalizations can be derived from research, but nearly all of these must be carefully placed into the environmental and physiological contexts in which they apply. This section deals with effects of flooding only as they relate to mineral relations of plants. Other chapters in this volume discuss in depth the effects of flooding on soils (Ponnamperuma, Chapter 2), hormone relations (Reid and Bradford, Chapter 6), disease (Stolzy and Sojka, Chapter 7), and plant adaptation to flooding (Hook, Chapter 8, and Wainwright, Chapter 9).

Many studies have employed actual waterlogging treatments to ascertain nutrient responses of plants. However, because much evidence indicates that dysfunction in nutrient absorption by roots under flooding conditions is largely caused by a lack of O_2 and attendant deleterious metabolic effects, some investigators have simulated flooding by imposing anaerobic conditions either by altering the composition of gas supplied to the soil or by bubbling O_2-free or -deficient gases through solution cultures. The latter systems have reduced secondary effects of flooding, particularly those related to disease and accumulation of phytotoxic products in the soil. Moreover, because nutrient availabilities may change measurably with soil submergence, soil-free, anaerobic systems such as solution cultures afford a convenient means of studying root absorption of minerals free of confounding and, to a large extent, uncontrollable changes in substrate nutrient supply to roots.

With respect to nutrient relations, responses of flood-intolerant species often differ from those tolerant of flooding. Although there is no sharp dichotomy in responses of the two types, they provide a convenient structure for discussion.

A. Flood-Intolerant Plants

1. *Macronutrient Elements and Sodium*

a. Nitrogen. Macronutrient response to flooding or anaerobiosis in flood-intolerant plants is summarized from the literature in Table II, and typical effects of reduced O_2 availability on shoot mineral concentrations for several elements are shown in Fig. 7. Nitrogen concentrations in tissues are generally reduced by flooding, and in nearly every case total nitrogen content of plant tissues declines. Several factors contribute to this response. In waterlogged soils there is a rapid depletion of nitrate nitrogen as free O_2 is quickly consumed by soil biota and anaerobic conditions develop. As a result, volatilization and loss of nitrogen are promoted through denitrification in which nitrate serves as a terminal electron acceptor for anaerobic microbes (Ponnamperuma, 1972). Singh and Ghildyal (1980) attributed part of the reduction in nitrogen concentration in *Zea mays* shoots to reduced nitrate availability in the soil.

Reduced uptake of nitrate as a result of effects of low O_2 tension on root metabolism also appears to play an important role in reducing nitrogen levels in flooded plants. In a series of experiments Drew and co-workers studied growth and nitrogen uptake patterns of wheat (*Triticum aestivum*) subjected to several anaerobic regimes (Drew and Lynch, 1980). In waterlogged soils, slowing of root and shoot growth was more closely related to declining O_2 concentration in the soil solution than to concentration of dissolved inorganic nitrogen (Trought and Drew, 1980a). Hence in this experiment nitrogen availability to the plants was apparently not limiting growth. Waterlogged wheat plants exhibited reduced nitrogen concentrations, chlorosis, and generally accelerated senescence of older leaves, with the latter two occurring with the onset of remobilization of nitrogen from old to young leaves (Trought and Drew, 1980b). Applications of nitrate or ammonium ions to the relatively well-aerated soil surface or by foliar sprays with urea counteracted the effect of waterlogging on chlorosis and senescence. When the soil was replaced by solution cultures rendered anaerobic by bubbling nitrogen gas through them, uptake of nitrogen (and phosphorus and potassium) was inhibited as it was in waterlogged soil (Trought and Drew, 1980c). From these data it appears that inhibition of nitrate uptake, rather than availability of nitrogen, is responsible for early nitrogen-deficiency symptoms in wheat. Similar patterns of chlorosis, accelerated senescence, internal nitrogen redistribution, and inhibited nitrogen uptake have been observed in barley (Drew and Sisworo, 1977; Drew *et al.*, 1979). Hence initial soil-nitrogen status, denitrification, and uptake responses of plants to flooding interact to determine the amount and distribution of nitrogen in tissues of flooded plants.

b. Potassium. In general, inhibitory effects of flooding on potassium uptake are similar to those for nitrogen (Table II and Fig. 7). Severe inhibition of potassium uptake characteristically follows soil submergence, and this response may limit plant growth in certain flooded crops (Lawton, 1945). Reduction in potassium absorption is most likely attributable to the effects of anaerobiosis on uptake mechanisms of roots (Hopkins *et al.*, 1950; Hammond *et al.*, 1955; Singh and Ghildyal, 1980; Trought and Drew, 1980b,c). If organic matter is available and cation exchange capacity of the soil is low, submergence may increase soluble potassium somewhat in the soil solution through displacement of exchangeable potassium from the exchange complex by competing ions (most likely newly dissolved reduced iron and manganese) (Jones and Etherington, 1970; Jones, 1975). However, flood-associated increases in potassium available to the plant are apparently too small to overcome the large inhibitory effects of anaerobic conditions on potassium uptake.

c. Phosphorus. Phosphorus composition of flooded plants, like nitrogen composition, is influenced by both soil conditions and plant uptake responses to soil inundation. Where amounts of soluble phosphorus available in the soil are adequate, flooding of intolerant plants generally lowers both tissue concentration and total content of phosphorus (Table II). These declines are attributable to inhibited uptake under anaerobiosis (Labanauskas *et al.*, 1965, 1972; Leyson and Sheard, 1974; Reyes *et al.*, 1977; Drew and Lynch, 1980). However, the situa-

TABLE II

Responses of Mineral Element Composition of Flooding-Intolerant Plants to Waterlogging or Induced Anaerobiosis

Species	Tissue	Measurement basis[a]	Treatment	Element: N	P	K	Ca	Mg	Cl	Na	Fe	Mn	Cu	Zn	Mo	B	Reference
Hordeum vulgare	Shoot	a	Waterlogging, low soil O_2	↓[b]	↓	↓	→	→		→							Drew and Sisworo (1977); Leyshon and Sheard (1974); Letey *et al.* (1962)
Triticum aestivum	Grain	a	Low soil O_2	→							→	↑	→	↓			Labanauskas *et al.* (1975)
		b		↓							↓	↓	↓	↓			Labanauskas *et al.* (1975)
	Shoot	a,b	Waterlogging, solution culture, anaerobiosis	↓	↓	↓	↓	↓									Trought and Drew (1980b,c)
Zea mays	Ear leaf	a	Waterlogging	↓	↓	↓					↑	↑	↓	↓	↑	↓	Lal and Taylor (1970)
	Shoot	a	Waterlogging	↓	↓	↓	↓	↓									Lawton (1945); Shapiro *et al.* (1956)
	Roots	a	Low soil O_2	↑	↑	↑		↑									Shapiro *et al.* (1956)
5 varieties	Whole plant	a	Waterlogging	↓	↑	↓	→	→									Singh and Ghildyal (1980)
Trifolium subterraneum	Leaves	a,b	Waterlogging	↓	↑	↓	↓	↓		↓	↑	→	→	↓			Devitt and Francis (1972)
Pisum sativum	Stem and leaves	a	Waterlogging	↓	↑	→	→					→					Jackson (1979)
	Fruit	a		↓	→	→						→					
Persea americana	Leaves	a	Low soil O_2	↓	↓	↓	↓	↓	→	→	↑	↓	↓	↓			Slowik *et al.* (1979)
	Stems	a		→	→	↓	→	→	→	↑	→	↓	↓	→			
	Roots	a		↓	→	↓	→	↓	↓	↑	→	→	→	↑			
	Total plant	b		↓	↓	↓	↓	↓	↓	↓	↓	↓	↓	↓			
Liquidambar styraciflua	Whole seedling	b	Waterlogging	↓	↓	↓	→	↓									Hosner and Leaf (1962)
Celtis laevigata	Whole seedling	b		↓	↑	↓	↓	↓									Hosner and Leaf (1962)
Pinus elliottii		c	Low O_2 supply to solution culture	↑	↓	↓	↓	↓									Shoulders and Ralston (1975)
Citrus sinensis	Leaves	a	Low soil O_2	↓	↓	↓	↓	↓	→	→	↓	↓	↓	↓		↓	Labanauskas *et al.* (1972)
	Stems	a		↓	→	↓	↓	↓	↑	↑	→	→	→	→		→	
	Roots	a		↑	↑	↑	↑	↑	→	→	↑	↑	↑	↑		↑	
	Whole plant	b		↓	↓	↓	↓	↓	↓	→	→	↓	↓	↓		↓	
	Tops	b	Low soil O_2	↓	↓	↓	↓	↓	↑	↑	↓	→	→	→		↓	Labanauskas *et al.* (1965)

	Roots	b		↓	↓	↓	↓	↓	↓	↓	↓	↓	↓	↓		↓	
	Total plant	b		↓	↓	↓	↓	↓	↓	→	↓	↓	↓	↓		↓	
Gossypium hirsutum	Shoots	a	Low soil O_2 solution culture, anaerobiosis	↓	↓	↓	↓			↑							Letey *et al.* (1961a)
	Leaves	a				↓	↓	↓		↑							Sowell and Rouse (1958)
Helianthus annuus	Shoots	a	Solution culture, anaerobiosis	↓	↓	↓	↓			↑							Letey *et al.* (1961a)
Simmondsia chinensis	Leaves	a	Low soil O_2	↑	→	↓	↓	↓		↑	↑	↓	→	↓		↓	Reyes *et al.* (1977)
	Stems	a		↑	→	↓	↑	↑		↑	↑	↑	→	→		↑	
	Roots	a		↓	↓	↓	→	↓		↓	↓	↑	↑	→		↑	
	Whole plant	b		↓	↓	↓	↓	↓		→	↓	↓	↓	↓		↓	
Agrostis stolonifera	Shoots	a	Waterlogging			→											Jones (1972)
	Roots	a				↓											
Festuca rubra	Shoots	a				→											
	Roots	a				↓											
Erica cinerea	Leaves	a	Waterlogging								↑	↑					Jones and Etherington (1972)
	Roots	a									↑	→					
Erica tetralix	Leaves	a									↑	→					
	Roots	a									→	→					
Dactylis glomerata	Shoots	a	Waterlogging	↓	↓	↓	↑	↓									Rogers and Davies (1973)
Phleum pratense	Shoots	a		↓	↓	↓	→	↓									
Festuca arundinacea	Shoots	a		↓	↓	↓	↓	↓									
Lolium perenne	Shoots	a		↓	↓	↓	→	↓									
Antirrhinum sp. (snapdragon)	Shoots	a	Low soil O_2		↓	↓				↑							Letey *et al.* (1961b)
SUMMARY	Number																
Concentration	↓			18	13	25	12	14	0	2	3	5	4	6	0	3	
	→			2	6	4	8	3	2	4	5	8	6	4	0	1	
	↑			5	5	2	3	3	1	10	9	6	2	2	1	3	
Total content	↓			11	7	10	8	10	3	3	6	6	6	6	3	5	
	→			0	0	0	1	0	0	3	1	2	2	1	0	0	
	↑			0	2	0	0	0	1	1	1	0	0	0	1	0	

[a]a, Concentration of element in plant part; b, total element content in plant part; c, uptake of element per unit volume of root per hour.
[b]↑, increase; ↓, decrease; →, no significant change in element concentration or content.

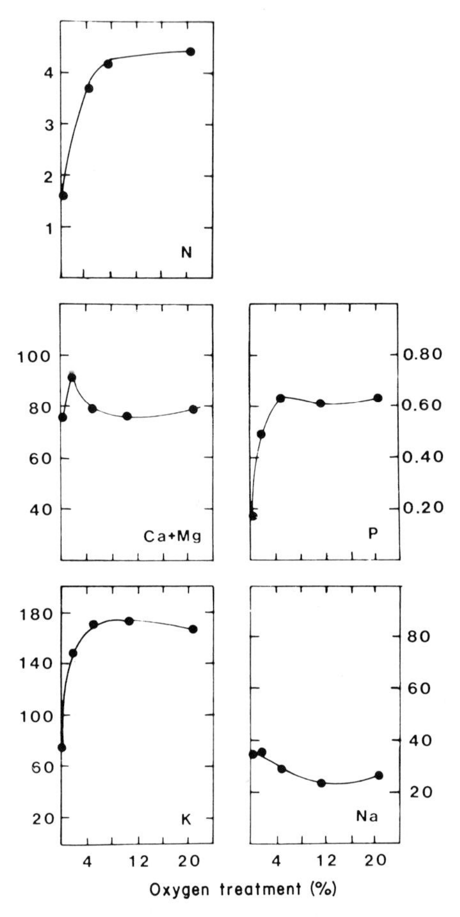

Fig. 7. Concentrations of various minerals in the shoots of cotton (nitrogen only) and leaves of barley (other elements) for several soil O_2-supply treatments. Vertical scales: N, P, %; Ca + Mg, K, Na, milliequivalents per 100 g. After Letey *et al.* (1961a, 1962). Reproduced in part from *Agron. J.* **54,** 538–540, 1962, by permission of the American Society of Agronomy.

tion is more complex for soils moderately or severely deficient in phosphorus. In such well-aerated soils much phosphorus can be held in unavailable forms. For example, in acid soils phosphorus can be tied up in iron and aluminum phosphates and tightly held on anion-exchange sites of clays and hydrous oxides of ferric iron and aluminum. In alkaline soils much phosphorus is present as very sparingly soluble hydroxyapatite. When soil is flooded, soil pH moves toward neutrality and soil reduction levels increase; as a result, phosphorus can be released from insoluble, adsorbed, and bound forms (Shapiro, 1958; Ponnamperuma, 1972), thereby becoming more available for uptake by roots. Hence, if phosphorus uptake is not severely inhibited by the imposed level of anaerobiosis and preflooding-available soil phosphorus levels are not inordinately high, flooding can result in temporarily increased plant phosphorus content (Devitt and Francis, 1972; Singh and Ghildyal, 1980). Hence the balance between increased phosphorus availability and inhibition of phosphorus uptake determines the response of plant phosphorus levels to brief floods. Plants subjected to prolonged flooding generally have reduced tissue phosphorus concentration and total content, as increased phosphorus availability cannot compensate for severe degeneration of the root system.

d. Calcium and Magnesium. Plant-tissue contents of calcium and magnesium often respond to flooding in similar fashion (Table II and Fig. 7). Flooding appears to have a much less pronounced inhibitory effect on accumulation of calcium and magnesium than on nitrogen, phosphorus, or potassium. Hence calcium and magnesium concentrations are not altered as much by flooding as are those of nitrogen, potassium, and phosphorus; however, concentrations may decrease slightly and total contents decline appreciably because of severely reduced growth. Absorption of calcium and magnesium may be metabolically mediated, and therefore be dependent on an adequate supply of O_2, but some data suggest that calcium and magnesium ions are actively extruded from the plasmalemma (Läuchli, 1979). Moreover, based on comparisons of mineral-element analyses of xylem exudate and culture solutions, Trought and Drew (1980c) suggested that calcium and magnesium were excluded relative to water movement in anaerobically cultured wheat root systems and that calcium and magnesium contents of the exudate could be accounted for by simple mass flow (see also Drew and Biddulph, 1971). Therefore, the lack of close coupling between active uptake mechanisms and calcium and magnesium accumulation by plants may explain the reduced effect of flooding on tissue concentrations of these elements.

e. Sodium. In the limited number of studies that have examined sodium levels in plants subjected to flooding or reduced O_2 supply to the root system, plant sodium content generally increased under anaerobiosis (Table II and Fig. 7). This response is consistent with current understanding of mineral metabolism in that plant roots are thought to extrude sodium ions at the plasmalemma (Läuchli, 1979). It is possible that, with O_2 depletion, exclusion of sodium becomes less efficient and tissue sodium concentrations rise. Some evidence, however, shows that sodium uptake by root systems occurs under both aerobic and anaerobic conditions, and hence that anaerobiosis itself cannot account for observed changes of sodium in flooded plants (Leggett and Stolzy, 1961). In any case, the tendency for sodium to accumulate in flooded plants suggests the possibility of sodium toxicity, particularly in sodium-sensitive species (Labanauskas *et al.*, 1965).

2. *Micronutrients*

a. Iron and Manganese. In submerged soils iron and manganese availabilities increase as first ferric and then manganic forms are converted to the more reduced and soluble ferrous and manganous forms (Ponnamperuma, 1972). The change in availabilities of these two elements is reflected in increased tissue concentrations (Table II); however, total content of iron and manganese most often declines because of severely inhibited growth. Reduced forms of iron and manganese are more readily taken up by roots, and in the case of iron, conversion of the ferric to the ferrous form at the root surface is required for uptake and is normally accomplished by internally produced reductant (Brown, 1978).

The abundance of ferrous and manganous ions in flooded soils may result in acute phytotoxic effects in certain plant species. Differential responses of plants to buildups of soluble ferrous and manganous ions have been suggested as influencing factors in species ecology and habitat distribution. Jones and Etherington (1970) observed that cut shoots of a dune-slack species (*Erica cinerea*) characteristically found on waterlogged soils were much less injured by iron sulfate introduced into the water supply than were those of a drier-site dune species (*E. tetralix*). These results suggest that high soluble-iron concentrations in wet habitats might exclude or reduce the abundance of the latter species. Similarly, Jones (1972) observed that whereas the dune species *Agrostis stolonifera* and *Festuca rubra* exhibited reduced growth or toxicity symptoms as manganese concentration was increased in solution cultures, the dune-slack species *Carex flacca* and *C. nigra* grew larger and displayed no or much less severe toxicity at higher manganese concentrations. Thus, increasing soluble manganese contents in waterlogged soils would favor the carices and exclude the grasses.

b. Copper, Zinc, Molybdenum, and Boron. In submerged soils availability of copper, zinc, molybdenum, and boron changes with time. With flooding, iron and manganese reduction and production of organic complexing compounds tend to increase solubility of copper and zinc. Concurrently, in acid soils increased pH and formation of sulfides associated with flooding tend to lower copper and zinc solubilities. Overall, for noncalcareous soils, long-term flooding tends to increase availability of copper and molybdenum and depress that of zinc (Ponnamperuma, 1972). Tissue concentrations and total contents of zinc generally decline under anaerobiosis (Table II), whereas concentrations of copper and boron may remain constant, decrease, or increase. Total content of copper and boron usually decreases as growth is greatly inhibited. The few data available for molybdenum

(Table II) indicate that tissue concentration of this element increased in the ear leaf of flooded *Zea mays* (Lal and Taylor, 1970).

3. Plant Tissue–Mineral Element Interactions

There is some evidence that flooding effects on nutrient content vary among plant organs. Although not apparent in all studies, higher concentrations of elements in the roots of flooded plants may be coupled with decreased shoot concentrations of the same minerals (Table II; Shapiro *et al.*, 1956; Labanauskas *et al.*, 1972; Reyes *et al.*, 1977). To account for these responses, Shapiro *et al.* (1956) suggested that reduced O_2 availability to roots inhibited translocation of ions from roots to shoots more than it decreased ion uptake. In contrast, sodium concentration in flooded plants was sometimes increased in shoots and decreased in roots (Labanauskas *et al.*, 1965; Reyes *et al.*, 1977). As root systems of unflooded plants in these studies had higher sodium concentrations compared with shoots, a metabolically related reduction in the efficiency with which sodium is excluded from the shoot may occur in flooded plants.

In one study that examined the effects of reduced aeration and flooding on nutrient concentrations and contents of grain of the wheat cultivar Yecora, manganese concentration was elevated and zinc concentration depressed in response to flooding (Table II; Labanauskas *et al.*, 1975); concentration of other nutrients was not significantly affected by either flooding or exclusion of O_2 from the soil. Total contents of all elements examined decreased significantly with anaerobiosis, however, because of substantial inhibitory effects of both flooding and O_2 exclusion on kernel growth and yield.

B. Flood-Tolerant Plants

In contrast with the reduced growth and total nutrient uptake observed in flood-intolerant species, plants tolerant of flooding often grow better and take up more nutrients in response to flooding when compared with well-watered (but not waterlogged) controls. Such responses have been reported for several species of flood-tolerant woody angiosperms (*Fraxinus pennsylvanica, F. profunda, Nyssa aquatica, Salix nigra, Acer negundo, A. rubrum, A. saccharinum, Populus deltoides,* and *Platanus occidentalis*), conifers (*Taxodium distichum*), and rice (*Oryza* spp.) (Hosner and Leaf, 1962; Dickson *et al.*, 1972; Weeraratna, 1975; Chahal *et al.*, 1980). Response of tissue nutrient concentrations of these plants to flooding varies. For example, flooding of rice increased tissue concentrations of nitrogen, phosphorus, calcium and iron, whereas those of manganese and magnesium were not significantly altered (Islam and Islam, 1973). In certain cases, such as was observed for zinc in rice shoots (Jugsujinda and Patrick, 1977), concentrations of minerals may even decline somewhat in flooded plants. However, as growth is uniformly greater in flooded plants, such a response is most likely ascribable to dilution effects and is not a sign of nutrient-absorption dysfunction (Hosner *et al.*, 1965).

Several morphological adaptations of flood-tolerant species allow for continued nutrient absorption under submerged conditions. Whereas root systems of flood-intolerant plants undergo large reductions in growth and can decay as anaerobic pathogens, such as *Phytophthora* spp., invade (Stolzy *et al.*, 1975; Kozlowski, 1982), many flood-tolerant species initiate vigorous adventitous roots that proliferate most abundantly in the upper, well-aerated portion of submerged soil and in the water layer above (Hook *et al.*, 1971; Sena Gomes and Kozlowski, 1980a; see also Kozlowski, Chapter 4 in this volume). Hence the optimal environment for uptake of minerals is thoroughly exploited. In response to soil inundation, many tolerant plants also develop abundant aerenchymatic tissues in roots that facilitate O_2 movement from aerial portions of the plant into the rhizosphere (Armstrong, 1968; Coutts and Armstrong, 1976; Stelzer and Läuchli, 1977). In flooded rice and *Eriophorum angustifolium* plants, reddish brown streaks of oxidized iron were found in soil adjacent to roots, emphasizing the capacity for O_2 transfer into the soil by these species (Armstrong, 1967; Ponnamperuma, 1972).

These adaptations permit normal root metabolic function despite intensely reducing and anaerobic surroundings. Given adequate (but not excessive) quantities of available minerals in submerged soils, flood-tolerant plants are able to absorb sufficient quantities of nutrients to sustain rapid growth. However, because of particular soil conditions, severe nutritional problems can arise as a result of submergence. Several nutritional deficiency and toxicity problems have been identified in rice, and most of these have been correlated with the chemical properties of various submerged soils (Tanaka and Yoshida, 1970). These studies have been particularly thorough and will be discussed in some detail because they illustrate well the large influence that soils and submergence have on mineral nutrition of flood-tolerant species.

1. Zinc Deficiency

On calcareous soils rice plants frequently exhibit zinc deficiency, presumably as a result of zinc fixation (Buckman and Brady, 1969). Deficiency occurs primarily during early growth of the crop and during this period may be exacerbated by immobilization of zinc in roots by bicarbonate ions that are produced in alkaline soils soon after submergence (Tanaka and Yoshida, 1970). Alkaline soils subjected to prolonged flooding exhibit increased zinc availability (and rice zinc-deficiency symptoms decrease) as pH declines with increasing soil reduction (Ponnamperuma, 1972; Pavanasasivam and Axley, 1980). These changes in available zinc are reflected in increased zinc concentration in plant tissues (Table

TABLE III

Zinc Content of Rice Roots and Shoots Grown under Flooded and Unflooded Conditions on an Othello Silt Loam Treated with Sulfate[a,b]

	Zinc (ppm)			
	Unlimed (pH 4.8)		Limed (pH 7.6)	
$SO_4 - S$ (kg ha^{-1})	Unflooded	Flooded	Unflooded	Flooded
Roots				
0	51.9_a	48.9_a	22.8_c	37.2_b
50	56.1_a	20.3_c	30.4_c	19.1_c
150	56.7_a	33.5_c	28.5_c	21.1_c
Shoots				
0	55.5_a	27.3_c	14.2_e	21.2_c
50	56.3_a	34.4_b	19.7_{cd}	14.9_{de}
150	58.8_a	35.8_b	25.9_c	12.5_e

[a] Adapted from Pavanasasivam and Axley (1980). *Commun. Soil Sci. Plant Anal.* **11,** 163–174, by courtesy of Marcel Dekker, Inc.

[b] Means followed by different subscripts are significantly different at the 5% level. Separate analyses for root and shoot data were performed after a square-root transformation.

III; Pavanasasivam and Axley, 1980). Concurrent increase in sulfides produced by biological reduction of sulfate under anaerobic conditions may lower tissue concentrations of zinc by precipitation of very insoluble zinc sulfide in the soil (Table III; Tanaka and Yoshida, 1970).

2. *Iron Toxicity*

Generally, high levels of iron are found in the soil solution of submerged soil. In dry soil that has low pH and abundant sulfate (''acid sulfate soil''), extremely high amounts of soluble ferrous iron are found soon after submergence, causing ''bronzing'' of rice leaves (Tanaka and Yoshida, 1970). As pH rises toward neutrality with prolonged flooding, iron availability decreases, as does iron toxicity. Excessive iron can also interfere with uptake of other nutrients. For example, high concentrations of substrate iron can induce phosphorus-deficiency symptoms in rice plants. Iron toxicity and associated nutrient-interference problems can be ameliorated by several means, including continuous submergence of the soil (which eliminates the postinundation peak in soluble iron), applications of lime, compost, phosphorus, and potassium, and by application of urea instead of ammonium sulfate as a nitrogen fertilizer (Tanaka and Yoshida, 1970).

3. *Manganese Toxicity*

Manganese concentrations of rice plants grown on certain soils may reach 3000 ppm, but visible toxicity symptoms are unusual. Rice is apparently quite tolerant of manganese, as tissue levels that may reduce yields and induce toxicity symptoms (>2500 ppm) are an order of magnitude greater than those for other crop plants [e.g., 200 ppm for *Medicago sativa* (Ouellette and Dessureaux, 1958; Tanaka and Yoshida, 1970)].

4. *Sulfide Toxicity*

As the redox potential of the soil declines after inundation, sulfate is biologically reduced, generally by bacterial obligate anaerobes of the genus *Desulfovibrio* (Armstrong, 1975). Although sulfide is quite phytotoxic, healthy rice roots have sufficient capacity to oxidize proximal sulfide so that direct injury from sulfide is minimal. Further, rice plants interact with a bacterium (*Beggiotoa* sp.) to accomplish a reduction of sulfide content in flooded paddy soil [Pitts (1969), cited in Etherington (1975)]. *Beggiotoa* oxidizes hydrogen sulfide to molecular sulfur intracellularly with biologically toxic hydrogen peroxide accumulating as a by-product. Peroxide concentrations eventually increase to injurious levels unless an external source of catalase is available. Pitts (1969) showed that rice roots release catalase to the soil and that *Beggiotoa* isolates from a paddy soil grow well in the presence of rice roots. In contrast, rice plants of low vigor may undergo sulfide damage because they lack sufficient O_2-transporting capacity. Moreover, sulfide oxidation may consume so many oxidizing equivalents from healthy rice roots that iron oxidation is inhibited and iron toxicity results (Tanaka and Yoshida, 1970).

REFERENCES

Allaway, W. G., and Milthorpe, F. L. (1976). Structure and functioning of stomata. *In* "Water Deficits and Plant Growth" (T. T. Kozlowski, ed.), Vol. 4, pp. 57–102. Academic Press, New York.

Armstrong, W. (1967). The oxidizing activity of roots in waterlogged soil. *Physiol. Plant.* **20,** 920–926.

Armstrong, W. (1968). Oxygen diffusion from the roots of woody species. *Physiol. Plant.* **21,** 539–543.

Armstrong, W. (1975). Waterlogged soils. *In* "Environment and Plant Ecology" (J. R. Etherington, ed.), pp. 181–218. Academic Press, New York and London.

Beadle, C. L., Jarvis, P. G., and Neilson, R. E. (1979). Leaf conductance as related to xylem water potential and carbon dioxide concentration in Sitka spruce. *Physiol. Plant.* **45,** 158–166.

Bradford, K. J. (1982). Regulation of shoot responses to root stress by ethylene, abscisic acid and cytokinin. In "Plant Growth Substances 1982" (P. F. Wareing, ed.), pp. 599–608. Academic Press, New York and London.

Brown, J. C. (1978). Mechanism for iron uptake by plants. *Plant Cell Environ.* **1,** 249–257.

Buckman, H. O., and Brady, N. C. (1969). "The Nature and Properties of Soils," 7th ed. Macmillan, New York.

Caughey, M. G. (1945). Water relations of pocosin or bog shrubs. *Plant Physiol.* **20,** 671–689.

Chahal, R. S., Tulla, P. S., and Khanna, S. S. (1980). Nutrient uptake by rice crop under submerged and upland conditions. *J. Indian Soc. Soil Sci.* **28,** 254–255.

Chang, H. T., and Loomis, W. E. (1945). Effect of carbon dioxide on absorption of water and nutrients by roots. *Plant Physiol.* **20,** 221–232.

Childers, N. E., and White, D. G. (1942). Influence of submersion of the roots on transpiration, apparent photosynthesis, and respiration of young apple trees. *Plant Physiol.* **17,** 603–618.

Cooper, R. B., Blaser, E. E., and Brown, R. H. (1967). Potassium nutrition effects on net photosynthesis and morphology of alfalfa. *Soil Sci. Soc. Am. Proc.* **31,** 231–235.

Coulsen, C. L., Christy, A. L., Cataldo, D. A., and Swanson, C. A. (1972). Carbohydrate translocation in sugar beet petioles in relation to petiolar respiration and adenosine-5-triphosphate. *Plant Physiol.* **49,** 919–923.

Coutts, M. P. (1981). Effects of waterlogging on water relations of actively growing and dormant Sitka spruce seedlings. *Ann. Bot. (London)* **47,** 747–753.

Coutts, M. P., and Armstrong, W. (1976). Role of oxygen transport in the tolerance of trees to waterlogging. *In* "Tree Physiology and Yield Improvement" (M. G. R. Cannell and F. T. Last, eds.), pp. 361–385. Academic Press, New York and London.

Curtis, O. F., and Clark, D. G. (1950). "An Introduction to Plant Physiology." McGraw-Hill, New York.

Davies, W. J., and Kozlowski, T. T. (1975a). Effects of applied abscisic acid and plant water stress on transpiration of woody angiosperms. *For. Sci.* **22,** 191–195.

Davies, W. J., and Kozlowski, T. T. (1975b). Effect of applied abscisic acid and silicone on water relations and photosynthesis of woody plants. *Can. J. For. Res.* **5,** 90–96.

Devitt, A. C., and Francis, C. M. (1972). The effect of waterlogging on the mineral nutrient content of *Trifolium subterraneum. Aust. J. Exp. Agric. Anim. Husb.* **12,** 614–617.

Dickson, R. E., Broyer, T. C., and Johnson, C. M. (1972). Nutrient uptake by tupelo gum and bald cypress from saturated or unsaturated soil. *Plant Soil* **37,** 297–308.

Drew, M. C., and Biddulph, O. (1971). Effect of metabolic inhibitors and temperature on uptake and translocation of ^{45}Ca and ^{42}K by intact bean plants. *Plant Physiol.* **48,** 426–432.

Drew, M. C., and Lynch, J. M. (1980). Soil anaerobiosis, microorganisms and root function. *Annu. Rev. Phytopathol.* **18,** 37–66.

Drew, M. C., and Sisworo, E. J. (1977). Early effects of flooding on nitrogen deficiency and leaf chlorosis in barley. *New Phytol.* **79,** 567–571.

Drew, M. C., Sisworo, E. J., and Saker, L. R. (1979). Alleviation of water-logging damage to young barley plants by application of nitrate and a synthetic cytokinin, and comparison between the effects of waterlogging, nitrogen deficiency and root excision. *New Phytol.* **82,** 315–329.

Etherington, J. R. (1975). "Environment and Plant Ecology." Wiley, London.

Geiger, D. R., and Savonick, S. A. (1975). Effect of temperature, anoxia, and other metabolic inhibitors on translocation. *Encycl. Plant Physiol. New Ser.* **1,** 256–286.

Hammond, L. C., Allaway, W. H., and Loomis, W. E. (1955). Effects of oxygen and carbon dioxide levels upon absorption of potassium by plants. *Plant Physiol.* **30,** 155–161.

Harris, D. G., and van Bavel, C. H. M. (1957). Nutrient uptake and chemical composition of tobacco plants as affected by the composition of the root atmosphere. *Agron. J.* **49,** 176–181.

Hiron, R. W. P., and Wright, S. T. C. (1973). The role of endogenous abscisic acid in the responses of plants to stress. *J. Exp. Bot.* **24,** 769–781.

Holmgren, P., Jarvis, P. G., and Jarvis, M. S. (1965). Resistances to carbon dioxide and water-vapor transfer in leaves of different plant species. *Physiol. Plant.* **18,** 557–573.

Hook, D. D., Brown, C. L., and Kormanik, P. O. (1971). Inductive flood tolerance in swamp tupelo [*Nyssa sylvatica* var. *biflora* (Walt.)Sarg.]. *J. Exp. Bot.* **22,** 78–79.
Hopkins, H. T., Specht, A. W., and Hendricks, S. B. (1950). Growth and nutrient accumulation as controlled by oxygen supply to plant roots. *Plant Physiol.* **25,** 193–209.
Hosner, J. F., and Leaf, A. L. (1962). The effect of soil saturation upon the dry weight, ash content and nutrient absorption of various bottomland species. *Soil Sci. Soc. Am. Proc.* **26,** 401–404.
Hosner, J. F., Leaf, A. L., Dickson, R., and Hart, J. B. (1965). Effects of varying soil moisture upon the nutrient uptake of four bottomland species. *Soil Sci. Soc. Am. Proc.* **29,** 313–316.
Humble, G. D., and Raschke, K. (1971). Stomatal opening quantitatively related to potassium transport. Evidence from electron probe analysis. *Plant Physiol.* **48,** 447–453.
Islam, A., and Islam, W. (1973). Chemistry of submerged soils and growth and yield of rice. I. Benefits from submergence. *Plant Soil* **39,** 555–565.
Jackson, M. B. (1979). Rapid injury to peas by waterlogging. *J. Sci. Food Agric.* **30,** 143–152.
Johnson, F. L., and Bell, D. T. (1976). Plant biomass and net primary production along a flood-frequency gradient in the streamside forest. *Castanea* **41,** 156–165.
Jones, H. E., and Etherington, J. R. (1970). Comparative studies of growth and distribution in relation to waterlogging. I. The survival of *Erica cinerea* L. and *E. tetralix* L. and its apparent relationship to iron and manganese uptake in waterlogged soil. *J. Ecol.* **58,** 487–496.
Jones, R. (1972). Comparative studies of plant growth and distribution in relation to waterlogging. VI. The effect of manganese on the growth of dune and slack plants. *J. Ecol.* **60,** 141–145.
Jones, R. (1975). Comparative studies of plant growth and distribution in relation to waterlogging. IX. The uptake of potassium by dune and dune slack plants. *J. Ecol.* **63,** 859–866.
Jones, R. J., and Mansfield, T. A. (1970). Suppression of stomatal opening in leaves treated with abscisic acid. *J. Exp. Bot.* **21,** 714–719.
Jugsujinda, A., and Patrick, W. H. (1977). Growth and nutrient uptake by rice in a flooded soil under controlled aerobic–anaerobic and pH conditions. *Agron. J.* **69,** 705–710.
Keller, T., and Koch, W. (1962). Der Einfluss der Mineralstoffernährung auf CO_2—Gaswechsel und Blattpigment gehalt der Papperl. II. Eisen, *Mitt. Schweiz. Anstalt Forstliche Versuchs.* **38,** 283–318.
Keller, T., and Koch, W. (1964). The effect of iron chelate fertilization of poplar upon CO_2 uptake, leaf size, and content of leaf pigments and iron. *Plant Soil* **20,** 116–126.
Kozlowski, T. T. (1982). Water supply and tree growth. II. Flooding. *For. Abstr.* **43,** 145–161.
Kozlowski, T. T., and Pallardy, S. G. (1979). Stomatal responses of *Fraxinus pennsylvanica* seedlings during and after flooding. *Physiol. Plant.* **46,** 155–158.
Kramer, P. J. (1940). Causes of decreased absorption of water by plants in poorly aerated media. *Am. J. Bot.* **27,** 216–220.
Kramer, P. J. (1951). Causes of injury to plants resulting from flooding of the soil. *Plant Physiol.* **26,** 722–736.
Kramer, P. J., and Jackson, W. T. (1954). Causes of injury to flooded tobacco plants. *Plant Physiol.* **29,** 241–245.
Kramer, P. J., and Kozlowski, T. T. (1979). Physiology of Woody Plants. Academic Press, New York.
Kriedemann, P. E. (1971). Photosynthesis and transpiration as a function of gaseous diffusive resistances in orange leaves. *Physiol. Plant,* **24,** 218–225.
Kriedemann, P. E., Loveys, B. R., Fuller, G. L., and Leopold, A. C. (1972). Abscisic acid and stomatal regulation. *Plant Physiol.* **49,** 842–847.
Labanauskas, C. K., Stolzy, L. H., Klotz, L. J., and DeWolfe, T. A. (1965). Effects of soil temperature and oxygen on the amounts of macronutrients and micronutrients in citrus seedlings (*Citrus sinensis* var. 'Bessie'). *Soil Sci. Soc. Am. Proc.* **29,** 60–64.
Labanauskas, C. K., Stolzy, L. H., Klotz, L. J., and DeWolfe, T. A. (1971). Soil oxygen diffusion

rates and mineral accumulations in citrus seedlings (*Citrus sinensis* var. Bessie). *Soil Sci.* **111,** 386–392.

Labanauskas, C. K., Stolzy, L. H., and Handy, M. F. (1972). Concentrations and total amounts of nutrients in citrus seedlings (*Citrus sinensis* Osbeck) and in soil as influenced by differential soil oxygen treatments. *Soil Sci. Soc. Am. Proc.* **36,** 454–457.

Labanauskas, C. K., Stolzy, L. H., and Luxmoore, R. J. (1975). Soil temperature and soil aeration effects on concentrations and total amounts of nutrients in 'Yecora' wheat grain. *Soil Sci.* **120,** 450–454.

Lal, R., and Taylor, G. S. (1970). Drainage and nutrient effects in a field lysimeter study. II. Mineral uptake by corn. *Soil Sci. Soc. Am. Proc.* **34,** 246–248.

Läuchli, A. (1979). Absorption and translocation of mineral ions. *Prog. Bot.* **41,** 44–54.

Lawton, K. (1945). The influence of soil aeration on the growth and absorption of nutrients by corn plants. *Soil Sci. Soc. Am. Proc.* **10,** 263–268.

Leggett, J. E., and Stolzy, L. H. (1961). Anaerobiosis and sodium accumulation. *Nature (London)* **192,** 991–992.

Letey, J., Stolzy, L. H., Blank, G. B., and Lunt, O. R. (1961a). Effect of temperature on oxygen diffusion rates and subsequent shoot growth, root growth and mineral content of two plant species. *Soil Sci.* **92,** 314–321.

Letey, J., Lunt, O. R., Stolzy, L. H., and Szuszkiewicz, T. E. (1961b). Plant growth, water use and nutritional responses to rhizosphere differentials of oxygen concentration. *Soil Sci. Soc. Am. Proc.* **25,** 183–186.

Letey, J., Stolzy, L. H., Valoras, N., and Szuszkiewicz, T. E. (1962). Influence of soil oxygen on growth and mineral composition of barley. *Agron. J.* **54,** 538–540.

Leyson, A. J., and Sheard, R. W. (1974). Influence of short term flooding on the growth and plant nutrient composition of barley. *Can. J. Soil Sci.* **54,** 463–473.

Loustalot, H. J. (1945). Influence of soil moisture conditions on apparent photosynthesis and transpiration of pecan leaves. *J. Agric. Res.* **71,** 519–532.

Mansfield, T. A., and Davies, W. J. (1981). Stomata and stomatal mechanisms. *In* "The Physiology and Biochemistry of Drought Resistance in Plants" (L. G. Paleg and D. Aspinall, eds.), pp. 315–346. Academic Press, Sydney.

Mansfield, T. A., and Jones, R. J. (1971). Effects of abscisic acid on potassium uptake and starch content of stomatal guard cells. *Planta* **101,** 147–158.

McGee, A. B., Schmierbach, M. R., and Bazzaz, F. A. (1981). Photosynthesis and growth of populations of *Populus deltoides* from contrasting habitats. *Am. Midl. Nat.* **105,** 305–311.

Mikhailova, A. V. (1977). Some characteristics of the work of barley leaf photosynthetic apparatus under the effect of soil inundation with water and vitamin PP. *Biol. Nauki (Moscow)* **20,** 104–105.

Moldau, H. (1973). Effects of various water regimes on stomatal and mesophyll conductances of bean leaves. *Photosynthetica* **7,** 1–7.

Newsome, R. D., Kozlowski, T. T., and Tang, Z. C. (1982). Responses of *Ulmus americana* seedlings to flooding of soil. *Can. J. Bot.* **60,** 1688–1695.

Noland, T. L., and Kozlowski, T. T. (1979). Influence of potassium nutrition on susceptibility of silver maple to ozone. *Can. J. For. Res.* **9,** 501–503.

Ouellette, G. C., and Dessureaux, L. (1958). Chemical composition of alfalfa as related to degree of tolerance to manganese and aluminum. *Can. J. Plant Sci.* **38,** 206–214.

Pallaghy, C. K., and Raschke, K. (1972). No stomatal response to ethylene. *Plant Physiol.* **49,** 275–276.

Pallas, J. E., and Kays, S. J. (1982). Inhibition of photosynthesis by ethylene—a stomatal effect. *Plant Physiol.* **70,** 598–601.

Parker, J. (1950). The effect of flooding on the transpiration and survival of some southeastern forest tree species. *Plant Physiol.* **25,** 453–460.

Pavanasasivam, V., and Axley, J. H. (1980). Influence of flooding on the availability of soil zinc. *Commun. Soil Sci. Plant Anal.* **11,** 163–174.

Pereira, J. S., and Kozlowski, T. T. (1977). Variation among woody angiosperms in response to flooding. *Physiol. Plant.* **41,** 184–192.

Phung, H. T., and Knipling, E. B. (1976). Photosynthesis and transpiration of citrus seedlings under flooded conditions. *HortScience* **11,** 131–133.

Pitts, K. G. (1969). Explorations in the chemistry and microbiology of Louisiana rice plant–soil relations. Ph.D. Dissertation, Louisiana State Univ., Baton Rouge, Louisiana.

Ponnamperuma, F. N. (1972). The chemistry of submerged soils. *Adv. Agron.* **24,** 29–96.

Qureshi, F. A., and Spanner, D. C. (1973). The effect of nitrogen on the movement of tracers down the stolon of *Saxifraga sarmentosa* L. with some observations on the influence of light. *Planta* **110,** 131–144.

Raschke, K. (1975). Stomatal action. *Annu. Rev. Plant Physiol.* **26,** 309–340.

Regehr, D. L., Bazzaz, F. A., and Boggess, W. R. (1975). Photosynthesis, transpiration, and leaf conductance of *Populus deltoides* in relation to flooding and drought. *Photosynthetica* **9,** 52–61.

Reyes, D. M., Stolzy, L. H., and Labanauskas, C. K. (1977). Temperature and oxygen effects in soil on nutrient uptake in jojoba seedlings. *Agron. J.* **69,** 647–650.

Rodriguez, R. R., and Gausman, H. W. (1973). Flooding effects on light reflectance, transmittance and absorptance of cotton, *Gossypium hirsutum* leaves. *J. Rio Grande Val. Hortic. Soc.* **27,** 81–88.

Rogers, J. A., and Davies, G. E. (1973). The growth and chemical composition of four grass species in relation to soil moisture and aeration factors. *J. Ecol.* **61,** 455–472.

Sena Gomes, A. R., and Kozlowski, T. T. (1980a). Growth responses and adaptations of *Fraxinus pennsylvanica* seedlings to flooding. *Plant Physiol.* **66,** 267–271.

Sena Gomes, A. R., and Kozlowski, T. T. (1980b). Responses of *Melaleuca quinquenervia* seedlings to flooding. *Physiol. Plant.* **49,** 373–377.

Shapiro, R. E. (1958). Effect of flooding on availability of phosphorus and nitrogen. *Soil Sci.* **85,** 190–197.

Shapiro, R. E., Taylor, G. S., and Volk, G. W. (1956). Soil oxygen contents and ion uptake by corn. *Soil Sci. Soc. Am. Proc.* **20,** 193–197.

Shaybany, B., and Martin, G. C. (1977). Abscisic acid identification and its quantification in leaves of *Juglans* seedlings during waterlogging. *J. Am. Soc. Hortic. Sci.* **102,** 300–302.

Shoulders, E., and Ralston, C. W. (1975). Temperature, root aeration and light influence slash pine nutrient uptake rates. *For. Sci.* **21,** 401–410.

Sij, J. W., and Swanson, C. A. (1973). Effect of petiole anoxia on phloem transport in squash. *Plant Physiol.* **51,** 368–371.

Singh, R., and Ghildyal, B. P. (1980). Soil submergence effects on nutrient uptake, growth and yield of five corn cultivars. *Agron. J.* **72,** 737–741.

Sipp, S. K., and Bell, D. T. (1974). The response of net photosynthesis to flood conditions in seedlings of *Acer saccharinum* (silver maple). *Univ. Ill. For. Res. Rep.* No. 74–9.

Slatyer, R. O. (1967). "Plant–Water Relationships." Academic Press, New York and London.

Slowik, K., Labanauskas, C. K., Stolzy, L. H., and Zentmyer, G. A. (1979). Influence of rootstocks, soil oxygen, and soil moisture on the uptake and translocation of nutrients in young avocado plants. *J. Am. Soc. Hortic. Sci.* **104,** 172–175.

Sojka, R. E., and Stolzy, L. H. (1980). Soil-oxygen effects on stomatal response. *Soil Sci.* **130,** 350–358.

Sojka, R. E., and Stolzy, L. H. (1981). Stomatal response to soil oxygen. *Calif. Agric.* **35,** 18–19.

Sojka, R. E., Stolzy, L. H., and Kaufmann, M. R. (1975). Wheat growth related to rhizosphere temperature and oxygen levels. *Agron. J.* **67,** 591–596.

Sowell, W. F., and Rouse, R. D. (1958). Effect of Na on cation content of leaves and boll production of cotton plants grown in solution culture in growth chamber. *Soil. Sci.* **86,** 70–74.

Stelzer, R., and Läuchli, A. (1977). Salt and flooding tolerance of *Pucinellia peisonis*. II. Structural differentiation of the root in relation to function. *Z. Pflanzenphysiol.* **84,** 95–108.

Stolzy, L. H., Taylor, O. C., Letey, J., and Szuszkiewicz, T. E. (1961). Influence of soil-oxygen diffusion rates on susceptibility of tomato plants to air-borne oxidants. *Soil Sci.* **91,** 151–155.

Stolzy, L. H., Taylor, O. C., Dugger, W. M., Jr., and Mersereau, J. D. (1964). Physiological changes in and ozone susceptibility of the tomato plant after short periods of inadequate oxygen diffusion to the roots. *Soil Sci. Soc. Am. Proc.* **28,** 305–308.

Stolzy, L. H., Labanauskas, C. K., Klotz, L. J., and DeWolfe, T. A. (1975). Nutritional responses and root rot of *Citrus limon* and *Citrus sinensis* under high and low soil oxygen supplies in the presence and absence of *Phytophthora* spp. *Soil Sci.* **119,** 136–142.

Tanaka, A., and Yoshida, S. (1970). Nutritional disorders of the rice plant. *Int. Rice Res. Inst. Tech. Bull.* No. 10.

Tang, Z. C., and Kozlowski, T. T. (1982a). Some physiological and morphological responses of *Quercus macrocarpa* seedlings to flooding. *Can. J. For. Res.* **12,** 196–202.

Tang, Z. C., and Kozlowski, T. T. (1982b). Physiological, morphological, and growth responses of *Platanus occidentalis* seedlings to flooding. *Plant Soil* **66,** 243–255.

Tang, Z. C., and Kozlowski, T. T. (1982c). Some physiological and growth responses of *Betula papyrifera* seedlings to flooding. *Physiol. Plant.* **55,** 415–420.

Trought, M. C. T., and Drew, M. C. (1980a). The development of waterlogging damage in wheat seedlings (*Triticum aestivum* L.). I. Shoot and root growth in relation to changes in the concentrations of dissolved gases and solutes in the soil solution. *Plant Soil* **54,** 77–94.

Trought, M. C. T., and Drew, M. C. (1980b). The development of waterlogging damage in wheat seedlings (*Triticum aestivum* L.). II. Accumulation and redistribution of nutrients by the shoot. *Plant Soil* **56,** 187–199.

Trought, M. C. T., and Drew, M. C. (1980c). The development of waterlogging damage in young wheat plants in anaerobic solution cultures. *J. Exp. Bot.* **31,** 1573–1585.

Weeraratna, C. S. (1975). Effect of flooding on some chemical and morphological characteristics of rice. *Indian J. Agric. Sci.* **45,** 461–463.

Willey, C. R. (1970). Effects of short periods of anaerobic and near-anaerobic conditions on water uptake by tobacco roots. *Agron. J.* **62,** 224–229.

Yelenosky, G. (1964). Tolerance of trees to deficiencies of soil aeration. *Proc. Int. Shade Tree Conf.* **40,** 127–147.

Zaerr, J. B. (1983). Short-term flooding and net photosynthesis in seedlings of three conifers. *For. Sci.* **29,** 71–78.

Zvareva, E. G., and Bartkov, B. I. (1976). Distribution of assimilates in soybeans during fruit growth in flooded soil. *Fiziol. Biokhim. Kult. Rast.* **8,** 204–208.

CHAPTER 6

Effects of Flooding on Hormone Relations

D. M. REID
Plant Physiology Research Group
Biology Department
University of Calgary
Calgary, Alberta, Canada

K. J. BRADFORD
Department of Vegetable Crops
University of California
Davis, California

I. INTRODUCTION

When subjected to soil flooding, most mesophytes exhibit characteristic changes in their patterns of growth and development (Kawase, 1981; Hook, Chapter 8 in this volume). Many of these changes, such as stomatal closure, petiole epinasty, leaf senescence, hypocotyl swelling, aerenchyma formation, and reduced growth rate, can be either induced in drained plants or prevented in

ISBN 0-12-424120-4

flooded plants by applications of plant hormones, leading to the notion that flooding alters endogenous hormonal relations. There has, unfortunately, been much more speculation on this hypothesis than concrete experimental testing of it. In this chapter the experimental data on the synthesis, metabolism, and transport of the major plant growth regulators as influenced by soil flooding are first reviewed. Then, the characteristic developmental responses of plants to flooding in relation to the documented alterations in hormonal physiology are discussed. This approach should serve to point out areas of ignorance with respect to the physiology of the individual hormones and also allow discussion of hormonal interactions in the etiology of the developmental responses to flooding. Although the discussion is limited to the role of plant hormones in flooding responses, it is recognized that many other factors are also involved in flooding injury (Chapters 4, 5, 7–9 in this volume; Crawford, 1982).

II. EFFECTS OF FLOODING ON SYNTHESIS, METABOLISM, AND TRANSPORT OF HORMONES

A. Ethylene

Turkova (1944) was probably the first to point out that the appearance of waterlogged plants is similar to that of plants exposed to ethylene. Various studies subsequently confirmed that the ethylene levels in flooded plants generally exceed those found in drained plants (Kawase, 1972; El-Beltagy and Hall, 1974; Jackson and Campbell, 1975a). Several mechanisms have been proposed to account for the accumulation of ethylene in flooded plants. An environmental source of the gas was suggested by the discovery that ethylene is present in flooded or poorly aerated soils (Smith and Russell, 1969; Smith and Dowdell, 1974). The ethylene is of microbial origin, although the exact organisms and substrates involved are unclear (Drew and Lynch, 1980). A major paradox is that although ethylene accumulates in soils only after O_2 has been depleted (Sheard and Leyshon, 1976; Ioannou *et al.*, 1977), biological ethylene synthesis requires O_2 (see following discussion). Lynch and Harper (1974) originally proposed that anaerobic microorganisms might provide substrates that could then be converted to ethylene by aerobes, primarily fungi. However, this fails to explain the increase in ethylene evolution rate from soils incubated under continuously anaerobic conditions. In a more recent hypothesis (Lynch and Harper, 1980), nonbiological conversion of 2-keto-4-methylmercaptobutyric acid into ethylene has been suggested as the source of the gas. The synthesis and metabolism of ethylene in soils is complex and has been discussed more fully by Drew and Lynch (1980). For present purposes, it is sufficient to note that ethylene is often present in flooded soils and may represent an exogenous source of the hormone in flooded plants (Jackson and Campbell, 1975b).

Kawase (1972, 1976) found that ethylene accumulated in submerged tissues or in plant parts enclosed in a diffusion barrier such as plastic wrap. The diffusion coefficient of ethylene in water is 10,000-fold less than it is in air, and the low solubility of ethylene in water further limits the rate of ethylene escape from tissues (Drew, 1979). Kawase (1976) proposed that ethylene synthesized by the submerged roots accumulates and moves up the stem through the intercellular spaces. A drawback of this mechanism is that the low O_2 levels in flooded roots would virtually stop ethylene biosynthesis, precluding the accumulation required for diffusion up the stem. Experiments with wrapped stem segments (Kawase, 1976) or calculations of ethylene entrapment in roots resulting from a water layer (Konings and Jackson, 1979) are based on aerobic ethylene production rates and therefore overestimate the degree of ethylene accumulation expected under flooded conditions. Although it is likely that O_2 diffuses intercellularly from the aerial organs and supports some ethylene synthesis in submerged tissues, it is doubtful that ethylene production even at the normal rate in roots could account for the large increase in ethylene levels in the shoots of flooded plants solely by this passive accumulation mechanism (Bradford and Dilley, 1978; Bradford *et al.*, 1982). The water-jacketing effect has been clearly demonstrated for certain aquatic plants (Musgrave *et al.*, 1972), but as the tissues involved are photosynthetic, O_2 would not become limiting for ethylene synthesis.

The mechanisms just discussed share the expectation that ethylene will diffuse considerable distances within the shoots, because elevated ethylene concentrations can be measured even in the apices of waterlogged plants (Jackson and Campbell, 1975a). Movement of [^{14}C]ethylene from the roots to the shoots has been demonstrated, although the actual amount of ethylene transported was small relative to the high concentration applied (Jackson and Campbell, 1975b). Zeroni *et al.* (1977) found that unless diffusion from the plant was physically blocked, ethylene would move only a short distance intercellularly before escaping from the stem. Similarly, gas exchange between the atmosphere and the submerged roots of *Puccinellia peisonis* occurred largely through tissues immediately above the water level (Stelzer and Läuchli, 1980). Thus, it is unlikely that either microbial- or root-synthesized ethylene can account for the elevated ethylene concentrations in the shoots of flooded plants, although both mechanisms may contribute to the overall plant response to the stress.

The importance of anaerobiosis as a factor in the ethylene physiology of flooded plants was established by Jackson and Campbell (1976). When the roots of tomato plants growing in aerated solution culture were made anaerobic with N_2 gas, increased ethylene production was measured in the shoots. Bradford and Dilley (1978) performed similar experiments in which the ethylene production rates of both roots and shoots of individual plants could be simultaneously measured. When the root zones of tomato plants growing in porous media were made anaerobic with a flowing stream of N_2, the ethylene production rates of the shoots increased with a time course similar to that for flooded plants. Ethylene emanation from the roots of the same plants, however, declined with increasing duration of the stress. Anaerobiosis of the roots was sufficient to cause elevated ethylene synthesis in the shoot.

Other evidence suggested that the factor promoting shoot ethylene synthesis is transported in the transpiration stream. For example, if the stem of a tomato plant is steam-girdled to collapse the intercellular spaces and phloem, but leaving the xylem functional, the movement of ethylene gas from the root zone into the shoot is prevented (Jackson and Campbell, 1975b). Ethylene emanation from the shoots of steam-girdled plants will, however, increase in response to flooding (Bradford, 1981). In other experiments, tomato plants were induced to develop two parallel root systems on cuttings with a vertically split stem base (Jackson and Campbell, 1975a). When only one of the root systems was treated anaerobically, ethylene production increased only in those leaves with a direct vascular connection to the stressed roots. These observations led to the suggestion that some precursor or inducer of ethylene synthesis is transported from the anaerobic roots via the transpiration stream to the shoot (Jackson and Campbell, 1976; Jackson *et al.*, 1978; Drew *et al.*, 1979).

The possible identity of this transported factor was suggested by the elucidation of the ethylene biosynthetic pathway. It has been known for many years that methionine is the biological precursor for ethylene biosynthesis (Lieberman, 1979). Adams and Yang (1977) concluded from tracer studies that *S*-adenosylmethionine (SAM) is an intermediate between methionine and ethylene. When apple tissues are incubated anaerobically and then returned to air, a large burst of ethylene production occurs, as if a precursor accumulates under anaerobiosis that is then converted to ethylene aerobically (Burg and Thimann, 1959). Adams and Yang (1979) employed anaerobic incubation of apple tissue to identify this precursor as 1-aminocyclopropane-1-carboxylic acid (ACC) and proposed a pathway for ethylene biosynthesis from methionine to SAM to ACC to ethylene (Fig. 1; for reviews see Yang *et al.*, 1980, 1982). ACC accumulates to high levels in anaerobic tissue but is rapidly converted to ethylene in air, indicating that it is the final step in the pathway, requiring O_2. Tomato roots also show an inhibition of ethylene production under anaerobiosis, followed by a burst of ethylene synthesis when returned to air (Jackson *et al.*, 1978). It is logical, therefore, that ACC might accumulate in flooded roots and be transported to the shoots in the transpiration stream. Virtually all vegetative tissues can synthesize ethylene from ACC (Cameron *et al.*, 1979), so ACC carried into the shoot in the xylem would be quickly converted into ethylene.

This hypothesis was tested by measuring the amount of ACC appearing in the xylem sap of tomato plants as a function of time under flooding (Bradford and Yang, 1980a,b). ACC export from the roots increased to a maximum after 48 hr of flooding, then declined (Fig. 2A). No ACC was detected in the xylem sap of

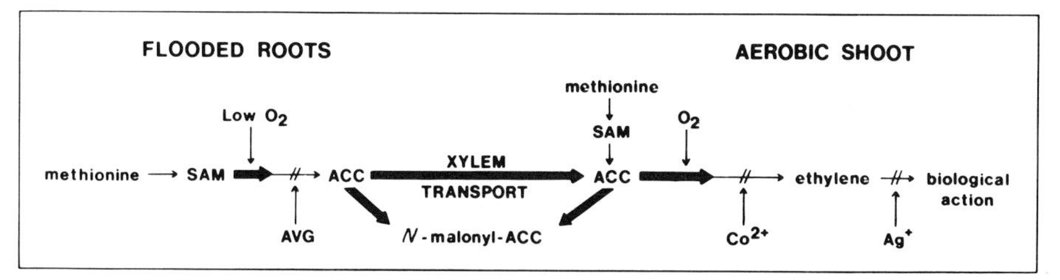

Fig. 1. Scheme for ethylene physiology of flooded plants. The heavy arrows indicate that the process is promoted, whereas the diagonal lines signify inhibition. ACC, 1-Aminocyclopropane-1-carboxylic acid; Ag^+, silver ion; AVG, aminoethoxyvinylglycine; Co^{2+}, cobaltous ion; SAM, *S*-adenosylmethionine.

control plants. Ethylene production by the petioles also increased following flooding but lagged behind the appearance of ACC by 12–24 hr (Fig. 2A). If plants were flooded for 30 hr and then drained, ACC flux in the xylem began to increase and then disappeared as O_2 was readmitted to the root zone (Fig. 2B). Petiolar ethylene production also declined to the control rate after draining (Fig. 2B). Supplying ACC to excised shoots at concentrations equal to those measured

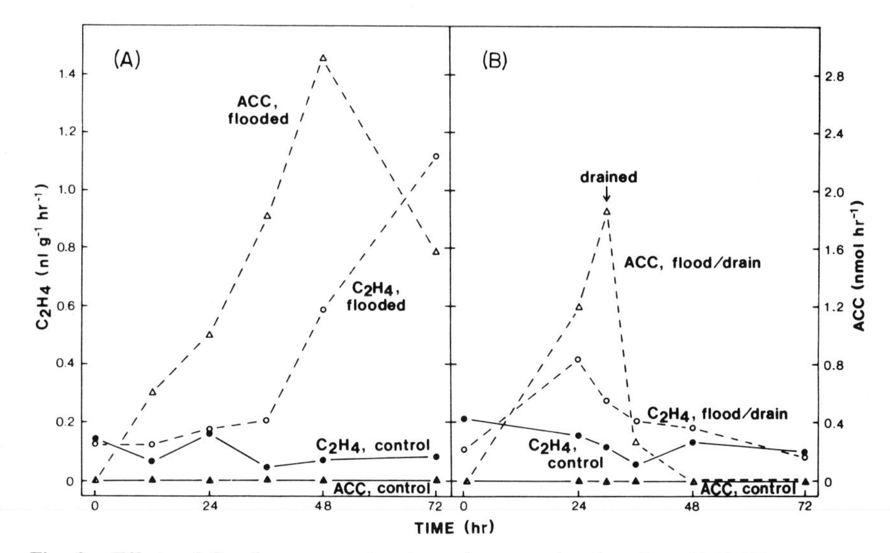

Fig. 2. Effects of flooding on root 1-aminocyclopropane-1-carboxylic acid (ACC) export and petiolar ethylene synthesis in tomato plants. Xylem sap was collected from plants detopped at the times indicated and was analyzed for ACC content. The ACC content was multiplied by the sap collection rate to give the ACC-transport rate (nmol hr^{-1}). The ethylene production rate was measured on excised petiolar segments within the first 0.5 hr following excision. (A) Continuous flooding for 72 hr. (B) Flooding for 30 hr followed by draining. Redrawn from Bradford and Yang (1980b).

in flooded plants stimulated shoot ethylene synthesis. These findings indicate that at least in tomato, root synthesis of ACC and its transport in the transpiration stream are involved in the acceleration of ethylene synthesis rates in the shoots of flooded plants. Furthermore, the synthesis of ACC in anaerobic roots could be inhibited by aminoethoxyvinylglycine (AVG), and conversion of ACC to ethylene in the shoots could be inhibited by CO^{2+} (Fig. 1; Bradford and Yang, 1981; Bradford *et al.*, 1982). Both inhibitors effectively prevented an increase in ethylene synthesis in the shoots of plants with anaerobically stressed roots.

In addition to preventing the conversion of ACC into ethylene, anaerobiosis actually stimulates the synthesis of ACC in the roots (Fig. 1; Bradford and Yang, 1980a; Bradford *et al.*, 1982). The mechanism of action of anaerobiosis is unknown, but it may be analogous to other types of chemical or mechanical injuries that induce the synthesis of 1-aminocyclopropanecarboxylate synthase, the ACC-forming enzyme (Boller and Kende, 1980; Yu and Yang, 1980). It has not yet been possible to isolate active enzyme from roots to test this hypothesis, although it has been demonstrated in mung bean hypocotyls that auxin can induce the formation of ACC synthase activity at O_2 levels too low to allow conversion of ACC to ethylene (Imaseki *et al.*, 1977; Yoshii *et al.*, 1980). On the other hand, Jackson (1982) found that hypoxia (5% O_2) stimulated ethylene production from apical segments of maize roots. This could result from sufficient stress in some parts of the tissue (resulting from respiratory consumption of O_2) to stimulate ACC synthesis, followed by movement of ACC to cells where adequate O_2 is available for the conversion of ACC to ethylene. Alternatively, the degree of hypoxia required to accelerate ACC synthesis may still permit ethylene formation. Such mechanisms may be operating in the hypocotyl and proximal root regions of flooded plants, where O_2 diffusion intercellularly will maintain hypoxic, rather than anaerobic, conditions. In floating rice (*Oryza sativa* cv. Habiganj Aman II), submersion of up to 80% of the plant promotes internode elongation by increasing both ethylene production and accumulation (Métraux and Kende, 1982). In this case as well, the relative O_2 sensitivities of the rates of ACC formation and of its conversion to ethylene may have important regulatory implications. More experimental information is needed on the O_2 requirements of both ACC and ethylene formation, particularly in tissues that respond developmentally to flooding.

It was discovered that ACC can also be metabolized to *N*-malonyl-ACC (Fig. 1; Amrhein *et al.*, 1981; Hoffman *et al.*, 1982). In flooded tomato plants, only the roots accumulated ACC, whereas both the roots and the shoots contained elevated levels of *N*-malonyl-ACC (Amrheim *et al.*, 1982). This suggests that much of the ACC exported from the roots to the shoot is diverted into this metabolic end product, perhaps explaining the relatively long lag between the appearance of ACC at the stem base and the rise in ethylene synthesis rate in the upper leaves (Fig. 2A). Other amino acids show an exponential decline in concentration as they move through the xylem because of uptake by the xylem parenchyma cells (van Bel *et al.*, 1979). Such uptake processes may result in xylem-to-phloem transfer of amino acids, and there is evidence that ACC can be transported in the phloem as well as in the xylem (Amrhein *et al.*, 1982). Many questions remain to be answered concerning the fate of root-synthesized ACC in flooded plants. One may predict that the xylem parenchyma will prove to be a useful tissue for studying ACC uptake, transport, and metabolism.

Other hormones can also modify the rate of ACC synthesis. The stimulation of ethylene production by auxin is well known (Section IIB). In addition, cytokinins promote ACC and ethylene synthesis, whereas ABA is inhibitory (Yoshii and Imaseki, 1981; McKeon *et al.*, 1982). Because the levels of these hormones are also affected by flooding (see following discussion), the rates of ACC synthesis and its conversion to *N*-malonyl-ACC and ethylene may depend in a complex manner on O_2 availability and hormonal interactions in the submerged tissues. Experimental data on this question are lacking.

The bulk of the literature on ethylene accumulation in flooded plants can be interpreted in a manner consistent with this ACC-transport hypothesis. Some apparent exceptions are cases where the concentrations or production rates of ethylene in anaerobic or submerged tissues have been reported to exceed those of the adjacent aerobic tissues (Kawase, 1978; Blake and Reid, 1982). However, in these studies the tissues were excised and handled in air before ethylene extraction or were incubated in air for estimates of ethylene production rates. Because the conversion of ACC to ethylene can occur very rapidly in the presence of O_2, the high levels of ethylene in previously flooded or anaerobic tissues may represent metabolism of accumulated ACC. Thus, the ethylene production rate in air following flooding should not be equated with the endogenous rate during flooding. The O_2 requirement for the synthesis of ethylene (but not ACC) should be recognized in future studies of the ethylene physiology of flooded plants.

B. Auxins

Despite the early suggestion by Kramer (1951) that anoxia and flooding might interfere with auxin transport and thus produce morphological effects such as adventitious rooting, few studies have been carried out that look at the relationship of auxins and flooding. Phillips (1964a,b) found that over a 14-day period of waterlogging, the concentration of auxins in sunflower shoots rose threefold. Flooding also seemed to delay a reduction in auxin content of roots. More recently, other workers have shown that flooding of roots acts more rapidly to raise the auxin levels of shoots (Hall *et al.*, 1977; Wample and Reid, 1979). In sunflower hypocotyls, auxin levels doubled within 24 hr of the onset of flooding (Wample and Reid, 1979).

The main flow of auxins is in a basipetal direction, although small quantities may move in the xylem and phloem (Hall and Medlow, 1974). Thus, flooding could affect auxin levels by interfering with auxin production in roots or with transport to and from the root. In the short term, elevated ethylene levels in shoots of flooded plants (Section II,A) could inhibit auxin transport (Burg and Burg, 1967; Beyer and Morgan, 1969) and slow the movement of auxin from shoots to roots. This is consistent with the finding of Hall *et al.* (1977) that the increase in indoleacetic acid (IAA) levels in leaves of *Vicia faba* occurred after the rise in ethylene evolution. The situation with regard to auxin–ethylene interactions is not simple, however, as auxin buildup may in turn promote ethylene production (Crocker *et al.*, 1935; Abeles and Rubinstein, 1964; Yu and Yang, 1979) or at least stimulate the ethylene-synthesizing enzyme system (Imaseki *et al.*, 1977). In the longer term, as in Phillips's (1964b) experiments, it is possible that dead and dying root tissue could produce auxins (Sheldrake, 1973), causing an overall increase in the whole plant.

Wample and Reid (1979) found that in addition to causing a rapid rise in auxin content of sunflower hypocotyls, flooding also inhibited the basipetal movement of [^{14}C]IAA and slowed its breakdown. Presumably, inhibition of both transport and metabolism caused the flood-induced auxin accumulation in the hypocotyls. Furthermore, aerated floodwater around the roots had the same effect as did nonaerated (anaerobic) flooded conditions in reducing the catabolism and transport of IAA (Wample and Reid, 1979). This raises two possibilities. The flood-induced changes in auxin transport and metabolism may be caused by only small reductions in O_2 supply to the tissues (aerated water being intermediate in O_2 concentration between nonflooded controls and totally anaerobic water). Alternatively, the presence of water itself may somehow interfere with root metabolism (the "water effect" of Wample and Reid, 1975), perhaps by leaching substances from roots (Drakeford and Reid, 1981, unpublished) or by acting as a water jacket and interfering with normal gas exchange of root tissue with the soil (Kawase, 1976).

The increases of abscisic acid (ABA) in waterlogged plants (Section IIC) might also slow the basipetal movement of auxins, as Little (1975) has found evidence that exogenous ABA could reduce the movement of [^{14}C]IAA in tree stems.

Much more work, using better techniques of auxin extraction, purification, and assay, is needed before we can get a clear picture of exactly how flooding of an anaerobic root system affects auxin metabolism and transport.

C. Abscisic Acid

The first report that flooding, like other stresses, promoted ABA accumulation was that of Wright and Hiron (1972), who found a fivefold increase of ABA-like inhibitors in leaves of *Phaseolus vulgaris* after 5 days of flooding. Four days of flooding raised ABA content by nine times in *Euphorbia lathyrus* (Sivakumaran and Hall, 1978), and 2 days produced a two- to threefold increase in *Vicia faba* leaves (Hall *et al.*, 1977). Shaybany and Martin (1977) found that ABA increased within 12–18 hr of flooding in seedlings of four *Juglans* species. The latter workers also positively identified ABA in their extracts using gas chromatography–mass spectrometry. Unfortunately, they did not measure the ABA content of unflooded controls over the entire time period of their experiment, making it difficult to be certain of the precise degree of effectiveness of the flooding treatment.

Exactly how flooding promotes ABA levels is not clear. Some species show symptoms of water stress (Kramer, 1940, 1951; Childers and White, 1942; Kozlowski and Pallardy, Chapter 5 in this volume) but this does not seem to be the general case (Section IIIA). Unfortunately, detailed measurements of plant water status have not been made in experiments where ABA contents were measured. It is possible that ABA may be responsible for a "feed forward" effect on stomata, preventing a drop in leaf water potential ψ (Bradford and Hsiao, 1982; Bradford, 1983a,b). However, measurements of the time course of accumulation of ABA in leaves of flooded plants in relation to alterations in stomatal behavior have not been made.

The source of the extra ABA in flooded plants is not known. It is possible that ABA might be supplied from the stressed roots via the xylem, as ABA has been found in xylem sap and root tissues (Davison and Young, 1973; Walton *et al.*, 1976). The elevated ABA levels could be the result of an effect of flooding on transport (direction and rate), rates of synthesis and destruction, or even binding and release. The experimental data are as yet too sparse to exclude any possibilities. It would be useful to investigate in more detail the effects of flooding separately on ABA production and turnover in roots and shoots and also the influence of flooding on the movement of ABA in the transpiration stream, with particular attention to the time course and plant water status. We are not aware of any detailed studies of this nature.

D. Gibberellins and Cytokinins

Within 1 day of flooding there is a reduction in the concentration of gibberellins (GA) in the stems, roots, and xylem of tomato plants (Reid *et al.*, 1969; Reid and Crozier, 1971) and of the cytokinins in sunflower xylem sap (Burrows and Carr, 1969). The reduced export from roots to shoots might be a partial cause of the lowered level of GA and cytokinins in shoots of flooded tomato plants. GA levels in shoots rise again at the time that new adventitious roots become visible. These data agree with the original postulation of Went (1938, 1943) that shoot and leaf growth is retarded during flooding because the anaerobic roots are unable to export "caulo- and phyllocalines" to the rest of the plant. Van Staden and Brown (1973) found that in seeds, elevated O_2 tensions increased cytokinin (CK) levels, suggesting that it is the reduced O_2 concentration in waterlogged roots that slows CK production.

Although roots are a major site of GA and CK synthesis and transformation (Phillips and Jones, 1964; Carr *et al.*, 1965; Kende and Sitton, 1967; Crozier and Reid, 1971; Short and Torrey, 1972), and flooding could directly block their production, at least GA can also be synthesized in vegetative shoot tissue (Jones and Phillips, 1966; Reid *et al.*, 1968; Kamienska and Reid, 1978). Thus, the possibility remains that signals from the flooded roots might indirectly interfere with GA and CK biosynthesis in shoots. As with auxins and ABA, little is known of the details of exactly how flooding interferes with GA or CK levels.

There has been one suggestion that the lowered GA levels found in flooded tomato plants might be the direct result of the reduced CK production. Reid and Railton (1974) found that benzyladenine (BA, a synthetic cytokinin) applied to shoots of waterlogged tomato plants greatly promoted endogenous GA levels.

III. INTERACTION OF ALTERED BALANCE OF HORMONES ON GROWTH AND DEVELOPMENT

A. Gas Exchange and Photosynthesis

1. Stomatal Behavior

The effects of flooding on water relations and photosynthesis are discussed by Kozlowski and Pallardy (Chapter 5 in this volume). The discussion here is therefore restricted to hormonal involvement in the alterations of gas-exchange characteristics during flooding. Partial stomatal closure is a common response to flooding, resulting in the maintenance of high leaf ψ's (Sojka and Stolzy, 1980; Bradford and Hsiao, 1982; Bradford, 1983a). The plant growth regulator commonly associated with stomatal closure is ABA, and ABA levels may increase in flooded plants even though turgor is not lost (see Section II,C). Bradford (1982, 1983b) found that spray applications of ABA to control tomato plants could reproduce the stomatal behavior of flooded plants. However, more detailed time courses of changes in leaf ABA content (or distribution) and stomatal behavior are needed to establish a causal relationship.

Ethylene concentrations also increase in the leaves of flooded plants, and the time course is very similar to that of stomatal closure (Bradford and Hsiao, 1982). Although ethylene has been reported to cause closure of peanut (*Arachis hypogaea*) stomata, the gas has no effect on stomatal conductance of tomato, maize, and a number of other species (Pallaghy and Raschke, 1972; Bradford, 1982, 1983b; Pallas and Kays, 1982). In general, it is doubtful that ethylene plays a role in altering stomatal conductance of flooded plants.

Other data suggest that processes in the roots affect stomatal behavior. Bradford and Hsiao (1982) found that stomatal adaptation to flooding required a period of transpiration following the imposition of the stress, suggesting that the root influence is transmitted via the transpiration stream. Root excision or stem girdling also results in stomatal closure (Xu and Lou, 1980; Bradford and Hsiao, 1982). Stomata of flooded *Fraxinus pennsylvanica* began to reopen coincident with the appearance of adventitious roots on the stem (Sena Gomes and Kozlowski, 1980). These observations suggest that aerobic roots in some way promote stomatal opening. Cytokinins and gibberellins are produced in the roots, are transported in the transpiration stream, and are synthesized in reduced amounts during flooding (see Section II,D). Spraying a combination of BA and GA_3 on flooded tomato plants increased their transpiration rates and frequently caused them to wilt (Jackson and Campbell, 1979). Bradford (1982, 1983b) confirmed that supplementary BA sprays would prevent the stomatal closure that normally occurs during flooding. A reduced supply of cytokinins and/or gibberellins from the flooded roots may therefore be involved in promoting stomatal closure, perhaps in conjunction with higher ABA levels. Much more detailed work on the synthesis, transport, and metabolism of these compounds in flooded plants is needed to confirm this hypothesis.

2. *Photosynthesis*

As would be expected following a reduction in stomatal conductance, photosynthetic rates also decline during flooding or root anaerobiosis (Stolzy *et al.*, 1964; Regehr *et al.*, 1975; Wiedenroth and Poskuta, 1981). However, at least part of the reduction in assimilation rates is the result of nonstomatal inhibition (Moldau, 1973; Bradford, 1982, 1983a). The photosynthetic capacity of tomato leaves was not affected by exposure to exogenous ethylene or ABA, making it unlikely that these compounds are responsible for the nonstomatal effects during flooding (Bradford, 1982, 1983b). However, supplementary cytokinins did at least partially maintain the assimilative capacity of flooded plants (Bradford, 1982, 1983b). Although these results are as yet preliminary and require further substantiation, they do indicate that cytokinins merit further examination in relation to stress effects on leaf assimilative capacity.

B. Epinasty

One of the earliest symptoms of flooding stress in many herbaceous plants is epinastic growth of petioles. Epinasty is caused by more rapid elongation of the cells on the upper side of the petiole relative to those on the lower side. The extreme epinasty induced by flooding is probably a variant of the normal movements that orient the leaf in relation to light and gravity (Crocker *et al.*, 1932; Kang, 1979). It has long been known that ethylene, even at extremely low concentrations (0.1 μl liter^{-1}), will promote epinasty in sensitive species (Doubt, 1917). In a survey of over 200 species and cultivars, 44% responded epinastically to applications of ethylene gas (Crocker *et al.*, 1932). The increased production of ethylene in flooded plants is discussed in Section II,A. Consequently, it is likely that ethylene is the primary regulator of flood-induced epinasty. This supposition has been supported by experiments with specific inhibitors of ethylene synthesis or action that can prevent epinasty in flooded plants (Fig. 1; Beyer, 1976; Bradford and Dilley, 1978; Bradford *et al.*, 1982). It was previously thought that redistribution of IAA, perhaps under the influence of ethylene, was responsible for the differential growth resulting in epinasty (Lyon, 1970; Kang and Burg, 1974). However, there is now evidence that ethylene can directly promote the elongation of adaxial cells of the petiole without affecting IAA distribution (Palmer, 1976). Many other examples of promotion of cell expansion by ethylene have also been identified (Jackson, 1982; Osborne, 1982). Application of cytokinins and gibberellins can prevent the epinastic response during flooding, possibly by reducing the sensitivity of the responding cells to the gas (Selman and Sandanam, 1972; Reid and Railton, 1974; Jackson and Campbell, 1979). Ethylene appears to be the hormonal signal inducing epinasty, although other factors can interact to influence the final extent of the growth response.

C. Chlorosis and Senescence

In many species grown in waterlogged soil conditions, premature chlorosis and senescence set in within a few days (Went, 1938, 1943; Kramer, 1951; Burrows and Carr, 1969). Because exogenous cytokinins promote retention of chlorophyll and delay leaf senescence (Osborne, 1967), and flooding reduces the flow of cytokinins to shoots (Section II,D), it is a reasonable hypothesis that flood-induced chlorosis may in part be due to CK starvation. Certainly, application of BA is effective in overcoming the deleterious effects of flooding in tomato (Selman and Sandanam, 1972; Railton and Reid, 1973).

However, a wide range of factors such as a reduced supply of root-synthesized nitrogenous and other compounds must also play a role. GA can inhibit senscence in *Rumex* leaf discs (Whyte and Luckwill, 1966). Thus the lowered concentration of GA found in flooded plants (Section II,D) may exacerbate the situation. Elevated levels of ethylene and ABA (Sections II,A and II,C) could further speed up the onset of senescence, because both hormones can promote chlorosis and senescence (see reviews by Hall, 1977; Thimann, 1980; Thomas and Stoddart, 1980). It is possible that among other effects such as influencing nutrient-sink strength (Thomas and Stoddart, 1980), lowered GA and CK levels allow for more rapid lipid peroxidation and thus promote membrane deterioration

(Dhindsa *et al.*, 1982). This in turn would influence almost all aspects of leaf metabolism such as protein synthesis, chlorophyll synthesis, etc. Our own work (R. S. Dhindsa, P. L. Plumb-Dhindsa, and D. M. Reid, unpublished) indicates that in flooded sunflower, higher levels of ethylene are not the primary factor in causing premature leaf senescence.

There are several reports of a rather unusual type of senescence in which waterlogging causes leaves to very rapidly and suddenly develop a "scalded" or brittle appearance (Richards, 1929; Van't Woud and Hagen, 1957; Jackson, 1979; Drakeford and Reid, 1981). This catastrophic type of senescence can occur within 24 hr after the onset of flooding (Drakeford and Reid, 1981), and leaf death does not show the normal slow progression of senescence symptoms. For example, leaves can remain green yet undergo extensive cellular collapse. A sudden uncontrolled loss of water would seem to be indicated, but whether alterations of hormone concentrations are responsible has not been reported.

D. Responses of Roots and Lower Portions of Stem–Hypocotyl

Roots and lower portions of the stem and hypocotyl show many responses to waterlogging, such as the formation of adventitious roots on the proximal parts of the root and basal parts of the stem, increased porosity of roots, and negative geotropism of growth of some roots. There is much evidence indicating that all of these, and other, responses aid in the movement of O_2 from aerated shoots to anoxic roots (Williams and Barber, 1961; Armstrong, 1978; Hook and Scholtens, 1978; de Witt, 1978; Hook, Chapter 8 in this volume).

1. Adventitious Roots

One of the most commonly observed morphological phenomena associated with waterlogging is the production of adventitious roots. It has been long argued that these roots alleviate some of the adverse symptoms of flooding (Bergmann, 1920; Jackson, 1955; Kramer, 1969). In essence, they partially replace the original anaerobic and sometimes dying root system.

There is a large literature on adventitious rooting in cuttings and the formation of new laterals on more or less intact root systems (Hess, 1969; Haissig, 1974a,b,c; Hartman and Kester, 1975). Much of the literature is difficult to interpret as some of the earlier reports did not adequately differentiate between *de novo* initiation and outgrowth of preexisting root primordia. Nevertheless, some general principles with regard to the controls of root initiation are clear. Flood-induced rooting is a fairly rapid process, the controls of which seem to be similar to those that influence rooting of cuttings. For example, new primordia can be seen on the upper portion of the roots and basipetal zone of hypocotyls or stems in less than 24 hr after the onset of flooding or similar root-initiating treatment (Wample and Reid, 1978; Fabijan *et al.*, 1981a). Rooting in most systems seems to be controlled by a complex of nutritive and hormonal factors (Hess, 1969; Wample and Reid, 1978; Fabijan *et al.*, 1981a,b). The original root system can produce hormonal factors inhibitory to rooting, whereas leaves and cotyledons supply factors that promote rooting.

It is possible that certain concentrations of cytokinins and GA might normally slow the production of new roots. For instance, exogenous GA (Brian *et al.*, 1960) and cytokinins (Heide, 1965) inhibit root formation. [This of course does not exclude the possibility that smaller quantities of these hormones might be required in rooting (Fabijan *et al.*, 1981b).] The flood-induced reduction in GA and CK supply from roots (Section II,D) may thus aid in the promotion of adventitious rooting. The removal of the original root to produce a cutting (which is the signal for root initiation) may be analogous to slowing root metabolism by waterlogging. Consistent with the idea that root-produced hormones normally block root initiation, are the observations that (1) BA application totally suppresses root formation in flooded tomato and sunflower (Selman and Sandanam, 1972; Railton and Reid, 1973); (2) GA levels rise again subsequent to the appearance of new roots in willow cuttings (Michniewicz *et al.*, 1970) and flooded tomato plants (Reid and Crozier, 1971); and (3) the rate of initiation of new roots in sunflower cuttings slows once an appreciable number of new roots are in a phase of rapid elongation (Fabijan *et al.*, 1981a).

Perhaps soon after an episode of flooding starts, and maybe even before O_2 concentration becomes severely limited (Wample and Reid, 1975), the original roots reduce GA and CK export to the shoots. This reduction in hormone level, along with other signals described subsequently, could trigger adventitious root initiation before all the older roots die of anoxia. As soon as an appreciable number of new roots form, GA and CK production resumes. This then acts as a negative feedback mechanism suppressing further root formation.

However, even ignoring nutritional factors, the situation obviously could not be so simple. Elevated levels of ethylene and auxin (Sections IIA and II,B) may also play an important role. Ethylene seems to act as a controlling factor in root formation in sunflower (Wample and Reid, 1979) and *Zea mays* (Drew *et al.*, 1979). Wample and Reid (1979) proposed that the increase in ethylene in sunflower hypocotyls might be the cause of the auxin buildup in that organ (Section II,B) and that it is auxin rather than ethylene that triggers rooting.

Because both water stress and ABA can, under certain circumstances, promote rooting formation (Chin *et al.*, 1969; Rajagopal and Anderson, 1980a,b), flood-induced increases in ABA might reinforce the aforementioned reductions in GA and CK and increases in auxins and ethylene.

2. *Geotropic and Other Responses of Roots*

In addition to producing a new root system, many species undergo other morphological modifications to roots such as negative or plageotropic growth and even occasionally a corkscrew growth pattern (Wample, 1976; Tjepkema, 1978). Also observed in roots and lower portions of stems are the development of lacunae, increases of porosity of tissues (McPherson, 1939; Luxmoore and Stolzy, 1969; de Witt, 1974; Das and Jat, 1977), and the development of larger cells (Pitman, 1969). Flood-induced alterations in hormones might be the cause of these changes. For instance, (1) the ABA increase in flooded plants (Section II,C) may be involved in the geotropic growth response of roots (Pilet, 1975), (2) Grable (1966) has advanced quite reasonable arguments proposing a role for IAA in lacunae formation, and (3) there is much work showing that alterations in the concentrations of GA, ABA, ethylene, and auxins influence cell size.

We do not think that it is profitable to follow up these ideas in any detail here, no matter how reasonable they might be, because of the lack of good data on the kinetics of alterations of hormone levels, metabolism, and transport in roots of flooded plants, with the possible exception of ethylene. Elevated ethylene levels in the uppermost roots and hypocotyl might influence cell enlargement and aerenchyma development by the mechanisms described by Kawase (1979; Kawase and Whitmoyer, 1980; see also Section III,E,2). Although there is some debate about the extent by which roots show increases in ethylene during flooding (Section II,A), some ethylene may diffuse down from the higher concentrations found in hypocotyls. In agreement with this idea, it is known that exogenous ethylene will promote aerenchyma formation in adventitious roots of maize (Drew *et al.*, 1979) and that the inhibitor of ethylene action, silver nitrate ($AgNO_3$), suppresses aerenchyma formation (Drew *et al.*, 1981).

3. *Root Elongation*

Root growth is usually slowed by anaerobiosis (Letey *et al.*, 1961; Stolzy *et al.*, 1981). Because only 1–2 hr of reduced O_2 are needed to severely reduce growth rate (Turner *et al.*, 1983), it may be that the main influence of anaerobiosis is simply to block aerobic respiration and thus starve the root of energy for growth (Pradet and Bomsel, 1978). Ethylene entering the roots from the soil could either promote or inhibit root growth, depending on the concentration (Smith and Robertson, 1971; Konings and Jackson, 1979). However, the effects of ethylene on root growth have been determined in the presence of atmospheric O_2 partial pressures, rather than in the low O_2 conditions that prevail when soil ethylene concentrations are high. Because ethylene becomes less effective as the O_2 partial pressure is decreased (Burg and Burg, 1967), the importance of ethylene in the regulation of physiological processes in submerged tissues is unclear.

E. Stem Growth and Morphology

1. Stem Elongation

Generally, waterlogging of roots causes a reduction of stem elongation. There are interesting exceptions to be found in some species of aquatic and marsh plants such as *Callitriche platycarpa* and lowland rice. In these cases, submergence in water promotes stem elongation, and ethylene may be one of the stimulatory factors (Ku *et al.*, 1970; Musgrave *et al.*, 1972; Métraux and Kende, 1982). Additional evidence indicating that this gas is involved in stem growth in these aquatic or semiaquatic species comes from experiments showing that $AgNO_3$ (an inhibitor of ethylene action) will slow the growth induced by submergence (Cookson and Osborne, 1978; Jackson, 1982), whereas ethylene will promote growth (Ku *et al.*, 1970; Jackson, 1982; Musgrave *et al.*, 1972; Métraux and Kende, 1982). The rachis of the semiaquatic fern *Regnellidium diphyllum* also elongates more rapidly in the submerged than in the nonsubmerged condition. This elongation is apparently stimulated, first, by an accumulation of ethylene, which softens the cell wall, and second, by the buoyant fronds stretching the elongating rachis (Musgrave and Walters, 1974).

The more commonly observed slowing of stem elongation in flooded plants is probably the result of many factors, including reduced aerobic respiration in roots, inhibition of transport of a wide range of metabolites to and from roots, and alterations in hormonal metabolism and transport. In flooded plants, those hormones normally thought to promote stem elongation (GA) decrease (Section II,D), and those that tend to inhibit this process (ABA and ethylene), increase (Sections II,A and II,C). The stems of waterlogged tomato plants, although deficient in GA, are more sensitive to exogenous GA's, supporting the idea that at least part of the dwarfing effect of flooding results from insufficient stem GA (Reid and Crozier, 1971). Whether or not the flood-induced elevations of auxin content (Section II,B) are at a concentration that is inhibiting to cell elongation has not been tested.

In tomato, Railton and Reid (1973) observed that flood-induced dwarfing could be partially overcome by application of BA. This was not found to be the case in the work of Selman and Sandanam (1972), who used the same species but a different cultivar. Reid and Railton (1974) further observed that this CK treatment more than overcame the inhibitory effect of flooding on levels of endogenous GA. They argued that the lowered CK production in flooded plants, as seen by Burrows and Carr (1969), might be partially responsible for reduced concentrations of endogenous GA. On the other hand, there might not have been any direct effect of BA on GA production. Rather, the high doses of BA used by Reid and Railton (1974) could have simply exerted some overall pharmacological effect on growth, such as diverting nutrient and hormonal flow toward the growing stem cells.

The data are presently insufficient to draw any firmer conclusions on the role of hormones in flood-induced dwarfing. Presumably, like adventitious rooting, stem elongation is under the control of a complex of interacting nutritional and hormonal factors. Detailed studies of hormone synthesis and metabolism at the localized zones of cell elongation are needed, rather than estimations of hormone levels in entire shoot systems.

2. *Aerenchyma and Hypocotyl–Stem Hypertrophy*

Aerenchyma is seen in stems as well as roots of waterlogged plants (Kawase, 1981), and it seems clear from the work of Kawase (1979) and Kawase and Whitmoyer (1980) that aerenchyma development is surprisingly rapid (within 2 days of the onset of flooding) and that an increase of ethylene-induced cellulase activity is one important factor. Swelling of the lower portions of the stem (hypertrophy) may be related to aerenchyma formation, in that the rapid lateral stem expansion can cause separation of cell walls and produce lacunae. Wample and Reid (1978) showed that hypocotyl hypertrophy in sunflower began within 24 hr after the onset of flooding and appeared to be promoted by factors moving from the illuminated leaves. Like adventitious rooting, hypertrophy occurred in plants flooded with either aerated water or deoxygenated water and thus was initiated by the presence of excess water rather than by anoxia (Wample and Reid, 1975, 1978).

Many studies have shown a positive correlation between hypertrophy and elevated concentrations of ethylene (see review by Abeles, 1973), and it seems likely that auxins interact with the gas in the regulation of cell size (Osborne, 1974). Wample and Reid (1979) suggested that because auxin and ethylene levels increased in sunflower hypocotyls, and Ethrel or flooding promoted hypocotyl hypertrophy, both hormones were involved. They also found that exogenous cytokinins partially blocked the buildup of endogenous auxin, but neither lowered ethylene levels nor influenced the extent of hypocotyl swelling. They thus concluded that auxin was of lesser importance than ethylene.

Tissue hypertrophy, lenticels, aerenchyma and stem buttresses have been observed during flooding of many species (Jackson and Drew, Chapter 3, and Kozlowski, Chapter 4 in this volume; Pereira and Kozlowski, 1977). The adaptive significance of these anatomical changes in terms of flooding survival is not totally clear, although Stelzer and Läuchli (1980), for example, found that in *Puccinellia* some O_2 might diffuse from aerated roots to the lower roots in anaerobic conditions. Blake and Reid (1982) suggested that such flood-induced morphological modifications might also be useful in the elimination of excess ethylene.

IV. CONCLUSIONS

In this chapter evidence for hormonal involvement in the responses of plants to waterlogging has been reviewed. In general, the shoot contents of auxin, ethylene, and abscisic acid (ABA) initially increase, whereas those of gibberellins (GA) and cytokinins decline. Although we and many others have put forward a number of proposals regarding the role of these changes in hormone levels on growth and development, it must be accepted that many of these ideas are still rather speculative. The most convincing case has been built around the effects of ethylene, and it is clear that this gas has a key role in many of the responses of plants to flooding. We do not have nearly enough good data on the flood-induced alterations in the biosynthesis, metabolism and transport of GA, auxins, ABA, or cytokinins. For example, the total number of studies in which reliable estimates have been made of endogenous GA, auxins, cytokinins, and ABA in flooded plants can be counted on two hands. There have been even fewer reports on the metabolism (as opposed to content or "level") of these hormones immediately after the onset of flooding and at a range of times thereafter. With the exception of ethylene, hormone biosynthesis and metabolism in the roots of waterlogged plants have been totally ignored. This absence of experimental interest is rather surprising, because what little information we do have shows substantial fluctuations in the levels of GA, ABA, auxins, and cytokinins, and many of the morphological changes occurring as a result of flooding are exactly what one would expect given those alterations in hormone concentrations. Filling the gaps in our knowledge will require information on alteration of hormone levels or metabolism at the actual site of the particular developmental change under investigation. The possibility of flood-induced changes in sensitivity of tissues to hormones (including ethylene) during flooding must also be considered. Finally, if we are to understand exactly how hormones influence the responses of plants to flooding, it is essential to examine the interactions of the various classes of hormones with each other and with metabolism in general.

REFERENCES

Abeles, F. B. (1973). "Ethylene in Plant Biology." Academic Press, New York and London.

Abeles, F. B., and Rubinstein, S. P. (1964). Regulation of ethylene evolution and leaf abscission by auxin. *Plant Physiol.* **39,** 963–969.

Adams, D. O., and Yang, S. F. (1977). Methionine metabolism in apple tissue. Implication of *S*-adenosylmethionine as an intermediate in the conversion of methionine to ethylene. *Plant Physiol.* **60,** 892–896.

Adams, D. O., and Yang, S. F. (1979). Ethylene biosynthesis: identification of 1-aminocyclopropane-1-carboxylic acid as an intermediate in the conversion of methionine to ethylene. *Proc. Natl. Acad. Sci. USA* **76,** 170–174.

Amrhein, N., Schneebeck, D., Skorupka, H., Tophof, S., and Stockigt, J. (1981). Identification of a major metabolite of the ethylene precursor 1-amino-cyclopropane-carboxylic acid in higher plants. *Naturwissenschaften* **68,** 619–620.

Amrhein, N., Breuing, F., Eberle, J., Skorupka, H., and Tophof, S. (1982). The metabolism of 1-amino-cyclopropane-1-carboxylic acid. *In* "Plant Growth Substances 1982" (P. F. Wareing, ed.), pp. 249–258. Academic Press, New York and London.

Armstrong, W. (1978). Root aeration in the wet-land condition. *In* "Plant Life in Anaerobic Environments" (D. D. Hook and R. M. M. Crawford, eds.), pp. 269–298. Ann Arbor Sci. Publ., Ann Arbor, Michigan.

Bergmann, H. F. (1920). The relation of aeration to the growth and activity of roots and its influence on the ecesis of plants in swamps. *Ann. Bot. (London)* **34,** 13–33.

Beyer, E. M. (1976). A potent inhibitor of ethylene action in plants. *Plant Physiol.* **58,** 268–271.

Beyer, E. M., and Morgan, P. W. (1969). Ethylene modification of an auxin pulse in cotton stem segments. *Plant Physiol.* **44,** 1690–1694.

Blake, T. J., and Reid, D. M. (1982). Ethylene, water relations and tolerance to waterlogging of three *Eucalyptus* species. *Aust. J. Plant Physiol.* **8,** 497–505.

Boller, T., and Kende, H. (1980). Regulation of wound ethylene synthesis in plants. *Nature (London)* **286,** 259–260.

Bradford, K. J. (1981). "Ethylene Physiology and Water Relations of Waterlogged Tomato Plants." Ph.D. Thesis, Univ. of California, Davis.

Bradford, K. J. (1982). Regulation of shoot responses to root stress by ethylene, abscisic acid, and cytokinin. *In* "Plant Growth Substances 1982" (P. F. Wareing, ed.), pp. 599–608. Academic Press, New York and London.

Bradford, K. J. (1983a). Effects of soil flooding on leaf gas exchange of tomato plants. *Plant Physiol.* **73,** 475–479.

Bradford, K. J. (1983b). Involvement of plant growth substances in the alteration of leaf gas exchange of flooded tomato plants. *Plant Physiol.*, **73,** 480–483.

Bradford, K. J., and Dilley, D. R. (1978). Effects of root anaerobiosis on ethylene production, epinasty and growth of tomato plants. *Plant Physiol.* **61,** 506–509.

Bradford, K. J., and Hsiao, T. C. (1982). Stomatal behavior and water relations of waterlogged tomato plants. *Plant Physiol.* **70,** 1508–1513.

Bradford, K. J., and Yang, S. F. (1980a). Stress-induced ethylene production in the ethylene-requiring tomato mutant diageotropica. *Plant Physiol.* **65,** 327–330.

Bradford, K. J., and Yang, S. F. (1980b). Xylem transport of 1-aminocyclopropane-1-carboxylic acid, an ethylene precursor, in waterlogged tomato plants. *Plant Physiol.* 65, 322–326.

Bradford, K. J., and Yang, S. F. (1981). Physiological responses of plants to waterlogging. *HortScience* **16,** 25–30.

Bradford, K. J., Hsiao, T. C., and Yang, S. F. (1982). Inhibition of ethylene synthesis in tomato plants subjected to anaerobic root stress. *Plant Physiol.* **70,** 1503–1507.

Brian, P. W., Hemming, H. G., and Lowe, D. (1960). Inhibition of rooting of cuttings by gibberellic acid. *Ann. Bot. (London)* **24,** 407–419.

Burg, S. P., and Burg, E. A. (1967). Molecular requirements for the biological activity of ethylene. *Plant Physiol.* **42,** 144–152.

Burg, S. P., and Thimann, K. V. (1959). The physiology of ethylene formation in apples. *Proc. Natl. Acad. Sci. USA* **45,** 335–344.

Burrows, W. J., and Carr, D. J. (1969). Effects of flooding the root system of sunflower plants on the cytokinin content of the xylem sap. *Physiol. Plant.* **22,** 1105–1112.

Cameron, A. C., Fenton, C. A. L., Yu, Y. B., Adams, D. O., and Yang, S. F. (1979). Increased production of ethylene by plant tissues treated with 1-amino-cyclopropane-1-carboxylic acid. *HortScience* **14,** 178–180.

Carr, D. J., Reid, D. M., and Skene, K. G. M. (1965). The supply of gibberellins from the root to the shoot. *Planta* **63,** 382–392.

Childers, N. F., and White, D. G. (1942). Influence of submersion of the roots on transpiration, apparent photosynthesis and respiration of young apple trees. *Plant Physiol.* **17,** 603–618.

Chin, T. Y., Meyer, M. M., Jr., and Beevers, L. (1969). Abscisic acid stimulated rooting of stem cuttings. *Planta* **88,** 192–196.

Cookson, C., and Osborne, D. J. (1978). The stimulation of cell extension by ethylene and auxin in aquatic plants. *Planta* **144,** 39–47.

Crawford, R. M. M. (1982). Physiological responses to flooding. *Encycl. Plant Physiol. New Ser.* **12B,** 453–477.

Crocker, W., Zimmerman, P. W., and Hitchcock, A. E. (1932). Ethylene-induced epinasty of leaves and the relation of gravity to it. *Contrib. Boyce Thompson Inst.* **4,** 177–218.

Crocker, W., Hitchcock, A. E., and Zimmerman, P. W. (1935). Similarities in the effects of ethylene and the plant auxins. *Contrib. Boyce Thompson Inst.* **7,** 235–248.

Crozier, A., and Reid, D. M. (1971). Do roots synthesize gibberellins? *Can. J. Bot.* **49,** 967–975.

Das, D. K., and Jat, R. L. (1977). Influence of three soil-water regimes on root porosity and growth of four rice varieties. *Agron. J.* **69,** 197–200.

Davison, R. M., and Young, H. (1973). Abscisic acid content of xylem sap. *Planta* **109,** 95–98.

de Witt, M. C. J. (1974). ''Reakties van Gerstwortels op een Tekort aan Zuurstof.'' Ph.D. Thesis, Univ. of Groningen.

de Witt, M. C. J. (1978). Morphology and function of roots and shoot growth of crop plants under oxygen deficiency. *In* ''Plant Life in Anaerobic Environments'' (D. D. Hook and R. M. M. Crawford, eds.), pp. 333–350. Ann Arbor Sci. Publ., Ann Arbor, Michigan.

Dhindsa, R. S., Plumb-Dhindsa, P. L., and Reid, D. M. (1982). Leaf senescence and lipid peroxidation. Effects of some phytohormones, and scavengers of free radicals and singlet oxygen. *Physiol. Plant.* **56,** 453–457.

Doubt, S. L. (1917). The response of plants to illuminating gas. *Bot. Gaz. (Chicago)* **63,** 209–224.

Drakeford, D. R., and Reid, D. M. (1981). The effects of prehistory on the responses of *Helianthus annuus* to flood stress. *Plant Physiol.* **67**(Suppl.), 121.

Drew, M. C. (1979). Plant responses to anaerobic conditions in soil and solution culture. *Curr. Adv. Plant. Sci.* **11,** 1–14.

Drew, M. C., and Lynch, J. M. (1980). Soil anaerobiosis, microorganisms, and root function. *Annu. Rev. Phytopathol.* **18,** 37–66.

Drew, M. C., Jackson, M. B., and Gifford, S. C. (1979). Ethylene-promoted adventitious rooting and development of cortical air spaces (aerenchyma) in roots may be adaptive responses to flooding in *Zea mays* L. *Planta* **147,** 83–88.

Drew, M. C., Jackson, M. B., Giffard, S. C., and Campbell, R. (1981). Inhibition by silver ions of gas space (aerenchyma) formation in adventitious roots of *Zea mays* L. subjected to exogenous ethylene or to oxygen deficiency. *Planta* **153,** 217–224.

El-Beltagy, A. S., and Hall, M. A. (1974). Effect of water stress upon endogenous ethylene levels in *Vicia faba. New Phytol.* **73,** 47–60.

Fabijan, D., Yeung, E., Mukherjee, I., and Reid, D. M. (1981a). Adventitious rooting in hypocotyls of sunflower (*Helianthus annuus*) seedlings. I. Correlative influence and development sequence. *Physiol. Plant.* **53,** 578–588.

Fabijan, D., Taylor, J. S., and Reid, D. M. (1981b). Adventitious rooting in hypocotyls of sunflower (*Helianthus annuus*) seedlings. II. Action of gibberellins, cytokinins, auxins and ethylene. *Physiol. Plant.* **53,** 589–597.

Grable, A. R. (1966). Soil aeration and plant growth. *Adv. Agron.* **18,** 57–105.

Haissig, B. E. (1974a). Origins of adventitious roots. *N.Z. J. For. Sci.* **4,** 299–310.

Haissig, B. E. (1974b). Influences of auxins and auxin synergists on adventitious root primordium initiation and development. *N.Z. J. For. Sci.* **4,** 311–323.

Haissig, B. E. (1974c). Metabolism during adventitious root primordium initiation and development. *N.Z. J. For. Sci.* **4,** 324–337.

Hall, M. A. (1977). Ethylene involvement in senescence processes. *Ann. Appl. Biol.* **85,** 424–428.

Hall, M. A., Kapuya, J. A., Sivakumaran, S., and John, A. (1977). The role of ethylene in the responses of plants to stress. *Pestic. Sci.* **8,** 217–223.

Hall, S. M., and Medlow, G. C. (1974). Identification of IAA in phloem and root pressure saps of Ricinus communis L. by mass spectrometry. *Planta* **119,** 257–261.

Hartman, H. T., and Kester, D. E. (1975). "Plant Propagation." Prentice-Hall, Englewood Cliffs, New Jersey.

Heide, O. M. (1965). Interaction of temperature auxins and kinins in the regeneration ability of *Begonia* leaf cuttings. *Physiol. Plant.* **18,** 891–920.

Hess, C. E. (1969). Internal and external factors regulating root initiation. *In* "Root Growth" (W. J. Whittington, ed.), pp. 42–53. Plenum, New York.

Hoffman, N. E., Yang, S. F., and McKeon, T. (1982). Identification of 1-(malonylamino)-cyclopropane-1-carboxylic acid as a major conjugate of 1-aminocyclopropane-1-carboxylic acid, and ethylene precursor in higher plants. *Biochem. Biophys. Res. Commun.* **104,** 765–770.

Hook, D. D., and Scholtens, J. R. (1978). Adaptations and flood tolerance of tree species. *In* "Plant Life in Anaerobic Environments" (D. D. Hook and R. M. M. Crawford, eds.), pp. 299–332. Ann Arbor Sci. Publ., Ann Arbor, Michigan.

Imaseki, H. A., Watanabe, A., and Odawara, S. (1977). The role of oxygen in auxin-induced ethylene production. *Plant Cell Physiol.* **18,** 577–586.

Ioannou, N., Schneider, R. W., and Grogan, R. G. (1977). Effect of flooding on the soil gas composition and the production of microsclerotica in *Venticillium dahlia* in the field. *Phytopathology* **67,** 651–656.

Jackson, M. B. (1979). Rapid injury to peas by soil waterlogging. *J. Sci. Food Agric.* **30,** 143–152.

Jackson, M. B. (1982). Ethylene as a growth promoting hormone under flooded conditions. *In* "Plant Growth Substances 1982" (P. F. Wareing, ed.), pp. 291–301. Academic Press, New York and London.

Jackson, M. B., and Campbell, D. J. (1975a). Ethylene and waterlogging effects in tomato. *Ann. Appl. Biol.* **81,** 102–105.

Jackson, M. B., and Campbell, D. J. (1975b). Movement of ethylene from roots to shoots a factor in the responses of tomato plants to waterlogged soil conditions. *New Phytol.* **74,** 397–406.

Jackson, M. B., and Campbell, D. J. (1976). Waterlogging and petiole epinasty in tomato: the role of ethylene and low oxygen. *New Phytol.* **76,** 21–29.

Jackson, M. B., and Campbell, D. J. (1979). Effects of benzyladenine and gibberellic acid on the responses of tomato plants to anaerobic root environments and to ethylene. *New Phytol.* **82,** 331–340.

Jackson, M. B., Gales, K., and Campbell, D. J. (1978). Effect of waterlogged soil conditions on the production of ethylene and on water relationships in tomato plants. *J. Exp. Bot.* **29,** 183–193.

Jackson, W. T. (1955). Role of adventitious roots in recovery of shoots following flooding of the original root systems. *Am. J. Bot.* **42,** 816–819.

Jones, R. L., and Phillips, I. D. J. (1966). Organs of gibberellin synthesis in light-grown sunflower plants. *Plant Physiol.* **41,** 1381–1386.

Kamienska, A., and Reid, D. M. (1978). The effects of stem girdling on levels of GA-like substances in sunflower plants. *Bot. Gaz. (Chicago)* **139,** 18–26.

Kang, B. G. (1979). Epinasty. *Encycl. Plant Physiol. New Ser.* **7,** 647–667.

Kang, B. G., and Burg, S. P. (1974). Ethylene action on lateral auxin transport in tropic responses, leaf epinasty, and horizontal nutation. *In* "Plant Growth Substances 1973," pp. 1090–1094. Hurokawa, Japan.

Kawase, M. (1972). Effect of flooding on ethylene concentration in horticultural plants. *J. Am. Soc. Hortic. Sci.* **97,** 584–588.

Kawase, M. (1976). Ethylene accumulation in flooded plants. *Physiol. Plant.* **36,** 236–241.

Kawase, M. (1978). Anaerobic elevation of ethylene concentration in waterlogged plants. *Am. J. Bot.* **65,** 736–740.

Kawase, M. (1979). Role of cellulose in aerenchyma development in sunflower. *Am. J. Bot.* **66,** 183–190.

Kawase, M. (1981). Anatomical and morphological adaptation of plants to waterlogging. *Hort-Science* **16,** 8–12.

Kawase, M., and Whitmoyer, R. E. (1980). Aerenchyma development in waterlogged plants. *Am. J. Bot.* **67,** 18–22.

Kende, M., and Sitton, D. (1967). The physiological significance of kinetin and gibberellin-like root hormones. *Ann. N.Y. Acad. Sci.* **144,** 235–243.

Konings, H., and Jakcson, M. B. (1979). A relationship between rates of ethylene production by roots and the promoting or inhibiting effects of exogenous ethylene and water on root elongation. *Z. Pflanzenphysiol.* **92,** 385–397.

Kramer, P. J. (1940). Causes of decreased absorption of water by plants in poorly aerated media. *Am. J. Bot.* **27,** 216–220.

Kramer, P. J. (1951). Causes of injury to plants resulting from flooding of the soil. *Plant Physiol.* **26,** 722–736.

Kramer, P. J. (1969). "Plant and Soil Water Relationships: A Modern Synthesis." McGraw-Hill, New York.

Ku, H. S., Suge, H., Rappaport, L., and Pratt, H. K. (1970). Stimulation of rice coleoptile growth by ethylene, *Planta* **90,** 333–339.

Letey, J., Lunt, O. R., Stolzy, L. M., and Szuszkiewicz, T. E. (1961). Plant growth, water use and nutritional response to rhizosphere differentials of oxygen concentration. *Soil Sci. Soc. Am. Proc.* **25,** 183–186.

Lieberman, M. (1979). Biosynthesis and action of ethylene. *Annu. Rev. Plant Physiol.* **30,** 533–591.

Little, C. H. A. (1975). Inhibition of cambial activity in *Abies balsamea* by internal water stress: role of abscisic acid. *Can. J. Bot.* **53,** 3041–3050.

Luxmoore, R. J., and Stolzy, L. M. (1969). Root porosity and growth responses of rice and maize to oxygen supply, *Agron. J.* **61,** 202–204.

Lynch, J. M., and Harper, S. H. T. (1974). Formation of ethylene by a soil fungus. *J. Gen. Microbiol.* **80,** 187–195.

Lynch, J. M., and Harper, S. H. T. (1980). Role of substrates and anoxia in the accumulation of soil ethylene. *Soil Biol. Biochem.* **12,** 363–367.

Lyon, C. L. (1970). Ethylene inhibition of auxin transport by gravity in leaves. *Plant Physiol.* **45,** 644–646.

McKeon, T. A., Hoffman, N. E., and Yang, S. F. (1982). The effect of plant-hormone pretreatments on ethylene production and synthesis of 1-aminocyclopropane-1-carboxylic acid in water-stressed wheat leaves. *Planta* **155,** 437–443.

McPherson, D. C. (1939). Cortical air spaces in the roots of *Zea mays* L. *New Phytol.* **38,** 190–202.

Métraux, J. P., and Kende, H. (1982). On the role of ethylene in the growth response of floating rice to submergence. *Abstr. Int. Conf. Plant Growth Subst. 11th,* p. 35.

Michniewicz, M., Kriesel, K., and Krassowska, J. (1970). Effect of adventitious roots on level of endogenous gibberelin-like substances in buds and newly formed shoots of willow cuttings (*Salix viminallis* L.). *Bull. Acad. Polon. Sci. Ser. Biol.* **18,** 355–359.

Moldau, H. (1973). Effects of various water regimes on stomatal and mesophyll conductances of bean leaves. *Photosynthetica* **7,** 1–7.

Musgrave, A., and Walters, J. (1974). Ethylene and buoyancy control rachis elongation of the semi-aquatic fern *Regnellidium diphyllum. Planta* **121,** 51–56.

Musgrave, A., Jackson, M. B., and Ling, E. (1972). *Callitriche* stem elongation is controlled by ethylene and gibberellin. *Nature (London) New Biol.* **238,** 93–96.

Osborne, D. J. (1967). Hormonal regulation of leaf senescence. *Symp. Soc. Exp. Biol.* **21,** 305–322.

Osborne, D. J. (1974). Auxin, ethylene and the growth of cells. *In* "Mechanisms of Regulation of Plant Growth" (R. L. Bieleski, A. R. Ferguson, and M. M. Creswell, eds.), pp. 645–654. Royal Soc. N.Z., Wellington.

Osborne, D. J. (1982). The ethylene regulation of cell growth in specific target tissues of plants. *In* "Plant Growth Substances 1982" (P. F. Wareing, ed.), pp. 279–290. Academic Press, New York and London.

Pallaghy, C. K., and Raschke, K. (1972). No stomatal response to ethylene. *Plant Physiol.* **49,** 275–276.

Pallas, J. E., Jr., and Kays, S. J. (1982). Inhibition of photosynthesis by ethylene—a stomatal effect. *Plant Physiol.* **70,** 598–601.

Palmer, J. H. (1976). Failure of ethylene to change the distribution of inholeacetic acid in the petiole of *Coleus blumei* and *frederici* during epinasty. *Plant Physiol.* **58,** 513–515.

Pereira, J. S., and Kozlowski, T. T. (1977). Variation among woody angiosperms in response to flooding. *Physiol. Plant.* **41,** 184–192.

Phillips, I. D. J. (1964a). Root–shoot hormone relations. I. The importance of an aerated root system in the regulation of growth hormone levels in the shoot of *Helianthus annuus. Ann. Bot. N. S.* **28,** 17–35.

Phillips, I. D. J. (1964b). Root–shoot hormone relations. II. Changes in endogenous auxin concentration produced by flooding of the root system in *Helianthus annuus. Ann. Bot. N.S.* **28,** 38–45.

Phillips, I. D. J., and Jones, R. L. (1964). Gibberellin-like activity in bleeding sap of root systems of *Helianthus annuus* detected by a new dwarf pea epicotyl assay and other methods. *Planta* **63,** 269–278.

Pilet, P. E. (1975). Abscisic acid as a root growth inhibitor: physiological analysis. *Planta* **122,** 299–302.

Pitman, M. G. (1969). Adaptation of barley roots to low oxygen supply and its relation to potassium and sodium uptake. *Plant Physiol.* **44,** 1233–1240.

Pradet, A., and Bomsel, J. L. (1978). Energy metabolism in plants under hypoxia and anoxia. *In* "Plant Life in Anaerobic Environments" (D. D. Hook and R. M. M. Crawford, eds.), pp. 89–118. Ann Arbor Sci. Publ., Ann Arbor, Michigan.

Railton, I. D., and Reid, D. M. (1973). Effects of benzyladenine on the growth of waterlogged tomato plants. *Planta* **111,** 261–266.

Rajagopal, V., and Anderson, A. S. (1980a). Water stress and root formation in pea cuttings. I. Influence of the degree and duration of water stress on stock plants grown under two levels of irradiance. *Physiol. Plant.* **48,** 144–149.

Rajagopal, V., and Anderson, A. S. (1980b). Water stress and root formation in pea cuttings. II. Effect of abscisic acid treatment of cuttings from stock plants grown under two levels of irradiance. *Physiol. Plant.* **48,** 150–154.

Regehr, D. L., Bazzaz, F. A., and Boggess, W. R. (1975). Photosynthesis, transpiration, and leaf conductance of *Populus deltoides* in relation to flooding and drought. *Photosynthetica* **9,** 52–61.

Reid, D. M., and Crozier, A. (1971). Effects of waterlogging on the gibberellin content and growth of tomato plants. *J. Exp. Bot.* **22,** 39–48.

Reid, D. M., and Railton, I. D. (1974). The influence of benzyladenine on the growth and gibberellin content of shoots of waterlogged tomato plants. *Plant Sci. Lett.* **2,** 151–156.

Reid, D. M., Clements, J. B., and Carr, D. J. (1968). Red light induction of gibberellin synthesis in leaves. *Nature (London)* **127,** 580–582.

Reid, D. M., Crozier, A., and Harvey, B. M. R. (1969). The effects of flooding on the export of gibberellins from the root to the shoot. *Planta* **89,** 376–379.

Richards, B. L. (1929). White-spot of alfalfa and its relation to irrigation. *Phytopathology* **19,** 125–141.

Selman, I. W., and Sandanam, S. (1972). Growth responses of tomato plants in non-aerated water culture to foliar sprays of gibberellic acid and benzyladenine. *Ann. Bot. (London)* **36,** 837–848.

Sena Gomes, A. R., and Kozlowski, T. T. (1980). Growth responses and adaptations of *Fraxinus pennsylvanica* seedlings to flooding. *Plant Physiol.* **66,** 267–271.

Shaybany, B., and Martin, G. C. (1977). Abscisic acid identification and its quantitation in leaves of *Juglans* seedlings during waterlogging. *J. Am. Soc. Hortic. Sci.* **102,** 300–302.

Sheard, R. W., and Leyshon, A. J. (1976). Short-term flooding of soil: its effect on the composition of gas and water phases of soil and on phosphorous uptake of corn. *Can. J. Soil Sci.* **56,** 9–20.

Sheldrake, A. R. (1973). The production of hormones in higher plants. *Biol. Rev.* **48,** 509–559.

Short, K. C., and Torrey, J. G. (1972). Cytokinins in seedling roots of pea. *Plant Physiol.* **49,** 155–160.

Sivakumaran, S., and Hall, M. A. (1978). Effects of age and water stress on endogenous levels of plant growth regulators in *Euphorbia lathyrus* L. *J. Exp. Bot.* **29,** 195–205.

Smith, K. A., and Dowdell, R. J. (1974). Field studies of the soil atmosphere. I. Relationships between ethylene, oxygen, soil moisture content, and temperature, *J. Soil. Sci.* **25,** 217–230.

Smith, K. A., and Robertson, P. D. (1971). Effect of ethylene on root extension in cereals. *Nature (London)* **234,** 148–149.

Smith, K. A., and Russell, R. S. (1969). Occurrence of ethylene, and its significance, in anaerobic soils. *Nature (London)* **222,** 769–771.

Sojka, R. E., and Stolzy, L. H. (1980). Soil-oxygen effects on stomatal response. *Soil Sci.* **130,** 350–358.

Stelzer, R., and Läuchli, A. (1980). Salt and flooding tolerance of *Puccinellia peisonis.* IV. Root respiration and the role of aerenchyma in providing atmospheric oxygen to the roots. *Z. Pflanzenphysiol.* **97,** 171–178.

Stolzy, L. H., Taylor, O. L., Dugger, W. M., Jr., and Mersereau, J. D. (1964). Physiological changes in and ozone susceptibility of the tomato plant after short periods of inadequate oxygen diffusion to the roots. *Soil Sci. Soc. Am. Proc.* **28,** 305–308.

Stolzy, L. J., Focht, D. D., and Fluhler, H. (1981). Indicators of soil aeration status. *Flora (Jena)* **171,** 236–265.

Thimann, K. V. (ed.) (1980). "Senescence in Plants," Vols. 1 and 2. C.R.C. Press, Boca Raton, Florida.

Thomas, H., and Stoddart, J. L. (1980). Leaf senescence. *Annu. Rev. Plant Physiol.* **31,** 83–111.

Tjepkema, J. (1978). The role of oxygen diffusion from the shoots and nodule roots in nitrogen fixation by root nodules of *Myrica gale. Can. J. Bot.* **56,** 1365–1371.

Turkova, N. S. (1944). Growth reactions in plants under excessive watering. *Dokl. Acad. Nauk. SSSR* **42,** 87–90.

Turner, F. T., Sij, J. W., MaCauley, G. N., and Chen, C. C. (1983). Soybean seedling response to anaerobiosis. *Crop. Sci.* **23,** 40–44.

van Bel, A. J. E., Mostert, E., and Borstlap, A. C. (1979). Kinetics of L-alanine escape from xylem vessels. *Plant Physiol.* **63,** 244–247.

Van Staden, J., and Brown, N. A. C. (1973). The effect of oxygen on endogenous cytokinin levels and germination of *Leucadendron daphnoides* seed. *Physiol. Plant.* **29,** 108–111.

Van't Woudt, B. D., and Hagen, R. M. (1957). Crop responses at excessively high soil moisture levels. *In* "Drainage of Agricultural Lands" (J. N. Luthin, ed.), pp. 514–611. Am. Soc. Agron, Madison, Wisconsin.

Walton, D. C., Harrison, M. A., and Cote, P. (1976). The effects of water stress on abscisic acid levels and metabolism in roots of *Phaseolus vulgaris* L. and other plants. *Planta* **131,** 141–144.

Wample, R. L. (1976). "Hormonal and Morphological Responses of *Helianthus annuus* L. to Flooding." Ph.D. Thesis, Univ. of Calgary, Canada.

Wample, R. L., and Reid, D. M. (1975). Effect of aeration on the flood-induced formation of adventitious roots and other changes in sunflower (*Helianthus annuus* L.). *Planta* **127,** 263–270.

Wample, R. L., and Reid, D. M. (1978). Control of adventitious root production and hypocotyl hypertrophy of sunflower (*Helianthus annuus*) in response to flooding. *Physiol. Plant.* **44,** 351–358.

Wample, R. L., and Reid, D. M. (1979). The role of endogenous auxins and ethylene in the formation of adventitious roots and hypocotyl hypertrophy in flooded sunflower plants. *Physiol. Plant.* **45,** 219–226.

Went, F. W. (1938). Specific factors other than auxins affecting growth and root formation. *Plant Physiol.* **13,** 55–58.

Went, F. W. (1943). Effect of the root systems on tomato stem growth. *Plant Physiol.* **18,** 51–65.

Whyte, P., and Luckwill, L. C. (1966). A sensitive bioassay for gibberellins based on retardation of leaf senescence in *Rumex obtusifolius* L. *Nature (London)* **210,** 1380.

Wiedenroth, E., and Poskuta, J. (1981). The influence of oxygen deficiency in roots on CO_2 exchange rates of shoots and distribution of ^{14}C-photoassimilates of wheat seedlings. *Z. Pflanzenphysiol.* **103,** 459–467.

Williams, W. T., and Barber, D. A. (1961). The functional significance of aerenchyma in plants. *Symp. Soc. Exp. Biol.* **15,** 132–144.

Wright, S. T. C., and Hiron, R. W. P. (1972). The accumulation of abscisic acid in plants during wilting and under other stress conditions. *In* "Plant Growth Substances, 1970" (D. J. Carr, ed.), pp. 291–298. Springer-Verlag, Berlin and New York.

Xu, X.-D., and Lou, C.-H. (1980). Presence of roots—a prerequisite for normal function of stomata on sweet potato leaves. *Bull. Beijing Agric. Univ. first issue,* 37–45 (in Chinese with English summary).

Yang, S. F., Adams, D. O., Lizada, C., Yu, Y., Bradford, K. J., and Cameron, A. C. (1980). Mechanism and regulation of ethylene biosynthesis. *In* "Plant Growth Substances 1979" (F. Skoog, ed.), pp. 219–229. Springer-Verlag, Berlin.

Yang, S. F., Hoffman, N. E., McKeon, T., Riov, J., Kao, C. H., and Yung, K. H. (1982). Mechanism and regulation of ethylene biosynthesis. *In* "Plant Growth Substances 1982" (P. F. Wareing, ed.), pp. 239–248. Academic Press, New York and London.

Yoshii, H., and Imaseki, H. (1981). Biosynthesis of auxin-induced ethylene. Effects of indole-3-acetic acid, benzyladenine and abscisic acid on endogenous levels of 1-aminocyclopropane-1-carboxylic acid (ACC) and ACC synthase. *Plant Cell Physiol.* **22,** 369–379.

Yoshii, H., Watanabe, A., and Imaseki, H. (1980). Biosynthesis of auxin-induced ethylene in mung bean hypocotyls. *Plant Cell Physiol.* **21,** 279–291.

Yu, Y. B., and Yang, S. F. (1979). Auxin-induced ethylene production and its inhibition by aminoethoxyvinylglycine and cobalt ion. *Plant Physiol.* **64,** 1074–1077.

Yu, Y. B., and Yang, S. F. (1980). Biosynthesis of wound ethylene. *Plant Physiol.* **66,** 281–285.

Zeroni, M., Jeri, P. H., and Hall, M. A. (1977). Studies on the movement and distribution of ethylene in *Vicia faba L. Planta* **134,** 119–125.

CHAPTER 7

Effects of Flooding on Plant Disease

L. H. STOLZY
Department of Soil and Environmental Sciences
University of California
Riverside, California

R. E. SOJKA
Coastal Plains Soil and Water Conservation Research Center
United States Department of Agriculture
Florence, South Carolina

ISBN 0-12-424120-4

I. INTRODUCTION

A. Oxygen in Soil Pores

In discussing flooding in relation to plant disease, one is faced immediately with problems in terminology. The concept of flooding encompasses a range of soil conditions, and the notion of disease incorporates a variety of plant dysfunctions. The severity and duration of flooding influence disease incidence, and each disease manifests itself at a particular threshold of flooding severity. Finally, the diverse nature of soils renders difficult the universal application of unique definitions, empiricisms, and theories. Certain principles can, nonetheless, be established. The following discussion of flooding and disease interaction is presented from a soils-oriented perspective.

The most negative effect of soil flooding derives from the impact of high water content on soil oxygen (Durbin, 1978). The status of soil O_2 may be characterized as a capacity factor, an intensity factor, or a rate factor. Although each is frequently treated singly, to characterize soil O_2 status all three must be defined for the best characterization of the soil–O_2 system.

As a capacity factor soil O_2, which is a component of soil air, is inferred from simple descriptions of the three-phase soil physical model. In describing soil O_2 status with capacity factors alone, it is assumed that O_2 concentrations and O_2 diffusion rates in soil are not limiting. Some of the capacity factors most frequently calculated are

Porosity (or free space) f: $$f = (V_a + V_w)/(V_s + V_a + V_w) \quad (1)$$

Void ratio e: $$e = (V_a + V_w)/V_s \quad (2)$$

Saturation ratio θ_s: $$\theta_s = V_w/(V_a + V_w) \quad (3)$$

Air-filled porosity f_a: $$f_a = V_a/(V_s + V_a + V_w) \quad (4)$$

Volume wetness θ: $$\theta = V_w/(V_s + V_a + V_w) \quad (5)$$

where V_a, V_w, and V_s are the volumes of soil air, soil water, and soil solids, respectively.

An excellent discussion of these fundamental parameters and their relationship to one another was presented by Hillel (1980). Various investigations have related plant performance to capacity factors (Bushnell, 1953; Miller and Mazurak, 1958; Flocker *et al.*, 1959; Willhite *et al.*, 1965). Each of these parameters in some way describes the capacity of the soil to hold air, but reveals nothing about the nature of the air or its transport properties.

The major constituents of soil air and the ambient atmosphere are N_2, O_2, and CO_2. About 78% (by volume) of both is nearly constantly composed of N_2.

However, whereas atmospheric O_2 is nearly constant at 21%, soil air varies between nearly 21 and nearly 0% (Russell and Appleyard, 1915; Boynton and Reuther, 1938). When O_2 in soil falls below 21%, the drop usually corresponds closely to the increase in CO_2. Mean annual CO_2 concentration in the atmosphere is currently slightly above 0.03% (Revelle, 1982). In soil, however, CO_2 concentrations can exceed 12%. Furthermore, O_2 concentration generally declines and CO_2 concentration generally increases with depth in the soil profile.

Other gases (primarily by-products of anaerobic metabolism such as methane or ethylene) can also occur, from trace amounts to a few percent in soils as the redox potential of the soil environment becomes more negative. Although CO_2 and these trace gases usually occur in relatively small proportions to the other gases present, they frequently induce pronounced plant physiological responses, even at low concentrations.

Because soil air is composed of numerous constituents, the need to determine intensity factors (concentrations or partial pressures) of the particular gases present is important for three reasons: (1) If the concentration of O_2 is low (relative to ambient air), this indicates there are dynamic processes active in the soil profile consuming soil O_2; (2) determination of significant concentration of gases known to affect plant response can foreshadow or explain changes in growth, physiology, and production of plants; and (3) O_2 concentration indicates the level of chemical activity of soil O_2, in essence providing an indirect assay of the oxidation–reduction status of the soil environment.

Although intensity factors describe the relative amounts of gases present in the soil, they do not indicate if the rate of O_2 consumption in the rhizosphere can be satisfied by the rate of O_2 movement through the air-filled soil pores. Understanding the balance of these two rate factors provides the basis of modern conceptual models of soil aeration. It is from these models that understanding of flooding phenomena is derived.

Although O_2 movement into the soil profile can be attributed to changes in soil temperature, surface turbulence, variation in barometric pressure, and physical displacement by and dissolution from infiltrating water, the most active mechanisms by far is diffusion from the ambient atmosphere (Russell, 1952). Similarly, gases produced in the soil are exhausted primarily by diffusion to the ambient atmosphere along concentration gradients.

Diffusion of O_2 through soil to sites of respiration by plant roots and microorganisms is governed by such factors as concentration differences along the diffusion pathway, length and tortuosity of the pathway, and diffusion coefficients of the media encountered in diffusing from source to sink. The last factor is particularly important, because O_2 diffuses 10^4 times more slowly through water than through air (Greenwood, 1961), and only one-fourth as rapidly through dense protoplasm as through water (Krogh, 1919; Warburg and Kubowitz, 1929). The physical principles governing these processes are described in Fick's law:

$$J = D_\epsilon \frac{dC_\epsilon}{dx} \tag{6}$$

where J is gas flux per unit cross section of soil, C_ϵ the concentration of the particular gas in the gas phase of the medium, and D_ϵ the apparent diffusion coefficient of the gas in the medium (Stolzy *et al.*, 1981).

Refinements and complex adaptations of Fick's law have been used to describe both gas movement and O_2 use by soil organisms. Expansion of this conceptual base has been reviewed by Armstrong (1978, 1979), Bouldin (1968), Cannell (1977), Currie (1970), Grable (1971), Greenwood (1971, 1975), Meek and Stolzy (1978), Stolzy (1974), Stolzy *et al.* (1981), Luxmoore and Stolzy (1972a), Vartapetian (1973), and Vartapetian *et al.*, (1978).

Aeration and flooding phenomena cannot, however, be easily dealt with in the confines of a single conceptual base. The same factors that complicate analysis of all biological phenomena confound our understanding of flooding and, more specifically, the interaction of flooding and disease. They include spatial, temporal, thermal, chemical, and biological variability. Finally, although aeration plays the single most dramatic role, numerous other phenomena associated with abundant soil water influence the flooding–disease interaction. These phenomena range from simple hydration effects to changes in active and passive motility of soil-borne pathogens.

B. Water in Soil Pores

Because of the dipolar nature of the water molecule and the predominantly negative charge of soil minerals, particularly clay minerals, water is readily adsorbed on soil particle surfaces. Furthermore, the smaller the interparticle distance, the greater the proportion of the water molecules in soil pores affected by surface attraction. As water is pulled across particle surfaces by surface attraction during adsorption, a drop in pressure is created behind the advancing meniscus, in much the same way that pressure drops behind a piston drawn out of a closed cylinder. Water behind the meniscus advances under the pressure gradient until the attractive forces at the wetting front are in equilibrium with the hydrostatic pressure and cohesive forces of the advancing water. This phenomenon, capillarity, is the foundation on which much of the physics of soil flooding is based.

The principles of capillarity explain the difference in water retention among soils of like texture with differing pore geometry and among soils with different pore geometries caused by textural differences. Soil water potential ψ (the chemical potential of soil water) expresses the energy status (capacity or "potential" to do work) of water in soil relative to pure free water in the same position. Soil water usually has a negative potential because it is held in soil by forces of adsorption, cohesion, and solution and therefore has less capacity to do work than free water at the same position. The less water there is in soil, the more tightly the remaining water is held, and therefore the lower (more negative) its ψ. The most easily determined component (and usually the largest in most agricultural soils without salinity problems) is the soil matric potential ψ_m. It is that part of the total ψ explainable by the surface attractions of soil particles and plant roots as arranged in the soil profile (Taylor and Ashcroft, 1972). As mean pore size decreases, θ_s increases for a given ψ_m, accounting for the difference in water-retention curves for soils of differing textures and levels of particle aggregation. This principle also dictates that on desorption, the smallest pores empty last.

Consequently, it is erroneous to consider soils as homogeneous bodies of uniformly distributed pore water. Particle surfaces in close proximity to macropores seldom lose contact with the gaseous phase even at ψ_m somewhat above field capacity (FC). Certain pores within the soil mass, however, are rarely exposed directly to soil air. Particle surfaces near the smaller micropores or near the center of aggregates may remain undrained even at potentials approaching 1 bar. Therefore, flooding of discrete pores occurs over a continuum of conditions. Some pores are nearly always water filled in almost all soils. This results in the simultaneous existence of regions of aerobic and anaerobic conditions throughout the soil profile, dependent on sink strength and limitations to diffusion (Currie, 1962; Letey and Stolzy, 1967). This situation also explains detection of small amounts of anaerobically derived chemicals even in soils that are otherwise well drained (Lynch, 1975; Primrose, 1976; Hunt *et al.*, 1980, 1982; Smith, 1980).

On inundation of the soil profile from surface flooding, the hierarchy (with respect to pore size) of water entry into the profile is the reverse of desorption. Because of the vastly greater saturated hydraulic conductivity of macropores, water rushes first into and through the large continuous pore spaces. As this occurs, large volumes of soil air are displaced from the macropores, while numerous small pores are left with trapped air that slows subsequent entry of water. Water continues to move into the profile until the limits of retention at FC are reached for the depth of water applied, or until a restricting plane is encountered, slowing the advance of water and resulting in saturation of the soil above the less permeable layer.

For a horizon of large pores overlying a horizon of smaller pores, if the rate of water application at the soil surface is lower than the sautrated hydraulic conductivity of the overlying horizon (but greater than that of the lower horizon), water flows downward in the unsaturated state until the interface is encountered. On penetrating the lower horizon, its flow is reduced to a rate equal to its saturated hydraulic conductivity. Because water is arriving at the interface faster than it can flow through the lower horizon, it begins to accumulate, saturating the soil above the interface. If application of water continues, the zone of saturation progresses upward, eventually ponding at the surface.

To understand why flooding of pores occurs when a horizon of relatively small pores overlies one of larger pores, one must turn again to capillary phenomena. This is all the more remarkable inasmuch as it can occur to a limited extent even if the rate of water application at the soil surface is less than the saturated hydraulic conductivity of either horizon. As explained previously, surface forces and capilarity lower soil water pressure to a value below atmospheric pressure. When a soil pore abruptly increases in diameter, the situation is analogous to the pore's draining into an open void. Because water in the pore is at a pressure lower than the air pressure in the void, the meniscus is curved inward, away from the side of higher pressure. Flow into the void cannot proceed until the pressure in the smaller pore exceeds atmospheric pressure. Sufficient pressure is finally achieved when, again, water ponds above the interface of the two horizons as a result of the temporary cessation of flow. This backing up of water results in a zone of flooded pores above the coarser-textured horizon. Eventually, sufficient pressure is achieved to counter the negative ψ_m. The water pressure in the small pores at the interface finally exceeds atmospheric pressure, and water flows into the larger pore.

Ironically, gravel is often placed at the bottom of pots or planting beds, ostensibly for drainage. This can result both in flooding on irrigation with less water for a given volume of soil, and in less water storage in the reduced soil volume for plant use between waterings.

On introducing consideration of living organisms in flooded profiles, the situation is further complicated regardless of the indicator selected to define aeration. Average profile O_2 status may be irrelevant to the particular O_2 status at a microsite near a root hair or single-celled microorganism. Virtually all current technology integrates O_2 status over soil volumes more representative of macropores than micropores. Biological sinks for O_2, and therefore soil O_2 status, may vary by one or two orders of magnitude over a few micrometers distance from respiring cells. The mean status of aeration in the soil mass is therefore less important than the likelihood of encountering sites where the rate of O_2 supply is insufficient to meet the potential respiratory demand of soil organisms. Differences in the sampling volumes of various aeration sensors and actual differences in soil O_2 status over short distances within the soil led Flühler *et al.* (1976) to conclude that soil aeration is most appropriately characterized by a statistical expression of microsite spatial heterogeneity. Application of this concept was developed in greater depth by Stolzy *et al.* (1981).

II. FLOOD-PRONE SOILS

A. Profile Characteristics

The frequency and duration of flooding have a marked impact on soil properties and plant response. Flooding problems on a field scale occur most commonly on relatively level land with restricted surface or internal drainage (due either to fine soil textures with low saturated hydraulic conductivity or to shallow water tables). Soils with these properties are characteristically found in low-lying river deltas or floodplains, on remnant or recessional lacustrian formations, and along low-lying maritime coastal plains. Parent materials in these soils (particularly the first two groups) are frequently dominated by expanding lattice 2 : 1 clay minerals. The pore geometry of these shrink–swell soils varies immensely with ψ_m. Often, the higher the clay content, the smaller the mean pore size at saturation and the greater the likelihood of extensive and deep cracking when ψ_m is only slightly lower than ψ_m at FC. The complexity and variability of the pore geometry in these soils make it difficult to determine the degree or extent of O_2 depletion in the soil profiles in the field at an acceptable level of statistical reliability.

In the newly adopted comprehensive soil taxonomy (Soil Survey Staff, 1975), there is no soil order in which flooding cannot occur, although the properties of certain soil orders are such that flooding can be expected to occur naturally more often. This distinction is apparent in the absence of an Aquic subgroup in the order Aridisol, for example. However, Histisols are also without an Aquic subgroup; in this case, flooding and its impedance of organic matter oxidation is a formative factor that occurs throughout the entire order. Soil flooding and its role in soil genesis and classification are discussed in greater detail by Ponnamperuma (Chapter 2 in this volume). As a diagnostic tool, several profile characteristics are ready indicators of frequent flooding in nature.

Where soil flooding is frequent and prolonged, the lower profile usually remains reduced. Iron and manganese are present in ferrous and manganous forms and impart a deep gray color, frequently grading into a bluish cast. This condition is referred to as gleying (Soil Survey Staff, 1951). Frequently, flooded soils are also relatively high in organic matter in their surface horizons, rendering them dark or black in color. Where high soil temperatures result in oxidation of surface organic matter between periods of inundation, the flooding proneness of the soil may still be evident in the stratified deposition of parent materials in lenses of various thickness. Horizons that are intermittently flooded for prolonged periods and that have irregular internal drainage are frequently mottled. In diagnosing site susceptibility to flooding or flood-induced diseases, all of these profile characteristics should be examined.

B. Physicochemical Characteristics

Where soils are persistently flooded, the physiochemical characteristics of the soils that have developed *in situ* are basically those associated with steady-state reducing environments. Free water is likely to be present at a shallow depth in the soil profile, and the remaining profile may be within the capillary fringe, close to saturation. Free O_2 is nearly absent even near the soil surface and is certainly extinct at some modest depth within the profile. Metallic cations such as iron and manganese exist in their lower valence states, and oxidation–reduction potential E_h, a measure of electron availability in soil, is low. In highly aerated (oxidized) soils, E_h is of the order of +0.6 V. In extremely anaerobic (reduced) soils, E_h can be as low as −0.2 V. The E_h is especially dependent on soil pH as well as on iron and manganese content. The concept of E_h has been developed in detail by Novozamsky *et al.* (1976) and by Bohn *et al.* (1979). Application of the concept of E_h has also been explained by Russell (1976) and is discussed in detail by Ponnamperuma, (Chapter 2 in this volume).

When a soil is flooded, characteristic shifts occur in populations of soil organisms. Algal populations may be large near the surface, and there is frequently a larger than normal microfauna component. Higher-order plants almost certainly are composed largely of flood-tolerant species.

Where soils are intermittently flooded for periods of days or weeks, the profile may never achieve a steady state. Depending on the initial organic and inorganic properties and the initial O_2 availability within the soil at the beginning of each flooding episode, the soil profile will pass through a range of transient intermediate stages. The duration and nature of each intermediate stage will depend largely on soil temperature, amount of fresh organic matter present, initial soil microflora component, and nature of the higher plants growing in the soil. Rice, for example, introduces significant amounts of O_2 into the soil profile in paddy culture by diffusion of atmospheric O_2 from shoots to the highly porous root systems and by release of some of the O_2 to surrounding soil. The balance of these inorganic, organic, and biological inputs can explain the falling E_h of the system for varying lengths of time at discrete redox potentials as the profile depletes each pool of successively less willing electron acceptors (Patrick and Mikkelson, 1971; Russell, 1976). These physiochemical and biochemical considerations are presented in detail by Ponnamperuma (Chapter 2 in this volume).

III. FLOODING EFFECTS ON THE HOST PLANT: MORPHOLOGY AND FUNCTION

Several reviews have detailed the morphological and physiological changes that occur in higher plants as a result of flooding (Williamson and Kriz, 1970; Stolzy, 1972; Drew and Lynch, 1980; Bradford and Yang, 1981; Kawase, 1981;

Wiedenroth, 1981; Campbell, 1981; Armstrong, 1981). These changes are in great part responsible for a given species' individual level of susceptibility or resistance to flood injury and related disease. Several specific predisposing factors are discussed later in this chapter, and most of the general consequences of flooding on plant morphology and physiology are discussed in other chapters in this volume by Jackson and Drew (Chapter 3), Kozlowski (Chapter 4), Kozlowski and Pallardy (Chapter 5), and Reid and Bradford (Chapter 6). A few specific effects are briefly emphasized here.

A. Roots

When a root channel is flooded, roots experience an immediate reduction in O_2 supply. One of several consequences may result. In the least severe scenario, root respiration, deprived of free O_2, proceeds along a fermentative pathway, rapidly consuming the available pool of stored carbohydrates—the Pasteur effect (Wiedenroth, 1981; Bertani *et al.*, 1981; Wiebe *et al.*, 1981). Instead of CO_2, alcohol is the predominant by-product released, and the relative amount of energy released has been estimated to be as low as 5% of that liberated for a given amount of substrate via the aerobic pathway (Widenroth, 1981). Wiebe *et al.* (1981) discussed numerous alternative pathways, including the reduction of inorganic compounds such as sulfur and production of other by-products, such as methane.

Active uptake of nutrients is greatly diminished as a result of slowing of energy conversion. In the early stages of anaerobiosis (Drew and Sisworo, 1979; Trought and Drew, 1980a,b), mineral requirements of immature aerial plant tissue are met by mobilization from more mature tissue. The substrate requirement of roots is now so elevated that it cannot be met by translocation from shoots alone. Consequently, less resistant cellular constituents are metabolized in place. The latter eventually results in development of lysigenous zones of intercellular voids. These voids formed in the roots allow the diffusion of O_2 from the shoots to active root tissue. This ultimately is the mechanism through which a certain level of normal plant functioning is able to resume (Jensen *et al.*, 1969; Luxmoore and Stolzy, 1966, 1972a,b; Yu *et al.*, 1969; Varade *et al.*, 1970, 1971; Luxmoore *et al.*, 1971, 1972; Papenhuijzen and Roos, 1979; Benjamin and Greenway, 1979; Konings and Verschuren, 1980; Stelzer and Laüchli, 1980; Drew *et al.*, 1980; Kawase and Whitmoyer, 1980; Konings, 1982). Oxygen diffuses more freely through the newly created root air spaces from shoots or better-aerated portions of the root system, internally aerating the submerged portions of the root system.

When flooding is complete and root systems are less adaptable, or respiratory demand is too large to be met by a shift in respiratory pathway, root systems become necrotic. The necrotic tissues lose physical integrity. Such root necrosis is often termed root pruning. In addition to providing a vector for pathogen entry, root pruning also impairs physiological recovery on drainage of water from the soil profile, by limiting the root volume. The resultant decrease in root-to-shoot ratio impairs soil-nutrient and soil-water extraction in the recovering plant, slowing its subsequent growth and, in most cases, decreasing crop yield.

B. Shoots

Until O_2 is depleted from the flooded soil profile, shoots continue to respond as though irrigated. When soil O_2 is depleted, the chain of events just described ensues. Eventually, shoot physiological activity becomes limited as explained by insufficient water- and nutrient-supplying capability. Other significant responses occur that result from less well understood consequences of flooding. Changes in root membrane permeability and alterations in hormonal composition of the plant produce wilting, leaf epinasty, and stomatal closure (Bradford and Dilley, 1978; Hunt *et al.*, 1981) and stimulate adventitious root development, emanating from stem nodal positions above the zone of saturation (Bose *et al.*, 1977; Wample and Reid, 1978; Drew *et al.*, 1979, 1980; Kozlowski, 1982).

Stomatal closure is not always a simple mechanical response to increased root resistance resulting in lowered xylem pressure potentials (Sojka and Stolzy, 1980). Aside from the physiological implications (decreased CO_2 fixation and decline in photosynthesis), stomatal closure is a negative and almost synergistic reaction to flooding. Once stomata are closed, transpiration declines, which has the net effect of prolonging the flooding episode. Furthermore, stomatal closure restricts gas exchange between leaves and the atmosphere, thus restricting the aerial pathway for internal diffusion of O_2 from the shoots to flooded root systems. Stomatal closure by flooding has been recognized as a means of limiting shoot damage during episodes of air pollution (Stolzy *et al.*, 1961, 1964).

Other, more direct effects of soil flooding on shoot growth also occur. The mechanical support of the root system is substantially reduced in saturated soil. This significantly predisposes the plant canopy to lodging. Lodging brings plant tissue from adjacent plants in contact with one another and, when severe, in contact with wet soil and/or floodwaters. Both effects increase the likelihood of inoculation and spread of disease in the plant canopy. Even in the absence of lodging, if the ground is wet for prolonged periods, the resulting elevation of relative humidity in the canopy favors shoot diseases such as rusts and mildews, which invade the leaf surface when moist.

IV. PREDISPOSITION EFFECT OF WATERLOGGING

A. Root Surface

The term *predisposition* has been used since the late 1800's in reference to plant stress in relation to disease (Schoeneweiss, 1975, 1978). The term refers to host disposition or "proneness" to disease prior to infection. Levitt (1980) divided excess water or flooding stress into primary and secondary stress and indicated that excess water and flooding, although not injurious directly, can cause a secondary O_2-deficient strain. In this section we discuss some studies that demonstrate predisposing effects of flooding on plant diseases and some hypotheses of predisposition mechanisms.

As emphasized by Cook and Papendick (1972), soil water may act on (1) the pathogen, (2) the host plant, and (3) soil microorganisms. The rhizosphere, a narrow zone of soil around the root, is of most interest because of leakage or exudation from plants of various compounds into this zone that promote or inhibit plant pathogens (Curl, 1982). In the rhizosphere, the root is surrounded by a waterfilm that varies in thickness, controlling the O_2 supply for root respiration (Letey and Stolzy, 1967). A second term, rhizoplane, refers to the root surface providing a very favorable nutrient base for a large number of micro organisms.

One hypothesis of the mechanism of flooding stress on root disease suggests that soil saturation changes root tissues, enhancing zoospore attraction and infection. Experiments with safflower (*Carthamus tinctorius* L.) suggested that disease enhancement by flooding was not due entirely to inoculum behavior, but that disease was enhanced by host predisposition (Duniway, 1977b). Occurrence of drought prior to irrigation may also increase the severity of *Phytophthora* root rot in safflower (Duniway, 1979). Therefore, drought could enhance disease development following irrigation, because water stress predisposes the host and water deficiency in the soil increases subsequent production of effective inoculum by the pathogen (Duniway, 1977b).

Phytophthora root rot of alfalfa (*Medicago sativa* L.) is closely associated with excessive rainfall and poorly drained or heavily irrigated soils. Kuan and Erwin (1980) cited "enhanced attraction" as the mechanism by which alfalfa roots were predisposed to infection when subjected to saturation. Using scanning electron microscopy, they showed that after 1 day, breaks in the root epidermal cells occurred on the surfaces of roots grown in saturated soil, whereas the surfaces of roots grown in unsaturated soil were smooth and intact (Fig. 1). Papenhuijzen and Roos (1979) recognized that a recurring question in studies of root development in poorly aerated nutrient solutions is whether the low O_2 concentration is a primary inhibiting factor. In such studies, root growth is inhibited more than expected on the basis of the O_2 measured in the medium. However, O_2 concentration at the root surface is much lower than in the bulk medium because of consumption of O_2 by the root and low diffusion of O_2 inward through the water film around the root. Papenhuijzen and Roos (1979) found several effects after stopping aeration of the nutrient solution. In bean (*Phaseolus vulgaris* L. cv. Berna), starch grains disappeared from the amyloplasts of root-tip cells, small dense lipid bodies formed, and the rough endoplasmic reticulum of cortical cells showed concentric arrangements that

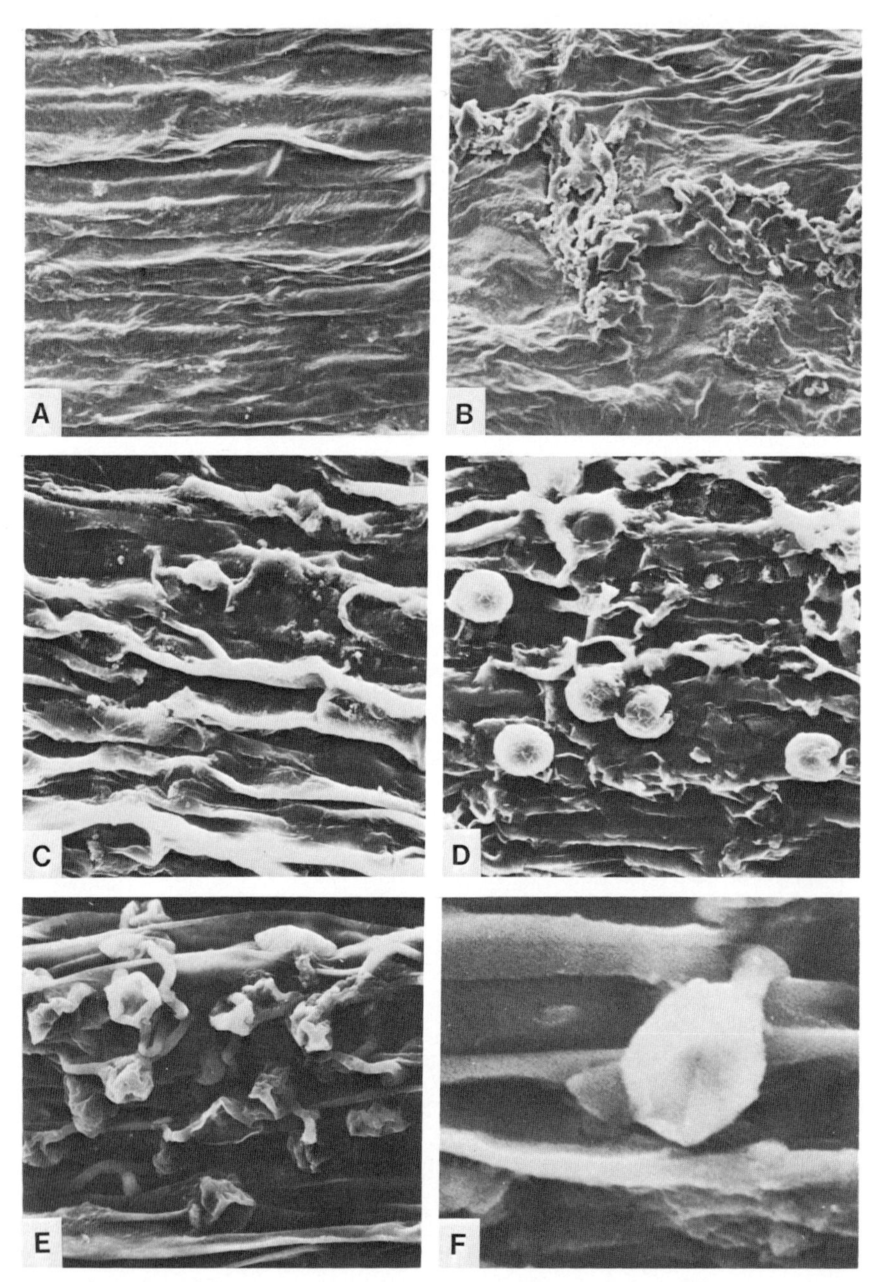

Fig. 1. Scanning electron micrographs of the surface of alfalfa roots (A) Alfalfa roots grown in unsaturated soil (1000×). (B) Alfalfa roots grown in saturated soil for 1 day (1000×). (C) For 5 days (1000×). (D) Zoospores of *Phytophthora megasperma* f. sp. *medicaginis* penetrating a root from saturated soil (1000×). (E) Germinated zoospores on a root from saturated soil (1000×). (F) Growth of the germinated zoospores into the break in the root surface (3000×). From Kuan and Erwin (1980).

were attributed to anoxia. One day after aeration was stopped, cells of the extreme tip of the root axis were dead. After 2 or 3 days, older cells 2–3 mm from the root tip were dead.

Knous and Maxfield (1976) hypothesized predisposition associated with root rot of alfalfa under O_2 stress. They reported that in a field irrigated every 4 days, increase of squalene concentration of roots in the susceptible cultivar Saranac was associated with root-tissue dissolution characteristic of *Phytophthora* root rot. The squalene increase may have resulted from root-tissue predisposition resulting from O_2 stress and not from flooding directly. In comparison, plants in plots irrigated every 10 days had no root rot and showed no increase in squalene.

Effects of waterlogging on concentrations of gases and solutes dissolved in soil water were investigated by Trought and Drew (1980a,b) to determine whether the early disruption in growth of wheat (*Triticum*) was most closely associated with depletion of dissolved O_2, accumulation of toxins, or changes in nutrient concentrations in the soil water. Injury to shoots and roots was attributed to a decrease in soil O_2 concentration rather than to concentration of inorganic nutrients or accumulation of toxins in the soil water.

Several studies have shown that the apical root meristem stops growing almost immediately in an anoxic soil environment. Using time-lapse photography, Klepper and Huck (1969) and Huck (1970) showed that in less than 3 hr, cells in the region of root elongation died without sufficient O_2. When the soil was reaerated, lateral roots branched from the undamaged part of the taproot.

Letey and Stolzy (1967) analyzed the soil and water geometry around the root in relation to O_2 supply at the root surface. Soil O_2 diffusion rate (ODR) from gas-filled soil pore spaces decreased as water-film thickness around the root increased. An earlier study showed that root growth was correlated with ODR (Stolzy and Letey, 1962). The limiting water-film thickness is dependent on the soil porosity around the root and O_2 concentration in the pores. For example, a root cannot grow into the center of a saturated aggregate if its porosity is 30% and its diameter greater than 0.60 mm (assuming pore O_2 concentration is 21%). Calculations can be made on thin sections of soil aggregates to determine the fraction of the area that can allow root growth at various soil water contents and pore O_2 concentrations.

Fawcett (1936) attributed root rot of citrus to *Phytophthora citrophthora* (R. E. Sm. & E. H. Sm.) Leonian and *P. parasitica* Dast. The disease was favored by abundant soil moisture and was severe in clay soils where water drainage was impeded. Klotz *et al.*, 1971) found that parasitism by *Phytophthora* spp. was severe in well-aerated, but thoroughly watered, soil. Stolzy *et al.* (1965a) reported little or no decay of citrus feeder roots unless the soil was saturated (Table I). Several watering methods that enhanced root decay were developed by Stolzy *et al.* (1965b). Soil saturation was the key factor that increased root decay, and duration of saturation was more critical than frequency of saturation. Saturating the soil three times a month resulted in more root decay than saturating soil twice

TABLE I

Effects of Irrigation and O_2 Partial Pressure on Plant Growth, Root Decay, and Root-Tip Initiation in *Citrus sinensis*[a]

Effects observed	Dry weight (per plant; g)[b] Leaves	Stems	Roots	Total plant	Root decay (%)	New white root tips (number)
Irrigation[c]						
Control	2.14	1.42	1.86_x	5.42	5_w	8
Water table	2.06	1.25	1.49_{xy}	4.81	40_x	11
Saturated every 4 days	2.15	1.34	1.29_y	4.77	52_y	9
Saturated 3 times a month	2.17	1.33	1.26_y	4.75	65_z	7
Saturated twice	2.15	1.37	1.65_x	5.18	9_w	15
Value	NS	NS	**	NS	**	NS
Aeration[d] (O_2 partial pressure)						
Air (152 mm Hg)	2.48_x	1.63_x	2.50_x	6.61_x	8_x	24_x
N_2 + air (10–13 mm Hg)	2.16_y	1.32_y	1.14_y	4.61_y	40_y	3_y
N_2 gas (0.3 mm Hg)	1.76_z	1.08_y	0.90_y	3.73_z	55_z	3_y
F value	**	**	**	**	**	**
Coefficient of variability (%)	16	25	20	15	23	70

[a]From Stolzy *et al.* (1965b).
[b]Subscript letters w, x, y, and z after mean values indicate statistical populations. Mean values are statistically significant only if they do not have a letter in common after values. NS, Differences of means not significant; **, *F* value significant at the 1% level.
[c]Each value is a mean of 12 internal replications.
[d]Each value is a mean of 20 internal replications.

a month, the latter a common practice among growers during summer months. Stolzy *et al.* (1965b) showed that low O_2 supply prevented growth and regeneration of citrus roots even in the absence of pathogens. Stolzy *et al.* (1967) studied root decay of avocado (*Persea americana* Mill. cv. Mexicola) in relation to O_2 diffusion, water regime, and *Phytophthora cinnamomi* Rands (Fig. 2). Significant root decay was found in saturated soils in the presence or absence of the fungus (Table II). The high water content also appeared to have an adverse effect on establishment of the fungus in the zoospore germ-tube stage. Oxygen deficiency in the root zone was the most important soil physical factor affecting growth and decay of roots. For plants growing in soils with O_2 diffusion rates of 0.17 $\mu g\ cm^{-2}\ min^{-1}$, 44–100% of the root systems were in decay.

Saturated soils commonly occur in nursery and landscape plantings, which can predispose normally resistant rhododendrons to root and crown rot caused by *Phytophthora cinnamomi* (Blaker and MacDonald, 1981). Blaker and Mac-

Fig. 2. One average plant from each treatment replication. The plants in the top row had O_2 concentration at the soil surface the same as that of air, whereas the bottom row had a reduced O_2 concentration at the surface. The letters refer to irrigation treatments: (A) control, (B) saturated every 10 days, (C) saturated every 4 days, and (D) water table. Soils were infested with a zoospore suspension of *Phytophthora cinnamomi*. From Stolzy *et al.* (1967).

Donald (1981) reported that root and crown rot of *Rhododendron* spp. was caused by *P. cinnamomi*. They subjected 1-year-old rhododendrons to various soil water regimes and inoculated them with motile zoospores of *P. cinnamomi*. In the absence of flooding, plants of the cultivar Purple Splendour developed severe root and crown rot following inoculation, whereas the cultivar Caroline remained free of symptoms and was relatively resistant. However, if Caroline roots were flooded for 48 hr before inoculation with *P. cinnamomi*, they developed severe symptoms of root and crown rot.

Armstrong (1981) listed four primary sequential stages of adaptive responses of higher plants to waterlogging: (1) soil waterlogging, (2) root O_2 stress, (3) decreased permeability of roots to water, and (4) plant water stress. Many of these effects were reviewed by Kozlowski (1976), Schoeneweiss (1975), Cannell (1977), and Levitt (1980). The effect of hormonal changes on plant response is an entire additional area of investigation of predisposing effects, but it also suggests the involvement of another important mechanism of plant adaptation, the release of root exudates (Fig. 3).

TABLE II

Effects of *Phytophthora cinnamomi*, Irrigation, and O_2 Supply on Plant Growth and Root Decay[a]

Treatments	Dry weight (per plant; g)			Root decay (%)	Stage of disease[b]
	Shoots	Roots	Total		
Phytophthora[c]					
Absent	5.6	2.0	7.6	42	1.0
Present	5.1	2.0	6.9	55	1.4
Significance	*	NS	NS	*	*
Irrigation[d]					
Check	5.4	2.0	7.4	39_x	0.4_x
Saturated every 10 days	5.1	1.9	6.7	46_x	0.8_x
Saturated every 4 days	5.8	2.1	7.9	37_x	0.9_x
Water table	5.0	1.9	7.0	72_y	2.6_y
Significance	NS	NS	NS	**	**
Aeration[c]					
High O_2 supply	6.5	2.7	9.2	18	0.1
Low O_2 supply	4.1	1.2	5.3	79	2.3
Significance	**	***	***	**	**
Interactions					
Phytophthora × irrigation	**	NS	*	NS	NS
Coefficient of variability (%)	19	25	20	44	32

[a]From Stolzy *et al.* (1967).

[b]Numerical rating of plant tops as to the degree of root damage on a scale of 0 to 5 (0, healthy; 5, dead). Significance equals *F* value at level: *, 5%; **, 1%; ***, 0.1%. Subscript letters x and y after mean values indicate statistical populations. Mean values are statistically significant only if they do not have a letter in common after values.

[c]Each value is a mean of 32 internal replications.

[d]Each value is a mean of 15 internal replications.

B. Release of Exudates

Root exudates include soluble sugars, amino acids, organic acids, etc.; and insoluble substances, for example, cells and tissue fragments from root caps, epidermis, and cortex, polysaccharides, volatile compounds, etc. (Hale *et al.*, 1978). Virtually every soluble compound found in plants can be found in root exudates, depending on the plant species. Root exudation occurs as a result of certain environmental conditions, insects, nematodes, mechanical injury, and growth of secondary or lateral roots. The region of meristematic cells behind the root tip and the region of root elongation are sites of major exudation (Rovira and Davey, 1974).

Nematodes, fungi, and bacteria enter root tissues by penetrating the walls of the epidermis or through wounds (Balandreau and Knowles, 1978). Such inva-

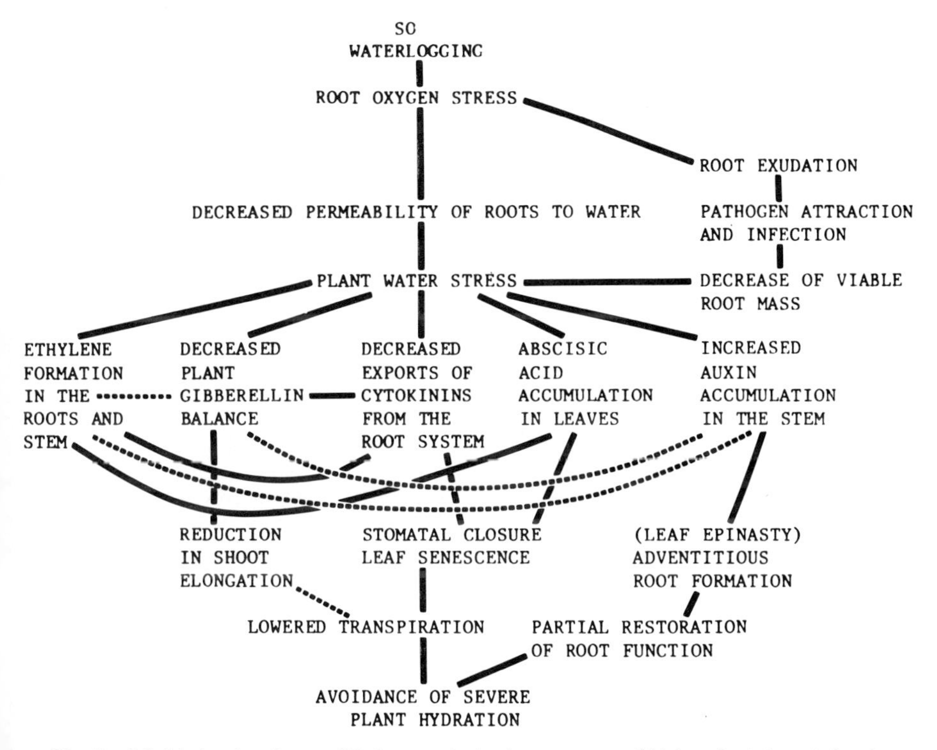

Fig. 3. Model showing the possible hormonal adaptive responses of higher plants to waterlogging stress. Unconfirmed interactions are represented by broken lines. Adapted from Armstrong (1981).

sion of roots affects the entire cortex of a living root, which provides invading microflora with a substrate equivalent to leaf litter, referred to as "root litter." Barber and Lynch (1977) supported the view that microorganisms stimulate the loss of soluble organic materials.

Hypoxia or anoxic conditions around the roots of flooded plants affect root exudation. Kuan and Erwin (1980) found lesions on the surface of alfalfa roots grown in soil saturated for 1 day. Also, the electrical conductivity of root exudate from the saturated soils increased from 15 to 22 mhos cm^{-1}, amino acids increased 30%, and sugar increased by 10% when compared to exudates from unsaturated soils. Labanauskas *et al.* (1974) measured amino acids in citrus leaves of plants growing in soils with normal and low soil-O_2 concentrations. The concentrations of the individual protein amino acids, the sum of individual protein amino acids, and percentage nitrogen in the leaves of low-O_2 plants were significantly lower than in leaves of plants grown with normal O_2 supply. Their study suggested a possible loss of amino acids from the plant to the soil under low soil O_2. However, there is no direct evidence to substantiate this assumption.

Grineva (1961) found that soil anaerobiosis increased the dry weight of root exudates and the proportions of oxidizing compounds in corn (*Zea mays* L.) and sunflower (*Helianthus* sp.). Grineva concluded that cessation of aerobic respiration induced a shift in metabolism, resulting in secretion of nonmetabolized compounds, and induced formation and excretion of ethanol at the expense of sugar (Grineva, 1963). According to Hale *et al.* (1978), information is lacking on the effects of anoxia on root exudation. The effects of reduced O_2 on respiration are well known. Reduction in aerobic root respiration reduces energy available for maintenance of the active transport system. The permeability of cellular membranes changes and some substances leak out. The effects of exudates on plant diseases are discussed in Section VII.

V. FLOODING EFFECTS ON PLANT PATHOGENS

A. Fungi

1. Sporulation and Zoospore Release

Root-infecting microorganisms include fungi, bacteria, nematodes, and viruses. The list of diseases caused by fungi is extensive. Many are associated with flooded soil conditions. Diseases of root-infecting fungi may be divided into pathogen-dominant diseases and host-dominant diseases (Kommedahl and Windels, 1979).

Pathogen-dominant fungi are unspecialized pathogens that attack plants having little or no disease resistance; the pathogen is dominant over the host. Examples are *Rhizoctonia solani* Kühn, *Pythium* spp., and *Phytophthora* spp. *Rhizoctonia solani* is a complex pathogen that can damage plant roots or shoots in semiarid to aquatic environments. There are at least 66 known species of *Pythium* that are distributed in agricultural and undisturbed soils worldwide. They are pioneers in fungal ecological successions and are poor competitors for substrate. They form resting oospores when other fungi appear, accounting for their extended survival in soil in the absence of a host. *Phytophthora* and *Pythium* are closely related but differ in pathogenicity. *Phytophthora* causes disease on most plant parts, whereas *Pythium* is mainly a soil-inhabiting, root-infecting fungus (Kommedahl and Windels, 1979).

Host-dominant diseases involve pathogen–host associations in which the host has greater influence than the pathogen on the course of disease. Disease severity is greatest when the environment stresses the host for some time. Some examples of fungi causing host-dominant diseases are *Armillaria mellea* Vahl ex Fr., which causes root rot on hundreds of woody plants, *Fusarium* and *Verticillium,* which cause vascular wilt, and *Thielaviopsis basicola* (Berk. & Br.) Ferr., a soil-borne pathogen that causes root rot of more than 100 plant species.

Root exudates are essential for host-dominant diseases. Both groups of fungi can persist in soil in an inactive state as resistant spores, sclerotia, or quiescent mycelia until activated by living roots (Curl, 1982). The influence of soil water on the reproductive structures of several *Phytophthora* species has been the subject of many studies. Soil water requirements for formation of sporangia in soil by *Phytophthora* are rather precise (MacDonald and Duniway, 1978a,b). The significance of ψ_m and osmotic ψ during the difficult developmental phases of selected mold oomycetes, including *Saprolegnia, Aphanomyces, Phythium,* and *Phytophthora* spp., was reviewed by Duniway (1979). Early studies recognized the association of saturated soil conditions with root decay. More precise methods for controlling ψ_m of soils have provided quantitative data on fungal water requirements. Duniway (1975b,c), MacDonald and Duniway, 1978a,b), Pfender *et al.* (1977), and Bernhardt and Grogan (1982) studied the formation of sporangia and release of zoospores by *Phytophthora cryptogea* Pethybr. & Laff, *P. megasperma* Drechs., *P. parasitica,* and *P. capsici* Leonian.

Gisi *et al.* (1979) studied production of sporangia by *Phytophthora cinnamomi* and *P. palmivora* (Butler) in sandy loam and clay soils at values of ψ_m between 0 and -15 bars. Number of sporangia produced was strongly correlated with ψ_m, but not with soil water content. Numbers of sporangia formed at a given ψ_m were similar for both soils. With buried mycelial inoculum, *P. cinnamomi* produced the most sporangia at -160 mbars with upper and lower limits between -10 and -250 mbars, whereas buried mycelial inoculum of *P. palmivora* produced the maximum numbers of sporangia in both soils at -10 mbars. Maximum numbers of sporangia were produced by *P. cinnamomi* on the soil surface only under flooded (1mbar) and saturated (0 mbar) soil conditions. When washed mycelium or mycelial discs were used, the optimal ψ_m was between -10 and -100 mbars. When infected radicles (Pfender *et al.*, 1977) and infected leaf discs (Sugar, 1977) were used, the optimum was 0 mbar. The difference was attributed to inter- and intraspecific differences in sensitivity to aeration and/or water requirements.

It was suggested that the optimum ψ_m for infection from mycelial mats was different when infected tissue was used as inoculum. Kuan and Erwin (1982) showed that optimal ψ_m for sporangium formation on infected roots was 0 mbar compared to -100 mbars on mycelial discs. They suggested that infected roots provided a substrate for the fungus that differed from mycelial mats in its nutrient and water status and interaction with other microorganisms. Duniway (1983) and Sterne and McCarver (1980) reported that fewer sporangia formed at 0 than at -150 mbars ψ_m, but production of sporangia was similar at ψ_m between -25 and -150 mbars (Fig. 4). This was expected from the behavior of mycelial discs in soil. The important differences were that sporangia of *Phytophthora cryptogea* arising on roots formed more rapidly and in greater abundance and persisted longer than those produced on mycelial discs. The stage in the life cycle of *Phytophthora* most sensitive to flooding is during the release of zoospores from

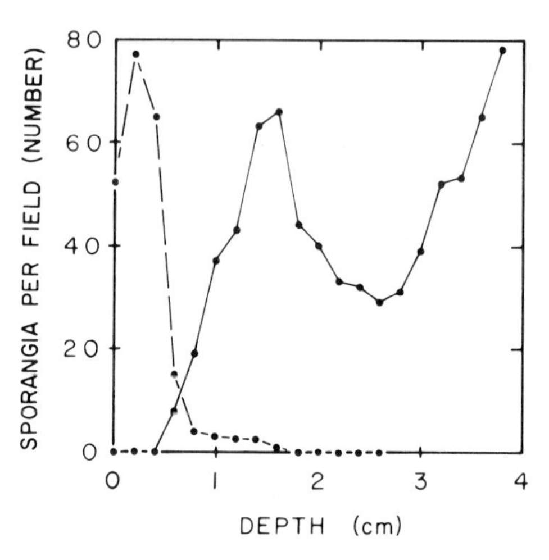

Fig. 4. Sporangium formation by *Phytophthora cryptogea* on roots of living safflower seedlings at various depths in soil maintained at ψ_m values of 0 (- - -) and −150 (——) mbars for 48 hr after inoculation. A microscope was used to count the sporangia in each field as one edge of a root was scanned from the level of the soil surface to the root tip. Representative data from 1 of 10 seedlings in each treatment are shown. From Duniway (1983).

sporangia (Duniway, 1979). The transition of soil from a drained to a saturated state stimulates zoospore release by sporangia. Prolonged flooding of soil helps in zoospore dispersal (Duniway, 1976). Many of the pathogenic host-dominant fungi infect roots by release of zoospores from sporangia.

Zoospore release occurs over a very narrow range of ψ_m near saturation. Indirect germination by release of zoospores may have a greater requirement for free water than does direct germination (Duniway, 1975a,c, 1979; Pfender *et al.*, 1977; Sugar, 1977; MacDonald and Duniway, 1978a,b).

Zoospore release from the sporangium of several species of *Phytophthora* can be divided into four intervals between swelling of the papilla: release of the second zoospore, rupture of the discharge vesicle, beginning of zoospore circling in the sporangium, and complete emptying of the sporangium (Gisi *et al.*, 1979; Gisi and Zentmyer, 1980). Most of the zoospore mass is forcibly expelled from the sporangium. The force of this cytoplasmic flow was attributed to changes in osmotic gradient and turgor pressure inside the sporangium (Gisi *et al.*, 1979).

2. *Germination and Mobility*

The effect of O_2 concentration on germination of zoospores of *Phytophthora parasitica, P. citrophthora,* and endoconidia of *Thielaviopsis basicola* was studied by Klotz *et al.* (1962, 1965). Liquid films around the zoospores were kept thin so O_2 diffusion for respiration would be high. The three fungi responded differently to O_2 concentration by failure to germinate or by varying germ-tube lengths. These studies indicate the importance of O_2 supply on distribution of root-rotting fungi and, indirectly, on their parasitism of citrus roots. Zoospores of both *P. parasitica* and *P. citrophthora* germinated at very low O_2 concentrations, as would be the case in flooded soils. Zoospores of *T. basicola* germinated at a much higher O_2 concentration, more nearly resembling soil conditions at FC.

Phytophthora cinnamomi has been isolated from a wide range of soils and roots of plants (Hwang and Ko, 1978). The fungus produces sporangia on root surfaces, and chlamydospores in root tissues. Chlamydospores and oospores are believed to be the primary survival propagules of *P. parasitica* in soils. Tsao and Bricker (1968) demonstrated that a higher percentage of chlamydospores germinated in moist, nonsterile soils than in dry soils. Feld (1982) showed that chlamydospores of *P. parasitica* germinated more readily in soil with ψ_m between 0 and -50 mbars than in soil with ψ_m between -100 and -700 mbars. Germination of chlamydospores of *P. cinnamomi* in sandy loam soil was significantly lower at soil ψ_m of -250 mbars than at 0 to -10 mbars (Sterne *et al.*, 1977). However, when glucose and asparagine were added, germination increased in the drier soils.

Duniway (1976) determined that 100% of *Phytophthora cryptogea* sporangia were expelled within 1 day after the soil was saturated. MacDonald and Duniway (1978a) found that maximum germination of *P. megasperma* and *P. cryptogea* sporangia occurred when they were wetted to 0 mbar, with much less germination at -10 mbars and none at -25 mbars ψ_m. Zoospore release began ~1 hr after the soil was saturated and was nearly completed within 4 hr.

Several factors influence germination of oospores of *Phytophthora* spp., including light, sterols, temperature, maturity of the oospores, and the genetically controlled capacity of particular isolates to germinate (Zentmyer and Ervin, 1970). Light enhances germination of oospores of most *Phytophthora* spp. Soaking oospores of *P. megasperma* f. sp. *glycinea* in water for 48 hr increased their subsequent germination (Jimenez and Lockwood, 1981). Oospores germinated in flooded soil smears, soybean root exudates, and other substrates. Germination in natural or autoclaved soil was 50–60% compared with 10% in deionized water. Sporangia produced in unsterilized field soil seldom germinated.

B. Nematodes

1. Survival and Reproduction

Flooding soil for long periods has been used in attempting to control nematodes in field soils. This method, however, has usually been unsuccessful. McElroy (1967) observed that field populations of *Hemicycliophora arenaria* Raski were reduced in proportion to the frequency and duration of irrigation and postulated that reduced aeration was the primary factor. Van Gundy *et al.* (1968) investigated ODR's in a flood-irrigated citrus orchard on sandy loam soil. Only a trace amount of dissolved O_2 was present to a 61-cm depth immediately following irrigation. After 12 hr, O_2 had diffused to a depth of 15 cm, but 7 days were required for restoration of normal ODR's to the entire soil depth. A profile of the ODR's following irrigation was determined (Fig. 5). Even short-duration irrigations slowed nematode reproduction. In this study, O_2 supply dropped to low levels during flood irrigation and gradually rose following water application, requiring 7–9 days for normal O_2 levels to return to a depth of 61 cm. Therefore, a large part of the soil profile, between 15 and 61 cm, experienced microaerobic conditions ($<5\%$ O_2) for several days. When irrigation frequency increased during summer months, the duration of adequate soil aeration periods for nematodes decreased, and at depths of 30–61 cm, nematodes were continuously exposed to low levels of O_2. Continuous aeration of the surface soil horizon, however, could maintain nematode populations. Such is not the case in many agricultural soils, however,

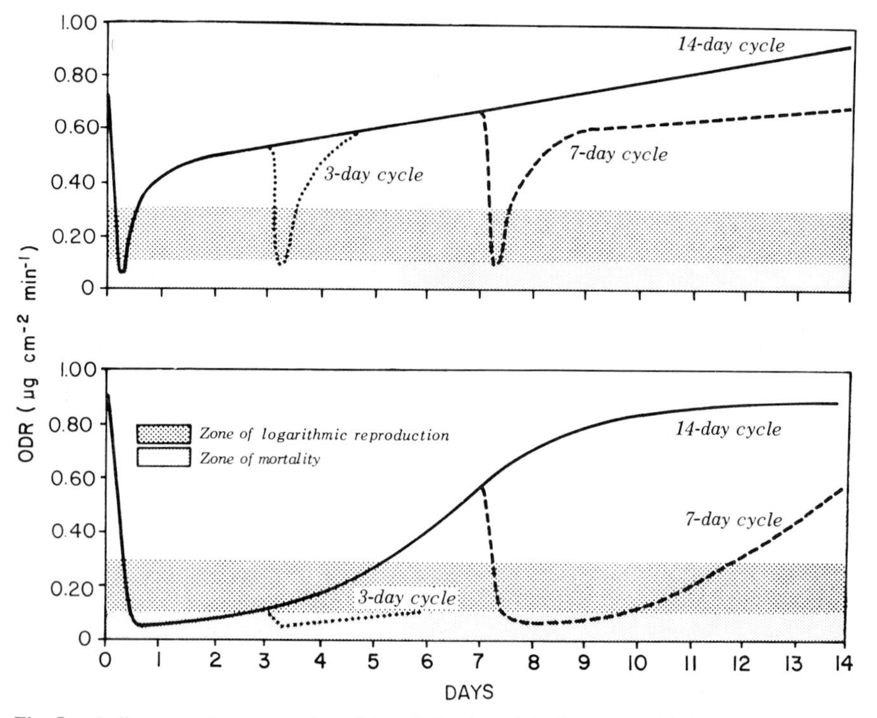

Fig. 5. A diagrammatic presentation of the relationship of the frequency of irrigation of ODR and to nematode mortality and reproduction at 12-in. (top) and 24-in. (bottom) depths. From Van Gundy *et al.* (1968). Copyright 1968 The Williams and Wilkins Co., Baltimore.

because tillage, partial drying, and high temperature in the surface horizon are not conducive to nematode growth and reproduction.

Oxygen in the soil atmosphere depends on the rate of diffusion in the gaseous phase. The most important factor controlling gaseous diffusion is the soil ψ_m. Furthermore, O_2 diffusion into water-saturated soil aggregates is not controlled by gas-filled macropore spaces, but by the water-filled micropores of the aggregates (Stolzy *et al.*, 1981). Because of their size, egg sacs and larvae of nematodes are found in these aggregates and are subjected to wide fluctuations in aeration. Both the hatch and mobility of *Meloidogyne javanica* (Treub) Chitwood are low when the pore spaces are filled with water (Wallace 1968b). Van Gundy and Stolzy (1961) showed that the lowest O_2 concentration that allowed development of the host and the nematode was 3.5%. Also, there was a linear relationship between movement of *M. javanica* larvae and the rate of O_2 diffusion in a porous medium (Van Gundy and Stolzy, 1963).

Cooper *et al.* (1970) pointed out that the study of nematodes in controlled environments involves constant conditions that are easily controlled (Overgaard-Nielson, 1949; Feder and Feldmesser, 1955; Wallace, 1956; Feldmesser *et al.*, 1958; Fairbairn, 1960; Nicholas and Jantunen, 1964, 1966). Their data indicate that this general group of nematodes consists of facultative anaerobes; growth and reproduction are dependent on a sufficient supply of O_2, but they can survive anaerobic conditions for varying periods. Soil-inhabiting nematodes are adapted to the changes in soil aeration. However, aeration increases in importance for nematode survival under irrigation and in high-rainfall areas.

Reproduction of a wide variety of soil-inhabiting nematodes is severely reduced in continuously microaerobic environments (Fairbairn, 1960; Nicholas and Jantunen, 1966; Bryant *et al.*, 1967; Saz, 1969). The rate of reproduction of *Aphelenchus avenae* Bastian, *Hemicycliophora arenaria,* and *Caenorhabditis* sp. was significantly decreased in a continuous environment of 5% O_2 and was inhibited at 4% (Cooper *et al.*, 1970). The physiological processes essential for reproduction and growth are apparently aerobic. Many researchers have assumed that short interruptions of O_2 are of little consequence to nematodes. However, Cooper *et al.* (1970) found that short-interval fluctuations between high and low O_2 can also be highly disruptive to nematode survival and reproduction. The closer the interval between the high and low O_2 concentrations, the greater influence on population number. Conversely, the longer the interval, the less the influence. Van Gundy *et al.* (1968) also showed that longer intervals between applications of irrigation water maintained larger nematode populations. Two of the nematodes (*A. avenae* and *Caenorhabditis* sp.) are capable of performing anaerobic glycolysis in microaerobic ($<5\%$ O_2) and anaerobic environments and easily survive short exposures to low O_2 levels (Cooper and Van Gundy, 1970). The shunting back and forth between oxidative and fermentative metabolism prevents nematodes from developing an adequate capacity for continuous ana-

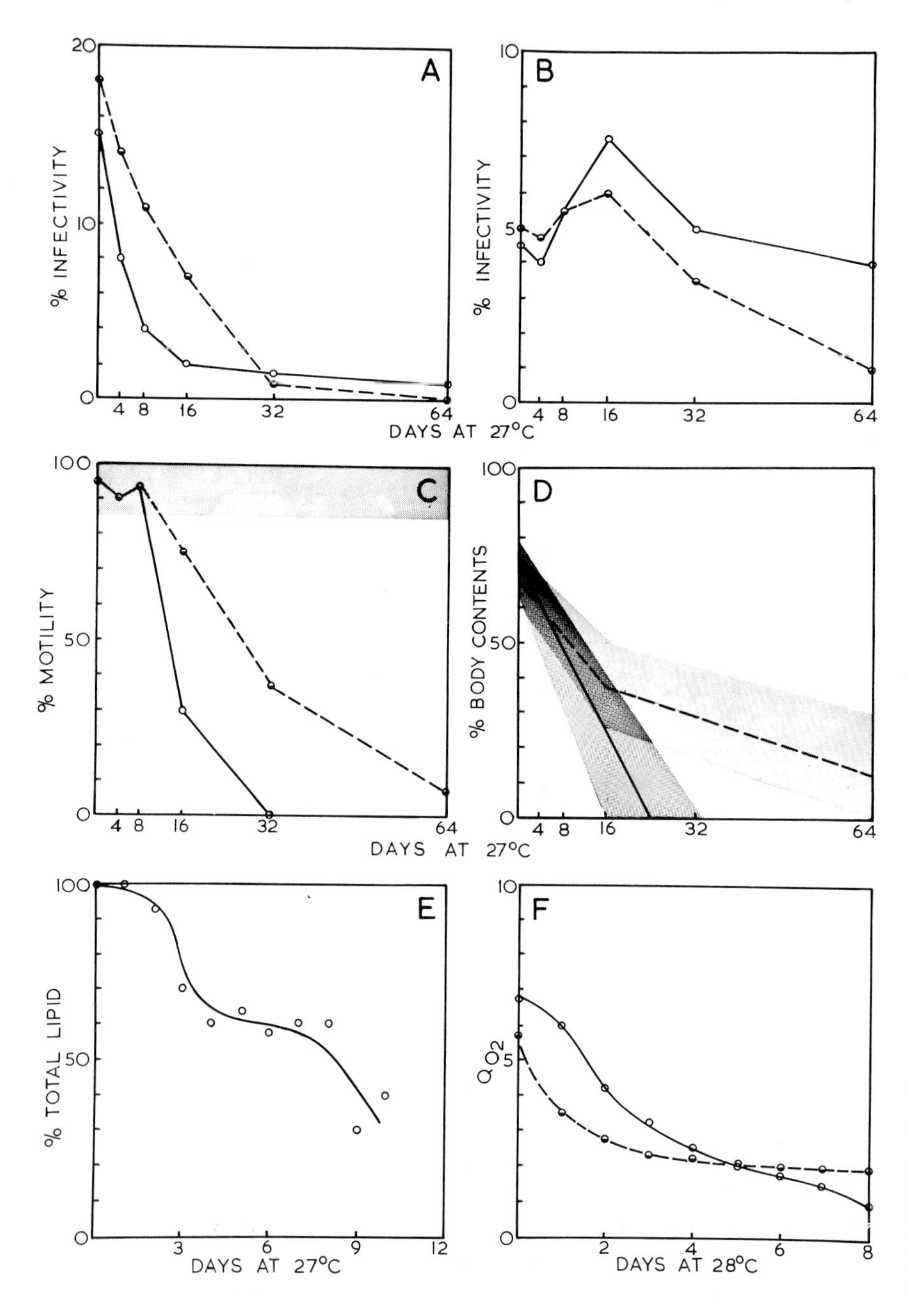

Fig. 6. (A) Mean percentages of *Meliodogyne javanica* larvae developing into adults on tomato and citrus seedlings, after being stored at 27°C *in vitro* (- - -) and in soil (——). (B) Mean percentages of *Tylenchulus semipenetrans* larvae developing into adults on citrus seedlings after being stored at 27°C

bolism and interferes with such processes as lipid metabolism in egg production, egg hatch, etc. Van Gundy and Stolzy (1961, 1963) observed that females of *H. arenaria, M. javanica,* and *Tylenchulus semipenetrans* Cobb did not lay as many eggs in continuously microaerobic environments as they did in continuously aerobic ones. Le Jambre and Whitlock (1967) reported similar results for some animal-parasitic nematodes. Behavioral responses to physiological stimuli caused by fluctuating environments affect the egg-laying stage in the life cycle.

During the first 12–16 hr of exposure to microaerobic and anaerobic environments, the principal glycolytic end product of *Aphelenchus avenae* and *Caenorhabditis* spp. was lactic acid. After 16 hr it was ethanol (Cooper and Van Gundy, 1971). On return to aerobiosis, [^{14}C]ethanol in the medium was utilized by the nematodes.

Wallace (1968b) showed a linear relationship between O_2 concentration and hatch rate of *Meloidogyne javanica*. Hatching did not occur in the absence of O_2. After 2 days without O_2, there was a marked decrease in hatch because of susceptibility of enbryonated eggs to anaerobic conditions. Wallace concluded that low O_2 resulting from waterlogging or soil depth may have a contrasting dual effect; it may kill embryos, but it may also maintain infectivity in larvae by inducing quiescence. Van Gundy *et al.* (1967) showed that a decrease in infectivity was associated with a corresponding decrease in mobility and body contents of second-stage larvae of *M. javanica* and *Tylenchulus semipenetrans* that were aged and starved in soil (Fig. 6). Body contents were consumed rapidly at high temperatures in dry soils and in oxygenated solutions. Conversely, body contents were conserved and motility and infectivity increased at low temperatures in wet soils and in low-O_2 solutions. Their observations explain why flooding of soils does not eradicate nematodes. Flooding in most agricultural soils does not produce completely anaerobic environments.

2. *Mobility*

Wallace (1968a) reviewed how nematodes, particularly plant-parasitic species, move in the various environments encountered during their life cycle. In most cases, nematodes move by undulatory propulsion, in which a train of dorsoventral waves passes from the head to tail. Because of their morphological

---

in vitro (- - -) and in soil (——). (C–F) ——, *Meliodogyne javanica;* - - -, *Tylenchulus semipenetrans.* (C) Mean percentages of larvae stored *in vitro* that migrated through a 1-cm sand column after 3 hr. Values within the shaded area are not statistically different ($p = 1\%$). (D) Mean percentages of body contents of larvae stored *in vitro*. The maximum variation of individuals in four experiments is represented by the shaded area for each nematode. The LSD between means is 3.37%. (E) Percentage loss of total lipid in *M. javanica* larvae when aged in shaking Warburg flasks at 27°C. (F) Mean respiration rates ($\mu l\ O_2\ mg^{-1}\ hr^{-1}$) of nematodes aged in a shaking Warburg flask at 28°C. From Van Gundy *et al.* (1967).

simplicity and lack of appendages, nematodes exert a force against external objects in *one* direction by undulatory propulsion. This mechanism is similar in all environments. The speed of larval migration and nematode hatch rate have the same relation to ψ_m (Collis-George and Wallace, 1968). In saturated or dry soil, the rate of hatch is low and migration of larvae low or negligible. When soil pores are filled with water, the nematodes are not attracted to soil particles because of water-film thickness. In dry soil, water films are thin and nematodes are tightly restricted to a small volume near the soil particles. As soil water content decreases from saturation to FC, the soil environment is most conducive to egg hatch, rapid migration of nematodes, and quick invasion of a host root.

VI. DISSEMINATION OF PATHOGENS BY FLOODING

A. Lateral Dispersal on the Soil Surface

Phytophthora citrophthora, P. parasitica, P. syringae Kleb, *P. hibernalis* Carne, and *P. megasperma* are pathogenic either to above- or below-ground tissues of commercially grown citrus in California. Klotz *et al.* (1959) extensively surveyed various surface water sources throughout the state over a year-long period and found that zoospores of *Phytophthora* spp. were effectively dispersed, both actively and passively, in all surface waters. Temperature and season affected the species prevalent at the time of sampling.

Southern Georgia is a major producer of certified vegetable and ornamental transplants for shipment to other areas. Fumigants are used to control soilborne plant pathogens (Shokes and McCarter, 1979). However, reinfestation of fumigated fields has limited the success of the control program. Transplants are irrigated with sprinkler systems drawing water from ponds that receive runoff from surrounding fields. Shokes and McCarter (1979) listed numerous studies conducted in many areas of the United States where pathogens were present in irrigation waters. Few of these studies show how these pathogens are disseminated or what their survival rate is in pond waters. Ponds were sampled regularly in southern Georgia for plant pathogens and at different depths and locations in the ponds as well as from filtered debris and bottom sediments. Plant-pathogenic fungi from samples included more than 20 species of *Pythium, Phytophthora, Fusarium,* and *Rhizoctonia.* One sample taken by filtering water from an irrigation line in the field yielded 5 *Pythium* species, 1 *Phytophthora* species, and 14 lance nematodes. Oospores of *Pythium aphanidermatum* (Edson) Fitzp. were recoverable for 185 days after submersion in a pond, whereas zoospores were not recovered after 12 days. Frequent sprinkler irrigation increased the incidence of avocado root rot (Zentmyer and Richards, 1952).

Hickman and English (1951) found that saturated soils with moving water were necessary to move zoospores of *Phytophthora fragariae* Hickman to the host, whereas excessive water in soybean [*Glycine max* (L.) Merr.] fields increased recovery of *P. megasperma* var. *sojae*, the amount of zoospore inoculum in runoff water, and incidence of disease (Kein, 1959). McIntosh (1964, 1966) detected *Phytophthora* spp. parasitic to apples (*Pyrus malus* L.) in irrigation waters.

B. Vertical Movement in the Soil

The movement of fungal zoospores vertically through soil profiles depends on soil-water conditions. Zoospores of *Phytophthora* spp. swim in a helical pattern (Ho and Hickman, 1967a,b; Hickman, 1970; Allen and Newhook, 1973). Stolzy *et al.* (1965b) calculated that soil pores with diameters between 40 and 60 μm are necessary to accommodate this type of motion of the zoospores with flagella extended. Water-filled pores of smaller diameter would lead to spore encystment when zoospores collided with soil particles. Soil pores of ~300 μm in diameter drain water at soil ψ_m of -10 mbars (Cook and Papendick, 1972; Griffin, 1972, Duniway, 1976). Chemotaxic and passive zoospore movements in soil generally occur only when soil ψ_m is higher than -10 mbars. Zoospore movement apparently requires continuous water-filled channels with a minimum of tortuosity. Movement of *P. cryptogea* zoospores in a coarse-textured soil mix was reduced at -100 mbars ψ_m, and in a finer loam, movement was reduced at -10 mbars. Duniway (1976, 1979) suggested that water-filled channels greater than 60 μm in diameter are required for prolonged zoospore movement. Duniway showed that zoospores readily swam 25–35 mm in the surface water over flooded soils or through coarse-textured soils with ψ_m of -0.5 mbar (Fig. 7).

Phytophthora megasperma zoospores and cysts and *Serratia marcescens* Bizio cells were infiltered from the outflow of horizontal soil columns during the establishment of a gradient of ψ_m (Wilkinson *et al.* 1981). In soil columns wetted a distance of 65 cm, zoospores moved 35 cm behind the wetting front in sand, 44 cm in sandy clay loam, 48 cm in loam, and did not move in silty loam soil. The soil ψ_m at the boundary between infested and noninfested soils ranged from -14 to -18 mbars. Zoospore cysts moved only half the distance that motile zoospores moved, whereas *S. marcescens* cells were recovered close behind the wetting front. The data of Wilkinson *et al.* indicate that infiltration of propugales occurred after soil pores with radii considerably larger than a priori estimates of the limiting pore radii were filled with water.

Pfender *et al.* (1977) found that zoospores originating from sporangia produced at various depths in two flooded soils were detected in surface water. Zoospores migrated upward through 65 mm of a sandy loam soil, but rarely moved 24 mm upward through a silt loam soil. In the flooded silt loam, the probability that zoospores would reach the ponded water at the soil surface depended on the number of sporangia and their depth in the soil.

The rate of active movement of five isolates of soil bacteria was measured in natural and artificial soils at a soil ψ_m of -50 and -150 mbars (Wong and

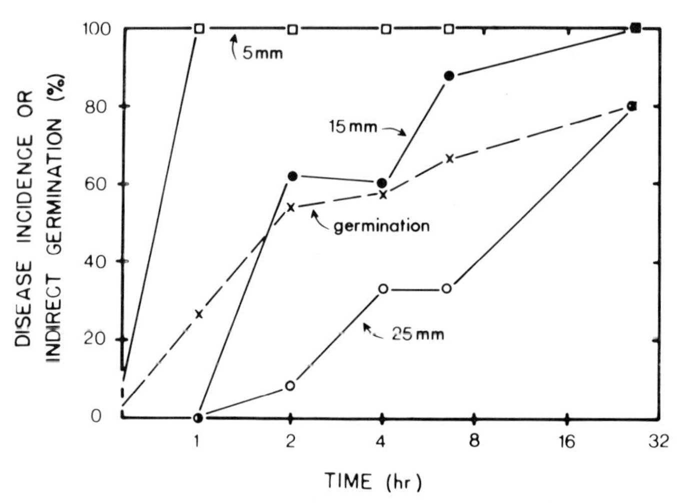

Fig. 7. Influence of the distance from sporangia on the rate at which *Phytophthora cryptogea* infected susceptible safflower seedlings when a coarse-textured soil mix was saturated. The soil was saturated to $\psi_m = 0$ without flooding at time zero, and tension plates were periodically used to drain the soil and stop indirect germination and zoospore movement. From Duniway (1983).

Griffin, 1976). At the same soil ψ_m, all organisms moved faster in the artificial soil was attributed to fewer surface charges to which bacteria became adsorbed. The studies concluded that bacterial movement is restricted in soil drier than FC. Generally, the distance of movement of microbial cells in the soil depends on adsorption of the cells on the soil solid phase, on water content, and on rate of water flow (Bitton *et al.*, 1974). Britton *et al.* found that passive infiltration of bacterial cells into dry soil columns was affected by soil type and that their upward movement stopped when the water content was at or below FC.

VII. ROOT–PATHOGEN INTERACTIONS

A. Root Attraction

The "rhizosphere effect" (Fig. 8) refers to the chemical or physical influence of living roots on microbial activity and the results of this activity on plant health and vigor (Curl, 1982). Root exudates are the primary contributing factor. Studies have shown that components of root exudates directly affect propagule germination, mycelial growth, and reproduction of pathogens.

Zoospores of the plant pathogenic fungi *Pythium* and *Phytophthora* are attracted to roots and to exudates from roots. Zoospore accumulation on roots is influenced by attraction, trapping, rapid encystment and germination, morphogenetic origin of zoospores, and directed growth of germ tubes (Hickman and Ho, 1966).

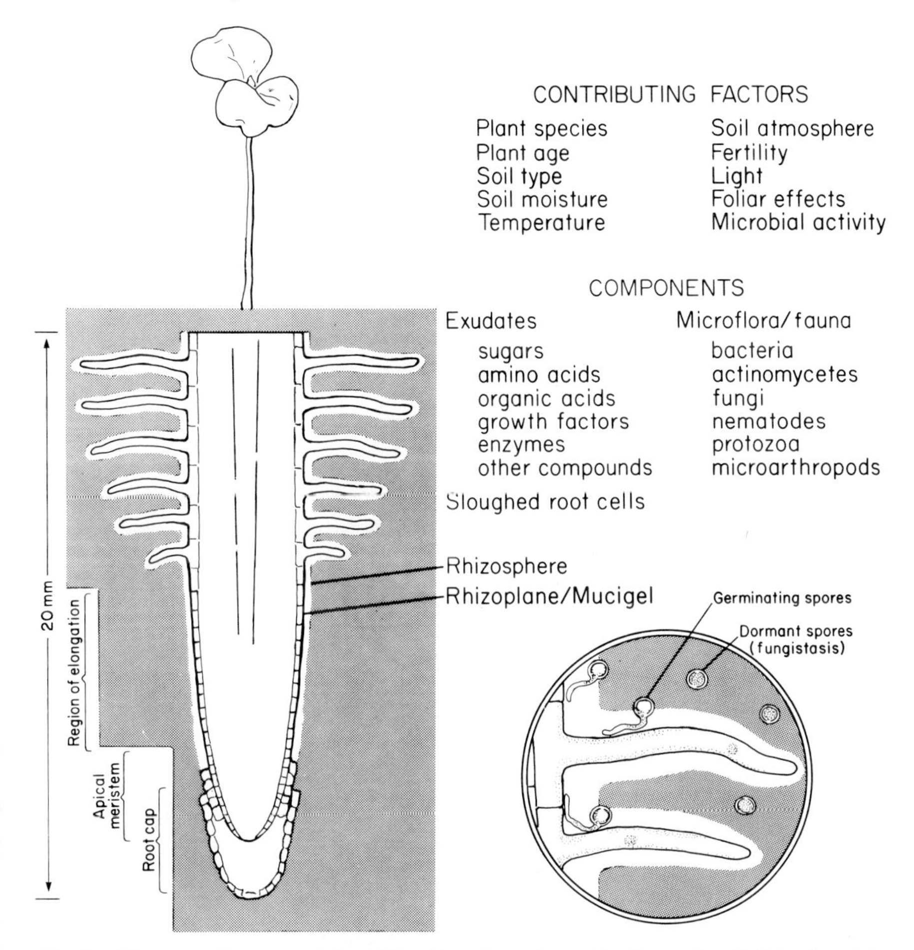

Fig. 8. Diagrammatic representation of the rhizosphere of a small cotton root, with lists of exudate and microbial components and influencing factors. Fungal spores within the rhizosphere (round inset) are stimulated to germinate. From Curl (1982).

According to Cameron and Carlisle (1978), zoospore attraction (positive chemotaxis) to amino acids, which occur in root exudates, and to ethanol produced by roots in waterlogged soils has been demonstrated. However, no single specific substance appears to be as strong an attractant as actual root exudate. Cameron and Carlisle (1978) emphasized that attractiveness of root exudates must be explained by additive or multiplicative effects between individual attractants or by powerful attractants that have not yet been detected, or both. Zentmyer (1961) confirmed that *Phytophthora palmivora* zoospores were attracted to certain amino acids and responded, as did *P. cinnamomi* zoospores, to ethanol. Allen and Newhook (1973) found that *P. cinnamomi* showed chemotaxis to solutions of methanol, *n*-propanol, *n*-butanol, and acetaldehyde. Cameron and Carlisle (1978), however, found that *P. palmivora* was attracted only to acetylaldehyde in the aforementioned chemical solutions. Several alcohols, aldehydes, and fatty acids also served as attractants, with thresholds considerably lower than those for ethanol. The most potent attractants had 4–6 carbon atoms, and iso compounds tended to be more effective than those with straight chains. Zoospores of *P. palmivora* and *P. cinnamomi* were repelled by H^+ and other monovalent ions.

Kuan and Erwin (1980) demonstrated that roots of plants grown in saturated soil attracted many more zoospores than did roots from plants in unsaturated soil. Zoospores were chemotactically attracted preferentially to the zone of cell elongation and to tissue from which secondary roots emerged (Fig. 9). More sugars and amino acids were found in exudate from roots grown in saturated soils than in exudates of roots grown in unsaturated soil. Zentmyer (1966) concluded that zoospores of *Phytophthora cinnamomi* were attracted to glutamic acid and aspartic acid and those of *P. palmivora* and *P. citrophthora* to glutamic acid, aspartic acid, asparagine, glycine, methionine, histidine, glutaric acid, and to several sugars.

The region of cell elongation of roots appears to produce more exudates and to attract more zoospores than other root tissues. Chi and Sabo (1978) showed that zoospores were strongly attracted to the region of cell elongation, immediately above the root cap or older regions of the roots, and none were attracted to root hairs in their study. This study also showed that susceptible seedlings of alfalfa roots pretreated in boiling water did not attract zoospores. Wounded parts of roots displayed a strong preferential attraction for zoospores. These responses were attributed to chemotaxis associated with chemicals in root exudates.

Zoospores of five species of *Phytophthora* responded positively to a wide range of chemicals, including vitamins, phenolic compounds, nitrogenous bases of nucleic acids, nucleotides, growth regulators, sugars, organic acids, and amino acids (Khew and Zentmyer, 1973); a distinct directionally oriented attraction to amino acids was noted. Several chemicals caused accumulation of zoospores by trapping and immobilization without evoking a directional movement of zoospores. Zoospores were more strongly attracted to positively charged molecules than to negatively charged ones. Ionic structure of the amino acid molecule is important in determining its chemotactic activity. A response to a possible concentration gradient of some diffusing stimulatory chemical in root exudates was demontrated by zoospores that had encysted at varying distances from avocado roots. Positive chemotropism of germ tubes for avocado roots was reported when zoospores settled on the bottom of the petri dish at distances of 2–3 mm from the root and then grew unidirectionally toward the root. However, ethanol in isotropic solutions did not affect the average speed, velocity, amplitude, frequency of the helical path, or rate of change of direction of zoospores of *P.*

Fig. 9. Attraction of zoospores of *Phytophthora megasperma* f. sp. *medicaginis* to alfalfa roots grown in saturated (S) and nonsaturated (NS) soil. Area between arrows is the zone of encysted zoospores. From Kuan and Erwin (1980).

cinnamomi. When moving through concentration gradients of ethanol, however, zoospores turned, moved to the higher concentration, and accumulated in numbers (Allen and Newhook, 1973). In the absence of ethanol, zoospores entering capillaries were disoriented, but zoospores germinated in gradients of ethanol had their germ tubes directed toward the source. Ethanol production by plant roots during brief periods of low soil aeration after flooding attracted zoospores to its site of exudation.

Zoospores of two *Pythium* species were attracted to the region of cell elongation of roots of cotton (*Gossypium hirsutum* L.) seedlings as well as to root-hair zones (Spencer and Cooper, 1967). Wounds in radicles greatly intensified the attraction.

Bird's (1962) experiments on orientation of *Meloidogyne javanica* larvae toward roots supported the root-attraction hypothesis. Bird concluded that attraction of larvae to roots is the result of several stimuli and not merely to CO_2 in the vicinity of roots, as had been suggested by several other investigators.

B. Root Infection by Zoospores

Zoospores of *Phytophthora megasperma* var. *sojae* were attracted to the root and then trapped in the vicinity of the root surface, after which they encysted in less than 1 hr (Ho and Hickman, 1967a,b; Chi and Sabo, 1978). The encysted zoospores formed a continuous sheath around the root behind the tip, and then germinated. Germ tubes were initiated on the cysts closest to the root, and all showed unidirectional growth toward the root. *Pythium* species penetrate and infect cotton-seedling radicles by germ tubes from single zoospores. These zoospores, as well as single mycelial tips, were attracted in aqueous media to plant roots (Spencer and Cooper, 1967). Infection of root radicles by a germ tube from single zoospores occurred within 2 hr and by single mycelial tips after the much longer time of 12 hr.

Studies by Klotz *et al.* (1962, 1963) on germination and germ-tube growth of zoospores of *Phytophthora* spp. and *Thielaviopsis basicola* showed that after swarming, *Phytophthora* spp. took only 1.5 hr to germinate even at low O_2 levels, whereas *T. basicola* took 7 hr at much higher O_2 concentrations.

In flooded fields, zoospores are generally the initial inoculant, as they are carried by moving water to plants in noninfested areas. Severity of root rot of avocado caused by *Phytophthora cinnamomi* increased when the water content of soil remained near saturation because of overirrigation or poor drainage (Zentmyer *et al.*, 1967). Disease severity on peach trees induced by *Pythium vexans* d. By. was correlated with an increase in zoospore production as a result of periodic excess water (Biesbrock and Hendrix, 1970). Encysted zoospores of *Phytophthora megasperma* on alfalfa roots germinated, and the germ tubes grew toward breaks in the roots caused by saturated soil conditions (Kuan and Irwin, 1980). Bacteria as well as germinated zoospores occupied the breaks on roots grown in saturated soil.

Root pathogens may damage the host primarily by decreasing effective root density and distribution of roots in the soil, thereby lessening the capacity of plants to extract water and nutrients from soil (Duniway, 1977a). The interrelationships of root-rotting fungi, soil O_2, and saturated soil were discussed by Stolzy *et al.* (1960, 1965b). Saturated soils limit soil aeration. Many orchards are saturated to different depths for a period of time after an irrigation. Root survival in a saturated zone depends on the drainage characteristics of the soil profile. If water moves out of the profile rapidly and O_2 is soon available, growth of older roots and regeneration of new roots will start soon after irrigation stops. If the water drains away slowly, root density will decrease with time. Good irrigation practices on poorly drained soils are more critical than on well-drained soils. The amount of irrigation water added at any time should be based on the amount needed to rewet the dry surface soil without saturating the subsoil. Orchards on coarse-textured soils, such as loamy sands and sands, lack feeder roots in the surface 1 m of profile between tree rows where water is applied. Water infiltration into and through the soil profiles is rapid because of large continuous pore spaces. The absence of feeder roots is the result of parasitism by zoospores produced at the soil surface that have moved to root surfaces in water. The running water at the soil surface is ideal for production of zoospores. The range of soil ψ_m that would drain soil pores between 40 and 60 μm would be -24 to -36 mbars, potentials achieved in these soils during furrow irrigation.

Feld (1982) conducted furrow- and drip-irrigation studies under controlled conditions in a lath house and greenhouse. Citrus root rot caused by *Phytophthora parasitica* was favored by overirrigated soil under furrow irrigation when soil ψ_m was between -0 and -150 mbars, and consistently moist soil under drip irrigation between -50 and -300 mbars. Very few new feeder root tips were produced by roots growing in soil under these two irrigation treatments. Feld (1982) showed that where citrus was grown under a furrow-irrigation system, and the soil was allowed to dry sufficiently between irrigations (to -700 mbars), new feeder roots were continually produced and plants were healthier than those allowed to dry to only -150 mbars or kept wet by drip irrigation.

Stolzy *et al.* (1959) flood-irrigated citrus seedlings growing in loamy fine sand contained in large concrete tiles with and without *Phytophthora* spp. Irrigation treatments consisted of watering the entire 1-m soil columns at values of soil ψ_m of -90 or -600 mbars. The amount of water added at each irrigation lowered the ψ_m at the bottom of the column to -20 mbars. Saturation of soil occurred only at the top of the soil column during irrigations. The -90-mbar treatment produced more total plant growth during 2 years, in both the presence and absence of the fungi, than did the dry treatments. However, presence of *Phytophthora* significantly reduced shoot and root growth. This and other experiments demonstrated that by proper management of irrigation waters, to prevent soil saturation in the profile, plants in hot, dry environments yielded more when watered at a high ψ_m (Stolzy *et al.*, 1959, 1960, 1965b; Lombard *et al.*, 1965).

VIII. PLANT WATER STRESS AS A PREDISPOSING FACTOR

Variations in precipitation and irrigation frequency differentially influence water balance in plants, often causing many adverse effects. One such effect is an increase in susceptibility of plants to attack by plant pathogens (Schoeneweiss, 1978). When the susceptibility of the plant is altered before an infection, the plant is predisposed to disease. The most common method of exposing plants to water stress is by varying watering regimes. For example, drought may enhance disease development in the plant following irrigation because of water-stress predisposition. Duniway (1977b) withheld water from safflower plants until leaf ψ was -13 or -17 bars before inoculation with zoospores of *Phytophthora cryptogea*. This treatment increased root rot by the fungi when compared to plants watered daily when leaf ψ was between -4 and -6 bars. With the exception of saturated soils, various soil water regimes after inoculation were almost equally suitable for development of severe root rot. Depending on methods used, saturated soil was more or less suitable than drier soil for disease development. Also, Zimmer and Urie (1967) demonstrated that irrigation frequency and intensity influenced root rot of safflower varieties grown in infected fields. Some varieties were killed by all irrigation treatments, whereas others were only slightly or moderately affected under frequent light irrigations and severely affected under prolonged irrigation. Water stress prior to irrigation increased the incidence of root rot.

Severity of alfalfa root rot in the seeding year increased through a combination of stress factors, including high soil-water levels, frequent cutting, high seeding rates, and late spring seeding (Pulli and Tesar, 1975).

The rhododendron cultivar Caroline was relatively resistant to crown rot of *Phytophthora cinnamomi* (Blaker and MacDonald, 1981). However, if plants were exposed to drought until leaf ψ dropped to -16 bars, or if their roots were flooded for 48 hr before inoculation with fungi, they developed severe symptoms of root and crown rot. Blaker and MacDonald's data showed that soil water extremes that commonly occur in nursery or landscape plantings can predispose normally resistant rhododendrons to root and crown rot. An understanding of how the host resists attack by a pathogen is necessary to determine how water stress causes a plant to become predisposed to disease (Schoeneweiss, 1978).

IX. CONTROL OF DISEASE BY FLOODING

Mulder (1979) and Schippers and Gams (1979) discussed soil management for control of soilborne pathogens. Their reviews indicate that nearly all aspects of soil management and soil-intrinsic properties may have a measurable impact on pathogen virulence. Such properties include soil texture, mineralogy, organic and biological composition, nutrient availability and other chemical effects, crop rotation, tillage practices, crop-residue management, soil temperature, soil ODR, and flooding. Infection and development of disease in plants can be reduced by changes in soil resulting from waterlogging. The waterlogged soil environment is more favorable to microorganisms anatagonistic to plant pathogens than at FC (Drew and Lynch, 1980). Stover (1979) reviewed the flooding of soil for disease control.

Ioannou *et al.* (1977a,c) studied production of microsclerotia in tomato [*Lycopersicon esculentum* (L.) Mill.] tissues infested with *Verticillium dahliae* Kleb. in soil subjected to different water treatments under field conditions. Equal numbers of microsclerotia were produced in dry or one-irrigation treatments, but none were produced during flooding. The inhibition of microsclerotia production was due in large part to decreased O_2 and increased CO_2 concentrations in the flooded soil. In another study, Ioannou *et al.* (1977b) showed that production of microsclerotia was completely inhibited at concentrations of 10% O_2 and 11% CO_2 in the soil.

A single rotation with paddy rice (*Oryza sativa* L.) controlled *Verticillium* wilt of cotton for 2–3 years and increased lint yield by an average of 31% over 3 years when compared to areas in which cotton was continuously grown (Pullman and DeVay, 1981). Four years after the single rice rotation, soil populations of *V. dahliae* were still much lower than the initial populations. Six weeks of soil flooding with or without rice was required before population densities began to decline. However, after 17 weeks of flooding in the presence of paddy rice, *V. dahliae* was not detectable, whereas with flooding alone 9% of the initial population was still present.

Fungi are favored by a relatively dry environment and bacteria by wet conditions (Cook and Papendick, 1972). Stover (1979) reported that only one species of bacteria, *Pseudomonas solanacearum,* on tobacco (*Nicotiana* and bananas (*Musa*) was controlled by flooding. The lower fungi are an exception and require more water for zoospore movement. Excessive soil water is often harmful to higher fungi, and waterlogging could eliminate some sensitive species (Mangenot and Diem, 1979). Stover (1979) pointed out that numerous practices used in long-term flooding helped control various pathogens. Because sufficient O_2 is available in the surface few millimeters of ponded or waterlogged soil to sustain organisms during flooding, the use of interflood deep plowing or postflood application of fungicide enhance control. Certain pathogens are more effectively controlled by alternating short-term flooding and drying. And, finally, because of the higher demand for O_2 by respiring microorganisms at high temperatures, pathogens can be better controlled by flooding when soil temperatures are high than when they are low. If crops are present during flooding, high soil temperatures during flooding will also cause O_2 stress to roots as well as to microorganisms.

REFERENCES

Allen, R. N., and Newhook, F. J. (1973). Chemotaxis of zoospores of *Phytophthora cinnamomi* to ethanol in capillaries of soil pore dimensions. *Trans. Br. Mycol. Soc.* **61,** 287–302.

Armstrong, W. (1978). Root aeration in the wetland conditions. *In* "Plant Life in Anaerobic Environments" (D. D. Hook and R. M. M. Crawford, eds.), pp. 269–297. Ann Arbor Sci. Publ., Ann Arbor, Michigan.

Armstrong, W. (1979). Aeration in higher plants. *Adv. Bot. Res.* **7,** 225–332.

Armstrong, W. (1981). The water relations of heathlands: general physiological effects of waterlogging. *In* "Heathlands and Related Shrublands" (R. L. Specht, ed.), pp. 111–121. Elsevier, New York.

Balandreau, J., and Knowles, R. (1978). The rhizosphere. *In* "Interactions between Non-Pathogenic Soil Microorganisms and Plants" (Y. R. Commerques and S. U. Krupa, eds.), pp. 243–268. Elsevier, New York.

Barber, D. A., and Lynch, J. M. (1977). Microbial growth in the rhizosphere. *Soil Biol. Biochem.* **9,** 305–308.

Benjamin, L. R., and Greenway, H. (1979). Effects of a range of O_2 concentrations on porosity of barley roots and on their sugar and protein concentrations. *Ann. Bot. (London)* **43,** 383–391.

Bernhardt, E. A., and Grogan, R. G. (1982). Effect of soil matric potential on the formation and indirect germination of sporangia of *Phytophthora parasitica, P. capiici,* and *P. cryptogea. Phytopathology* **72,** 507–511.

Bertani, A., Brambilla, I., and Menegus, F. (1981). Effect of anaerobiosis on carbohydrate content in rice roots. *Biochem. Physiol. Pflanz.* **176,** 835–840.

Biesbrock, J. A., and Hendrix, F. F. (1970). Influence of soil water and temperature on root necrosis of peach caused by *Pythium spp. Phytopathology* **60,** 880–882.

Bird, A. F. (1962). Orientation of the larvae of *Meliodognye javanica* relative to roots. *Nematologica* **8,** 275–287.

Bitton, G., Lahav, N., and Henis, Y. (1974). Movement and retention of *Klebsiella aerogenes* in soil columns. *Plant Soil* **40,** 373–380.

Blaker, N. S., and MacDonald, J. D. (1981). Predisposing effects of soil moisture extremes on the susceptibility of rhododendron to *Phytophthora* root and crown rot. *Phytopathology* **71,** 831–834.

Bohn, H. L., McNeal, B. L., and O'Connor, G A. (1979). Oxidation and reduction. *In* "Soil Chemistry" (H. L. Bohn, B. L. McNeal, and G. A. O'Connor, eds.), pp. 247–271. Wiley, New York.

Bose, T. K., Mukherjee, T. P., and Basu, R. N. (1977). Effects of ethylene and acetylene on the regeneration of adventitious roots. *Indian J. Plant Physiol.* **20,** 134–139.

Bouldin, D. R. (1968). Models for describing the diffusion of oxygen and other mobile constituents across the mud–water interface. *J. Ecol.* **56,** 77–87.

Boynton, D., and Reuther, W. (1938). A way of sampling soil gases in dense subsoils and some of its advantage and limitations. *Soil Sci. Soc. Am. Proc.* **3,** 37–42.

Bradford, K. J., and Dilley, D. R. (1978). Effects of root anaerobiosis on ethylene production, epinasty, and growth of tomato plants. *Plant Physiol.* **61,** 506–509.

Bradford, K. J., and Yang, S. F. (1981). Physiological responses of plants to waterlogging. *HortScience* **16,** 25–30.

Bryant, C., Nicholas, W. L., and Jantunen, R. (1967). Some aspects of the respiratory metabolism of *Caenorhabditis briggsae* (Rhabditidae). *Nematologica* **13,** 197–209.

Bushnell, J. (1953). Sensitivity of potatoes to soil porosity. *Ohio Agric. Exp. Stn. Res. Bull.* No. 726.

Cameron, J. N., and Carlisle, M. J. (1978). Fatty acids, aldehydes and alcohols as attractants for zoospores of *Phythophthora palmivora. Nature (London)* **271,** 448.

Campbell, E. O. (1981). The water relations of heathlands: morphological adaptations to waterlogging. *In* "Heathlands and Related Shrublands: Analytical Studies" (R. L. Sprecht, ed.), pp. 107–109. Elsevier, Amsterdam.

Cannell, R. Q. (1977). Soil aeration and compaction in relation to root growth and soil management. *Appl. Biol.* **2,** 1–86.

Chi, C. C., and Sabo, F. E. (1978). Chemotaxis of zoospores of *Phytophthora megasperma* to primary roots of alfalfa seedlings. *Can. J. Bot.* **56,** 795–800.

Collis-George, N., and Wallace, H. R. (1968). Supply of oxygen during hatching of the nematode *Meloidogyne javanica* under noncompetitive conditions. *Aust. J. Biol. Sci.* **21,** 21–35.

Cook, R. J., and Papendick, R. I. (1972). Influence of water potential of soils and plants on root disease. *Annu. Rev. Phytopathol.* **10,** 349–374.

Cooper, A. F., and Van Gundy, S. D. (1970). Metabolism of glycogen and neutral lipids by *Aphelenchus avenae* and *Caenorhabditis* sp. in aerobic, microaerobic, and anaerobic environments. *J. Nematol.* **2,** 305–315.

Cooper, A. F., and Van Gundy, S. D. (1971). Ethanol production and utilization by *Aphelenchus avenae* and *Caenorhabditis* sp. *J. Nematol.* **3,** 205–214.

Cooper, A. F., Van Gundy, S. D., and Stolzy, L. H. (1970). Nematode reproduction in environments in fluctuating aeration. *J. Nematol.* **2,** 182–188.

Curl, E. A. (1982). The rhizosphere: relation to pathogen behavior and root discasc. *Plant Dis.* **66,** 624–630.

Currie, J. A. (1962). The importance of aeration in providing the right conditions for plant growth. *J. Sci. Food Agric.* **7,** 380–385.

Currie, J. A. (1970). Movement of gases in soil respiration. *In* "Sorption and Transport Processes in Soils" (J. G. Gregory, ed.), pp. 152–171. Staples, Rochester, England.

Drew, M. C., and Lynch, J. M. (1980). Soil anerobiosis microorganisms and root function. *Annu. Rev. Phytopathol.* **18,** 37–66.

Drew, M. C., and Sisworo, E. J. (1979). The development of waterlogging damage in young barley plants in relation to plant nutrient status and changes in soil properties. *New Phytol.* **82,** 301–314.

Drew, M. C., Jackson, M. B., and Gifford, S. (1979). Ethylene-promoted adventitious rooting and development of cortical air spaces (aerenchyma) in roots may be adaptive responses to flooding in *Zea mays* L. *Planta* **14,** 83–88.

Drew, M. C., Chamel, A., Garrec, J. P., and Foucy, A. (1980). Cortical air spaces (aerenchyma) in roots of corn subjected to oxygen stress. *Plant Physiol.* **65,** 506–511.

Duniway, J. M. (1975a). Water relations in safflower during wilting induced by phytophthora root rot. *Phytopathology* **65,** 886–891.

Duniway, J. M. (1975b). Limiting influence of low water potential on the formation of sporangia by *Phytophthora drechsleri* in soil. *Phytopathology* **65,** 1089–1093.

Duniway, J. M. (1975c). Formation of sporangia by *Phytophthora drechsleri* in soil at high matric potentials. *Can. J. Bot.* **53,** 1270–1275.

Duniway, J. M. (1976). Movement of zoospores of *Phytophthora cryptogea* in soils of various textures and matric potentials. *Phytopathology* **66,** 877–882.

Duniway, J. M. (1977a). Changes in resistance to water transport in safflower during the development of *Phytophthora* root rot. *Phytopathology* **67,** 331–337.

Duniway, J. M. (1977b). Predisposing effect of water stress on the severity of *Phytophthora* root rot in safflower. *Phytopathology* **67,** 884–889.

Duniway, J. M. (1979). Water relations of water molds. *Annu. Rev. Phytopathol.* **17,** 431–460.

Duniway, J. M. (1983). Role of physical factors in the development of *Phytophthora* diseases. *In* "*Phytophthora:* Its Biology, Taxonomy, Ecology, and Pathology" (D. C. Erwin, S. Bartnicki-Garcia, and P. H. Tsao, eds.), pp. 175–187. Am. Phytopathol. Soc., St. Paul, Minnesota.

Durbin, R. D. (1978). Abiotic disease induced by unfavorable water relations. *In* "Water Deficits and Plant Growth" (T. T. Kozlowski, ed.), Vol. 5, pp. 101–117. Academic Press, New York.

Fairbairn, D. (1960). The physiology and biochemistry of nematodes. *In* "Nematology" (J. N. Sasser and W. R. Jenkins, eds.), pp. 267–296. Univ. of North Carolina Press, Chapel Hill.

Fawcett, H. D. (1936). "Citrus Diseases and Their Control." McGraw-Hill, New York.

Feder, W. A., and Feldmesser, J. W. (1955). Further studies of plant parasitic nematodes maintained in altered oxygen tension. *J. Parasitol.* **41,** 47.

Feld, S. J. (1982). "Studies on the Role of Irrigation and Soil Water Matric Potential on *Phytophthora parasitica* Root Rot of Citrus." Ph.D. Dissertation, Univ. of California, Riverside.

Feldmesser, J., Feder, W. A., and Rebois, R. W. (1958). Further effects of altered gas tension on plant parasitic nematodes. *J. Parasitol.* **44,** 31.

Flocker, W. J., Vomocil, J. A., and Howard, F. D. (1959). Some growth responses of tomatoes to soil compaction. *Soil Sci. Soc. Am. Proc.* **23,** 188–191.

Flühler, H., Stolzy, L. H., and Ardakani, M. S. (1976). A statistical approach to define soil aeration in respect to denitrification. *Soil Sci.* **122,** 115–123.

Gisi, U., and Zentmyer, G. A. (1980). Mechanism of zoospore release in *Phytophthora* and *Pythium. Exp. Mycol.* **4,** 362–377.

Gisi, U., Hemmes, D. E., and Zentmyer, G. A. (1979). Origin and significance of the discharge vesicle in *Phytophthora. Exp. Mycol.* **3,** 321–339.

Gisi, U., Zentmyer, G. A., and Klure, L. J. (1980). Production of sporangia by *Phytophthora cinnamomi* and *P. palmivora* in soils at different matric potentials. *Phytopathology* **70,** 301–306.

Grable, A. R. (1971). Effects of compaction on content and transmission of air in soils. *In* "Compaction of Agricultural Soils" (K. K. Barnes, W. M. Carleton, H. M. Taylor, T. I. Throckmorton, and G. E. VanderBerg, eds.), pp. 154-164. Am. Soc. Agric. Eng., St. Joseph, Michigan.

Greenwood, D. J. (1961). The effect of oxygen concentrations on decomposition of organic materials in soil. *Plant Soil* **14,** 360–376.

Greenwood, D. J. (1971). Soil aeration and plant growth. *Rep. Prog. Appl. Chem.* **55,** 423–431.

Greenwood, D. J. (1975). Measurement of soil aeration. *U.K. Min. Agric. Fish. Food Tech. Bull.* No. 29, pp. 261–272.

Griffin, D. M. (1972). "Ecology of Soil Fungi." Syracuse Univ. Press, Syracuse, New York.

Grineva, G. M. (1961). Excretion by plant root during brief periods of anaerobiosis. *Sov. Plant Physiol. (Engl. Transl.)* **8,** 549–552.

Grineva, G. M. (1963). Alcohol formation and excretion by plant roots under anaerobic conditions. *Sov. Plant Physiol. (Engl. Transl.)* **10,** 361–369.

Hale, M. G., Moore, L. D., and Griffin, G. J. (1978). Root exudates and exudation. *In* "Interactions between Non-Pathogenic Soil Microorganisms and Plants" (Y. R. Dommerques and S. V. Krupa, eds.), pp. 163–203. Elsevier, New York.

Hickman, C. J. (1970). Biology of *Phytophthora* zoospores. *Phytopathology* **60,** 1128–1135.

Hickman, C. J., and English, M. P. (1951). Factors influencing the development of red core in strawberries. *Trans. Br. Mycol. Soc.* **34,** 223–236.

Hickman, C. J., and Ho, H. H. (1966). Behavior of zoospores in plant-pathogenic Phycomycetes. *Annu. Rev. Phytopathol.* **4,** 195–220.

Hillel, D. (1980). "Fundamentals of Soil Physics." Academic Press, New York.

Ho, H. H., and Hickman, C. J. (1967a). Asexual reproduction and behavior of zoospores of *Phytophthora megasperma* var. *sojae*. *Can. J. Bot.* **45,** 1963–1981.

Ho, H. H., and Hickman, C. J. (1967b). Factors governing zoospore responses of *Phytophthora megasperma* var. *sojae* to plant roots. *Can. J. Bot.* **45,** 1983–1994.

Huck, M. G. (1970). Variation in taproot elongation rate as influenced by composition of the soil air. *Agron. J.* **62,** 815–818.

Hunt, P. G., Campbell, R. B., and Moreau, R. A. (1980). Factors affecting ethylene accumulation in a Norfolk sandy loam soil. *Soil Sci.* **129,** 22–27.

Hunt, P. G., Campbell, R. B., Sojka, R. E., and Parsons, J. E. (1981). Flooding-induced soil and plant ethylene accumulation and water status response of field-grown tobacco. *Plant Soil* **59,** 427–439.

Hunt, P. G., Matheny, T. A., Campbell, R. B., and Parsons, J. E. (1982). Ethylene accumulation in southeastern Coastal Plain soils: soil characteristics and oxidative–reductive involvement. *Commun. Soil Sci. Plant Anal.* **13,** 267–278.

Hwang, S. C., and Ko, W. H. (1978). Biology of chlamydospores, sporangia, and zoospores of *Phytophthora cinnamomi* in soil. *Phytopathology* **68,** 726–731.

Ioannou, N., Schneider, R. W., Grogan, R. G., and Duniway, J. M. (1977a). Effect of water potential and temperature on growth, sporulation, and production of microsclerotia by *Verticillium dahliae*. *Phytopathology* **67,** 637–644.

Ioannou, N., Schneider, R. W., and Grogan, R. G. (1977b). Effect of oxygen, carbon dioxide and ethylene on growth, sporulation, and production of microsclerotia by *Verticillium dahliae*. *Phytopathology* **67,** 645–650.

Ioannou, N., Schneider, R. W., and Grogan, R. G. (1977c). Effect of flooding on the soil gas composition and the production of microsclerotia by *Verticillium dahliae* in the field. *Phytopathology* **67,** 651–656.

Jensen, C. R., Luxmoore, R. J., Van Gundy, S. D., and Stolzy, L. H. (1969). Root air space measurements by a pycnometer method. *Agron. J.* **61,** 474–475.

Jiminez, B., and Lockwood, J. L. (1981). Germination of oospores of *Phytophthora megasperma* f. sp. *glycinea* in the presence of soil. *Phytopathology* **72,** 662–666.

Kawase, M. (1981). Anatomical and morphological adaptation of plants to waterlogging. *HortScience* **16,** 30–34.

Kawase, M., and Whitmoyer, R. E. (1980). Aerenchyma development in waterlogged plants. *Am. J. Bot.* **67,** 18–22.

Kein, H. H. (1959). Etiology of the *Phytophthora* disease of soybeans. *Phytopathology* **49,** 380–383.

Khew, K. L., and Zentmyer, G. A. (1973). Chemotactic response of zoospores of five species of *Phytophthora*. *Phytopathology* **63,** 1511–1517.

Klepper, E. L., and Huck, M. G. (1969). Time-lapse photography of root growth. Auburn T.V.

Klotz, L. J., Wong, P. P., and DeWolfe, T. A. (1959). Survey of irrigation water for the presence of *Phytophthora* spp. pathogenic to citrus. *Plant Dis. Rep.* **43,** 830–832.

Klotz, L. J., Stolzy, L. H., and DeWolfe, T. A. (1962). A method for determining the oxygen requirement of fungi in liquid media. *Plant Dis. Rep.* **46,** 606–608.

Klotz, L. J., Stolzy, L. H., and DeWolfe, T. A. (1963). Oxygen requirement of three root-rotting fungi in liquid medium. *Phytopathology* **53,** 302–305.

Klotz, L. J., Stolzy, L. H., DeWolfe, T. A., and Szuszkiewicz, T. E. (1965). Rate of oxygen supply and distribution of root-rotting fungi in soils. *Soil Sci.* **99,** 200–204.

Klotz, L. J., Stolzy, L. H., Labanauskas, C. K., and DeWolfe, T. A. (1971). Importance of *Phytophthora* spp. and aeration in root rot and growth inhibition of orange seedlings. *Phytopathology* **61,** 1342–1346.

Knous, T. R., and Maxfield, J. E. (1976). Amplication of metabolic squalene in alfalfa roots associated with *Phytophthora* root rot. *Proc. Am. Phytopathol. Soc.* **3,** 239 (Abstr.).

Kommedahl, T., and Windels, C. E. (1979). Fungi: pathogen or host dominance in disease? *In* "Ecology of Root Pathogens" (S. V. Krupa and Y. R. Dommerques, eds.), pp. 1–103. Elsevier, New York.

Konings, H. (1982). Ethylene promoted formation of aerenchyma in seedling roots of *Zea mays* L. under aerated and non-aerated conditions. *Physiol. Plant.* **54,** 119–124.

Konings, H., and Verschuren, G. (1980). Formation of aerenchyma in roots of *Zea mays* in aerated solutions and its relation to nutrient supply. *Physiol. Plant.* **49,** 265–270.

Kozlowski, T. T. (1976). Water supply and leaf shedding. *In* "Water Deficits and Plant Growth" (T. T. Kozlowski, ed.), Vol. 4, pp. 191–231. Acadamic Press, New York.

Kozlowski, T. T. (1982). Water supply and tree growth. II. Flooding. *For. Abstr.* **43,** 145–161.

Krogh, A. (1919). The rate of diffusion of gases through animal tissues with some remarks on the coefficient of invasion. *J. Physiol. (London)* **52,** 391–408.

Kuan, T. L., and Erwin, D. C. (1980). Predisposition effect of water saturation of soil on *Phytophthora* root rot of alfalfa. *Phytopathology* **70,** 981–986.

Kuan, T. L., and Erwin, D. C. (1982). Effect of soil matric potential on *Phytophthora* root rot of alfalfa. *Phytopathology* **72,** 543–548.

Labanauskas, C. K., Stolzy, L. H., and Handy, M. F. (1974). Soil oxygen and *Phytophthora* spp. root infestation effects on the protein and free amino acids in lemon and orange leaves. *J. Am. Soc. Hortic. Sci.* **99,** 497–500.

Le Jambre, L. F., and Whitlock, J. H. (1967). Oxygen influence on egg production by a parasitic nematode. *J. Parasitol.* **53,** 887.

Letey, J., and Stolzy, L. H. (1967). Limiting distances between root and gas phase for adequate oxygen supply. *Soil Sci.* **103,** 404–409.

Levitt, J. (1980). "Responses of Plants to Environmental Stresses," Vol. 2. Academic Press, New York.

Lombard, P. B., Stolzy, L. H., Garber, M. J., and Szuszkiewicz, T. E. (1965). Effects of climatic factors on fruit volume increase and leaf water deficit of citrus in relation to soil suction. *Soil Sci. Soc. Am. Proc.* **29,** 205–208.

Luxmoore, R. J., and Stolzy, L. H. (1969). Root porosity and growth responses of rice and maize to oxygen supply. *Agron. J.* **61,** 202–204.

Luxmoore, R. J., and Stolzy, L. H. (1972a). Oxygen diffusion in the soil–plant system. VI. A synopsis with commentary. *Agron. J.* **64,** 725–729.

Luxmoore, R. J., and Stolzy, L. H. (1972b). Oxygen consumption rates predicted from respiration, permeability and porosity measurements on excised wheat root segments. *Crop Sci.* **12,** 442–445.

Luxmoore, R. J., Stolzy, L. H., Joseph, H., and DeWolfe, T. A. (1971). Gas space porosity of citrus roots. *HortScience* **6,** 447–448.

Luxmoore, R. J., Sojka, R. E., and Stolzy, L. H. (1972). Root porosity and growth responses of wheat to aeration and light intensity. *Soil Sci.* **113,** 354–357.

Lynch, J. M. (1975). Ethylene in soil. *Nature (London)* **256,** 576–577.

MacDonald, J. D., and Duniway, J. M. (1978a). Influence of the matric and osmotic components of water potential on zoospore discharge in *Phytophthora*. *Phytopathology* **68,** 751–757.

MacDonald, J. D., and Duniway, J. M. (1978b). Temperature and water stress effects on sporangium viability and zoospore discharge in *Phytophthora cryptogea* and *P. megasperma*. *Phytopathology* **68,** 1449–1455.

Mangenot, F., and Diem, H. G. (1979). Fundamentals of biological control. *In* "Ecology of Root Pathogens" (S. V. Krupa and Y. R. Dommerques, eds.), pp. 207–265. Elsevier, New York.

McElroy, F. D. (1967). "Studies on the Biology and Host–Parasite Relations of *Hemicycliophora arenaria* Raski, 1958." Ph.D. Dissertation, Univ. of California, Riverside.

McIntosh, D. L. (1964). *Phytophthora* spp. in soils of the Okanagan and Similamean Valleys of British Columbia. *Can. J. Bot.* **42,** 1411–1415.

McIntosh, D. L. (1966). The occurrence of *Phytophthora* spp. in irrigation systems in British Columbia. *Can. J. Bot.* **44,** 1591–1596.

Meek, B. D., and Stolzy, L. H. (1978). Short-term flooding. *In* "Plant Life in Anaerobic Enviroments" (D. D. Hook and R. M. M. Crawford, eds.), pp. 351–373. Ann Arbor Sci. Publ., Ann Arbor, Michigan.

Miller, S. A., and Mazurak, A. P. (1958). Relationships of particle and pore sizes to the growth of sunflower (*Helianthus annuus,* L.). *Soil Sci. Soc. Am. Proc.* **22,** 275–277.

Mulder, D. (ed.) (1979). "Soil Disinfestation." Elsevier, Amsterdam.

Nicholas, W. L., and Jantumen, R. (1964). *Caenorhabditis briggsae* (Rhabditidae) under anaerobic conditions. *Nematologica* **10,** 409–418.

Nicholas, W. L., and Jantumen, R. (1966). The effect of different concentrations of oxygen and of carbon dioxide on the growth and reproduction of *Caenorhabditis briggsae* (Rahbditidae). *Nematologica* **12,** 328–336.

Novozamsky, S., Beek, J., and Bolt, G. H. (1976). Chemical equilibria. *In* "Soil Chemistry, A. Basic Elements" (G. H. Bolt and M. G. M. Bruggenwert, eds.), pp. 13–42. Elsevier, New York.

Overgaard-Nielson, C. (1949). Studies on the soil microfauna. II. Soil inhabiting nematodes. *Natura Jutlandica* **2,** 1–131.

Papenhuijzen, C., and Roos, M. H. (1979). Some changes in the subcellular structure of root cells of *Phaesseolus vulgaris* as a result of cessation of aeration in the root medium. *Acta Bot. Neerl.* **28,** 491–495.

Patrick, W. H., and Mikkelson, D. S. (1971). Plant nutrient behavior in flooded soil. *In* "Fertilizer Technology and Use" (R. A. Olson, T. J. Army, J. J. Hanway, and V. J. Kilmer, eds.), pp. 187–215. Soil Sci. Soc. Am., Madison, Wisconsin.

Pfender, W. F., Hine, R. B., and Stanghellini, M. E. (1977). Production of sporangia and release of zoospores by *Phytophthora megasperma* in soil. *Phytopathology* **67,** 657–663.

Primrose, S. B. (1976). Ethylene-forming bacteria from soil and water. *J. Gen. Microbiol.* **97,** 343–346.

Pulli, S. K., and Tesar, M. R. (1975). *Phytophthora* root rot in seedling-year alfalfa as affected by management practices inducing stress. *Crop Sci.* **15,** 861–864.

Pullman, G. S., and DeVay, J. E. (1982). Effect of soil flooding and paddy rice culture on the survival of *Verticilium dahliae* and incidence of vetticillium wilt in cotton. *Phytopathology* **71,** 1285–1289.

Revelle, R. (1982). Carbon dioxide and world climate. *Sci. Am.* **247,** 35–43.

Rovira, A. D., and Davey, C. B. (1974). Biology of the rhizosphere. *In* "The Plant Root and Its Environment" (E. W. Carson, ed.), pp. 154–208. Univ. Press of Virginia, Charlottesville.

Russell, E. J., and Appleyard, A. (1915). The atmosphere of the soil, its composition and the causes of variation. *J. Agric. Sci.* **7,** 1–48.

Russell, E. W. (1976). The chemistry of waterlogged soils. *In* "Soil Conditions and Plant Growth" (E. W. Russell, ed.), p. 849. Longmans, New York.

Russell, M. B. (1952). Soil aeration and plant growth. *Adv. Agron.* **2,** 253–301.

Saz, H. J. (1969). Carbohydrate and energy metabolism of nematodes and *Acanthocephala. In* "Chemical Zoology" (M. Florkin and B. J. Scheer, eds.), Vol. 3, pp. 329–359. Academic Press, New York.

Schippers, B., and Gams, W. (eds.). (1979). "Soil Borne Plant Pathogens." Academic Press, New York.

Schoeneweiss, D. F. (1975). Predisposition, stress, and plant disease. *Annu. Rev. Phytopathol.* **13,** 193–211.

Schoeneweiss, D. F. (1978). Water stress as a predisposing factor in plant disease. *In* "Water Deficits and Plant Growth" (T. T. Kozlowski, ed.), Vol. 5, pp. 61–99. Academic Press, New York.

Shokes, F. M., and McCarter, S. M. (1979). Occurrence, dissemination, and survival of plant pathogens in surface irrigation ponds in southern Georgia. *Phytopathology* **69,** 510–516.

Smith, K. A. (1980). A model of the extent of anaerobic zones in aggregated soils and its potential application to estimates of denitrification. *J. Soil Sci.* **31,** 263–277.

Soil Survey Staff (1951). "Soil Survey Manual." U.S. Dept. Agric. Handbook No. 18. U.S. Govt. Printing Office, Washington, D.C.

Soil Survey Staff (1975). "Soil Taxonomy." U.S. Dept. Agric. Handbook No. 436. U.S. Govt. Printing Office, Washington, D.C.

Sojka, R. E., and Stolzy, L. H. (1980). Soil-oxygen effects on stomatal response. *Soil Sci.* **130,** 350–358.

Spencer, J. A., and Cooper, W. E. (1967). Pathogenesis of cotton (*Gossypium hirsutum*) by Pythium species: zoospore and mycelium attraction and infectivity. *Phytopathology* **57,** 1332–1338.

Stelzer, R., and Laüchli, A. (1980). Salt and flooding tolerance of *Puccinellia peisonis.* IV. Root respiration and the role of aerenchyma in providing atmospheric oxygen to the roots. *Z. Pflanzenphysiol.* **97,** 171–178.

Sterne, R. E., and McCarver, T. H. (1980). Formation of sporangia by *Phytophthora cryptogea* and *P. parasitica* in artificial and natural soils. *Soil Biol. Biochem.* **12,** 441–442.

Sterne, R. E., Zentmyer, G. A., and Kaufmann, M. R. (1977). The influence of matric potential soil texture, and soil amendment on root disease caused by *Phytophthora cinnamomi. Phytopathology* **67,** 1495–1500.

Stolzy, L. H. (1972). Soil aeration and gas exchange in relation to grasses. *In* "The Biology and Utilization of Grasses" (V. B. Youngner and C. M. McKell, ed.), pp. 247–258. Academic Press, New York.

Stolzy, L. H. (1974). Soil atmosphere. *In* "The Plant Root and Its Environment" (E. W. Carson, ed.), pp. 335–361. Univ. Press of Virginia, Charlottesville.

Stolzy, L. H., and Letey, J. (1962). Measurement of oxygen diffusion rates with the platinum microelectrode. III. Correlation of plant responses to soil oxygen diffusion rates. *Hilgardia* **35,** 567–576.

Stolzy, L. H., Moore, P. W., Klotz, L. J., and DeWolfe, T. A. (1959). Growth rates of orange seedlings in the presence and absence of *Phytophthora* spp. at two levels of irrigation and two levels of soil nitrogen. *Plant Dis. Rep.* **43,** 1169–1173.

Stolzy, L. H., Moore, P. W., Klotz, L. J., DeWolfe, T. A., and Szuszkiewicz, T. E. (1960). Effects of *Phytophthora* spp. on water used by citrus seedlings. *J. Am. Soc. Hortic. Sci.* **76,** 240–244.

Stolzy, L. H., Taylor, O. C., Letey, J., and Szuszkiewicz, T. E. (1961). Influence of soil-oxygen diffusion rates on susceptibility of tomato plants to air-borne oxidants. *Soil Sci.* **91,** 151–155.

Stolzy, L. H., Taylor, O. C., Dugger, W. M., Jr., and Mersereau, D. (1964). Physiological changes in and ozone susceptibility of the tomato plant after short periods of inadequate oxygen diffusion to the roots. *Soil Sci. Soc. Am. Proc.* **28,** 305–308.

Stolzy, L. H., Letey, J., Klotz, L. J., and DeWolfe, T. A. (1965a). Soil aeration and root-rotting fungi as factors in decay of citrus feeder roots. *Soil Sci.* **99,** 403–406.

Stolzy, L. H., Letey, J., Klotz, L. J., and Labanauskas, C. K. (1965b). Water and aeration as factors in root decay of *Citrus sinensis. Phytopathology* **55,** 270–275.

Stolzy, L. H., Zentmyer, G. A., Klotz, L. J., and Labanauskas, C. K. (1967). Oxygen diffusion, water, and *Phytophthora cinnamomi* in root decay and nutrition of avocados. *Proc. Am. Soc. Hortic. Sci.* **90,** 67–76.

Stolzy, L. H., Focht, D. D., and Flühler, H. (1981). Indicators of soil aeration status. *Flora (Jena)* **171,** 236–265.

Stover, R. H. (1979). Flooding of soil for disease control. *In* "Soil Disinfestation" (D. Mulder, ed.), pp. 19–28. Elsevier, Amsterdam.

Sugar, D. (1977). "The Development of Sporangia of *Phytophthora cambivora, P. megasperma,* and *P. drechsleri* and Severity of Root and Crown Rot in *Prunus mahaleb* as Influenced by Soil Matric Potentials." M.S. Thesis, Univ. of California, Davis.

Taylor, S. A., and Ashcroft, G. L. (1972). The foundations of irrigation science. *In* "Physical Edaphology" (G. L. Ashcroft, ed.), pp. 1–28. Freeman, San Francisco, California.

Trought, M. C. T., and Drew, M. C. (1980a). The development of waterlogging damage in wheat seedlings (*Triticum aestivum* L.). I. Shoot and root growth in relation to changes in the concentrations of dissolved gases and solutes in the soil solution. *Plant Soil* **54,** 77–94.

Trought, M. C. T., and Drew, M. C. (1980b). The development of waterloggig damage in wheat seedlings (*Triticum aestivum* L.). II. Accumulation and redistribution of nutrients by the shoot. *Plant Soil* **56,** 187–199.

Tsao, P. H., and Bricker, J. L. (1968). Germination of chlamydospores of *Phytophthora parasitica* in soil. *Phytopathology* **58,** 1070 (Abstr.).

Van Gundy, S. D., and Stolzy, L. H. (1961). Influence of soil oxygen concentration on the development of *Meloidogyne javanica. Science (Washington, D.C.)* **134,** 665–666.

Van Gundy, S. D., and Stolzy, L. H. (1963). The relationship of oxygen diffusion rates to the survival, movement, and reproduction of *Hemicycliophora arenaria. Nematologica* **9,** 605–612.

Van Gundy, S. D., Bird, A. F., and Wallace, H. R. (1967). Aging and starvation in larvae of *Meloidogyne javanica* and *Tylenchulus semipenetrans. Phytopathology* **57,** 559–571.

Van Gundy, S. D., McElroy, F. D., Cooper, A. F., and Stolzy, L. H. (1968). Influence of soil temperature, irrigation and aeration on *Hemicycliophora arenaria. Soil Sci.* **106,** 270–274.

Varade, S. B., Stolzy, L. H., and Letey, J. (1970). Influence of temperature, light intensity, and aeration on growth and root porosity of wheat, *Triticum aestivum. Agron. J.* **62,** 505–507.

Varade, S. B., Letey, J., and Stolzy, L. H. (1971). Growth response and root porosity of rice in relation to temperature, light intensity, and aeration. *Plant Soil* **34,** 415–420.

Vartapetian, B. B. (1973). Aeration of roots in relation to molecular oxygen transport in plants. *In* "Plant Response to Climatic Factors" (R. O. Slatyer, ed.), pp. 259–265. UNESCO, Paris.

Vartapetian, B. B., Andreeva, I. N., and Nurtidinov, N. (1978). Plant cells under oxygen stress. *In* "Plant Life in Anaerobic Environments" (D. Hook and R. M. M. Crawford, eds.), pp. 13–88. Ann Arbor Sci. Publ., Ann Arbor, Michigan.

Wallace, H. R. (1956). Soil aeration and the emergence of larvae from cysts of the beet eelworm, *Heterodera schachtii* schm. *Ann. Appl. Biol.* **44,** 57–66.

Wallace, H. R. (1968a). The dynamics of nematode movement. *Annu. Rev. Phytopathol.* **6,** 91–114.

Wallace, H. R. (1968b). The influence of aeration on survival and hatch of *Meloidogyne javanica. Nematologica* **14,** 223–230.

Wample, R. L., and Reid, D. M. (1978). Control of adventitious root production and hypocotyl hypertrophy of sunflower (*Helianthus annuus*) in response to flooding. *Physiol. Plant.* **44,** 351–358.

Warburg, O., and Kubowitz, F. (1929). Atmung bei sehr kleinen Sauerstoffdrucken. *Biochem. Z.* **214,** 5–18.

Wiebe, W. J., Christian, R. R., Hansen, J. A., King, G., Sherr, B., and Skyring, G. (1981). Anaerobic respiration and fermentation. *In* "Ecological Studies: Analysis and Synthesis, Salt Marsh Soils and Sediments: Methane Production" (L. R. Pomeroy and R. G. Wiegert, eds.), Vol. 38, pp. 137–159. Springer-Verlag, New York.

Wiedenroth, E. M. (1981). Relations between phytosynthesis and root metabolism of cereal seedlings influenced by root anaerobiosis. *Photosynthetica* **15,** 575–591.

Wilkinson, H. T., Miller, R. D., and Millar, R. L. (1981). Infiltration of fungal and bacterial propagules into soil. *Soil Sci. Soc. Am. J.* **45,** 1034–1039.

Willhite, F. M., Grable, A. R., and Rouse, H. K. (1965). Interaction of nitrogen and soil moisture on the production and persistence of timothy in lysimeters. *Agron. J.* **57,** 479–481.

Williamson, R. E., and Kriz, G. J. (1970). Response of agricultural crops to flooding depth of water table and soil gaseous composition. *Trans. ASAE* **13,** 216–220.

Wong, P. T. W., and Griffin, D. M. (1976). Bacterial movement at high matric potentials. I. In artificial and natural soils. *Soil Biol. Biochem.* **8,** 215–218.

Yu, P., Stolzy, L. H., and Letey, J. (1969). Survival of plants under prolonged flooded conditions. *Agron. J.* **61,** 844–847.

Zentmyer, G. A. (1961). Chemotaxis of zoospores for root exudates. *Science (Washington, D.C.)* **133,** 1595–1596.

Zentmyer, G. A. (1966). The role of amino acids in chemotaxis of three species of *Phytophthora. Phytopathology* **56,** 907 (Abstr.).

Zentmyer, G. A., and Ervin, D. C. (1970). Development and reproduction of *Phytophthora. Phytopathology* **60,** 1120–1127.

Zentmyer, G. A., and Richards, S. J. (1952). Pathogenicity of *Phytophthora cinnamomi* to avocado trees, and the effects of irrigation on disease development. *Phytopathology* **42,** 35–37.

Zentmyer, G. A., Paulus, A. O., and Burns, R. M. (1967). Avocado root rot. *Bull. Calif. Agric. Exp. Stn.* No. 511.

Zimmer, D. E., and Urie, A. L. (1967). Influence of irrigation and soil infestation with strains of *Phytophthora drechsleri* on root rot resistance of safflower. *Phytopathology* **57,** 1056–1059.

CHAPTER 8

Adaptations to Flooding with Fresh Water

DONAL D. HOOK
Department of Forestry
Clemson University
Clemson, South Carolina

I. INTRODUCTION

Plants may adapt metabolically to tolerate anoxia (truly anoxia tolerant), adapt morphologically and physiologically to avoid anoxia (apparently anoxia tolerant), or they may not adapt and succumb very quickly to anoxia (anoxia intolerant) (Vartapetian, 1978). A few plant species that tolerate prolonged soil flooding exhibit both metabolic and avoidance traits (Hook *et al.*, 1971; John and Greenway, 1976; Vartapetian *et al.*, 1978; Hook and Scholtens, 1978; Davies, 1980; Kozlowski, 1982), which suggests that flood tolerance is not conveyed by a single adaptation but rather by a combination of adaptations. Hook and Brown

ISBN 0-12-424120-4

(1973) showed for five hardwood tree species that relative flood tolerance was a function of the number and degree of refinement of adaptations the species exhibited. Although classifying species into the three tolerance types may be helpful for ferreting out the basic mechanisms of flood tolerance, they have limited practical value for segregating species on the basis of relative flood tolerance. For instance, Vartapetian *et al.* (1978) showed that the mitochondria of roots of cotton (*Gossypium hirsutum*; relatively intolerant to flooding) maintained their ultrastructure longer under anoxia than those of rice (*Oryza sativa;* very tolerant to flooding); hence the former appeared to be the most flood tolerant. Apparently the tolerance of rice to flooding is strongly related to its capacity to transport oxygen to the rhizosphere, that is, to avoid anoxia (van Raalte, 1940; Armstrong, 1969; Kordan, 1974).

The literature is replete with descriptions of plant responses to flooding (Table I). Such a large number of different responses have contributed to confusion as to which of these are adaptations. However, adaptations that enable plants to tolerate or avoid anoxia are fairly common among vascular plants. This suggests there are only a few morphological, physiological, and biochemical solutions to the

TABLE I

Responses of Plants to Flooding[a]

Response	Selected references
Leaf	
Abscission	Yelenosky (1964); Hook (1968); Hook *et al.* (1971)
Epinasty of leaf and petiole	Kramer (1951); Jackson (1955); Wample and Reid (1975); Kawase (1981)
Petiole reorientation	Kramer (1951)
Stomatal closure	Pereira and Kozlowski (1977); Tang and Kozlowski (1982)
Reduced photosynthesis	Regehr *et al.* (1975)
Decreased transpiration	Bergman (1920); Parker (1949); Kramer (1951); Regehr *et al.* (1975)
Increased transpiration	Kramer (1951)
Chlorosis	Kramer (1951); Bergman (1959); Yelenosky (1964); Wample and Reid (1975)
Presence of anthocyanin	Parker (1949); Hook (1968)
Thickening of leaf	Hook (1968); Hook *et al.* (1971)
Decreased size of leaf	Heinicke (1932); Lindsey *et al.* (1961)
Wilting	Bergman (1920); Kramer and Jackson (1954); Dickson *et al.* (1965)
Flower and stem	
Flower abscission	Oskamp and Batjer (1932)

TABLE I *Continued*

Response	Selected references
Fruit abscission	Haas (1936)
Corky fruit	Heinicke *et al.* (1940)
Poor fruit set	Heinicke (1932)
Decreased growth and internode elongation	Bergman (1920); McDermott (1954); Hosner (1960); Hook (1968); Harms (1973); Wample and Reid (1975)
Increased diameter and height growth	Applequist (1959); Hook and Brown (1973); Langdon *et al.* (1978)
Spindly shoot	Heinicke (1932)
Development of aerenchyma	McPherson (1939); Sifton (1945); deWit (1978); Kawase (1981)
Transport of oxygen	van Raalte (1940); Armstrong (1979)
Hypertrophy of stem and lenticels	Penfound (1934); Sifton (1945); Hook *et al.* (1970b); Wample and Reid (1979)
Release of ethanol, ethylene, and acetaldehyde from lenticels	Chirkova and Gutman (1972)
Increased moisture stress	Sena Gomes and Kozlowski (1980); Wenkert *et al.* (1981)
Roots	
Death without replacement	Yu *et al.* (1969); Hook *et al.* (1971); Hook and Brown (1973); Wenkert *et al.* (1981)
Death with replacement by development of new roots	Hook *et al.* (1971); Coutts and Philipson (1978a,b); Sena Gomes and Kozlowski (1980)
Adventitious rooting	Kramer (1951); Hosner and Boyce (1962); Hook *et al.* (1971); Gill (1975); Drew and Lynch (1980); Kawase (1981)
Increased length of lateral roots	Schramm (1960)
More succulent roots	Kramer (1969); Hook *et al.* (1971); Keeley (1979)
Fewer root hairs	Snow (1904); Weaver and Himmell (1930)
Increased root diameter	Kramer (1969); Cochran (1972)
Decreased nutrient uptake	Kramer (1951); Greenwood (1967); Williamson and Splinter (1968)
Increased nutrient uptake	Valoras and Letey (1966); Jones and Etherington (1970); Hook *et al.* (1983)
Leaking of organic and inorganic compounds from roots	Crawford (1978); Smith and ap Rees (1979); Drew and Lynch (1980); Mendelssohn *et al.* (1981); Hook *et al.* (1983)
Accumulation of end products of anaerobic respiration	Crawford (1978)
Increased permeability of plasma membrane and loss of discriminatory uptake of nutrients	Crawford (1978); Smith and ap Rees (1979); Shadan (1980); Hook *et al.* (1983)
Decreased ectomycorrhizal development	Mikola (1973); Malajczuk and Lamont (1981)

[a]A portion of this list was taken from Whitlow and Harris (1979), but numerous other responses were added from the literature.

problem of plant life in a periodically flooded habitat. In this chapter the current level of understanding of these adaptations is reviewed.

II. SEED GERMINATION

It has long been known that seeds of the majority of land plants will not germinate under water and that they lose their vitality quickly under such conditions [Mazé, 1900 (after Crocker and Davis, 1914)].

Germination of seed under water may be beneficial or hazardous, depending on the depth and duration of inundation (Kramer and Kozlowski, 1979). Seeds of rice and some other lowland species germinate under water, and if the water is shallow enough for the shoot (coleoptile in rice) to extend above the water surface, root growth will be initiated [Nagai, 1916; Sasaki, 1930 (after Edwards, 1933)]. If the duration of flooding is not sufficient to cause injury to the germinating seedling, it may be in an advantageous position over competing vegetation whose seeds germinate only after the water level recedes. However, where flooding is deep and prolonged, as is the case in the swamps of the southern United States, seed germination under water is hazardous because few species can withstand prolonged submergence of the foliage. Seeds of species that dominate these swamps [bald cypress (*Taxodium distichum*), pond cypress (*T. dis-*

TABLE II

Selected Listing of Species with Regard to Capacity of Their Seeds to Germinate under Water[a]

Scientific name	Common name
Poor germination under water	
Herbaceous crop plants	
Festuca pratensis	Meadow fescue
Lolium perenne	English rye grass
Andropogon sorghum	Common sorghum
Zea mays	Field corn
Raphanus sativus	Radish
Allium cepa	Onion
Cucurbita maxima	Squash
Fagopyrum esculentum	Buckwheat
Pisum sativum	Alaska pea
Medicago sativa	Alfalfa
Festuca duriuscula	Hard fescue
Lycopersicon esculentum	Tomato
Thymus vulgaris	Thyme
Allium porrum	Leek
Ocymum basilicum	Basil

Prov. 1-184

Banco Crédito Agrícola de Cartago

DONAL CHARLES LIEBER KASS.

el mismo

75,oo SETENTA Y CINCO DOLARES.

BANCO CREDITO AGRICOLA
DE CARTAGO
8 ENE 1985
CAJERO
Orlando Hidalgo S

Compensado Ck.No 62524 C,C.No.460.

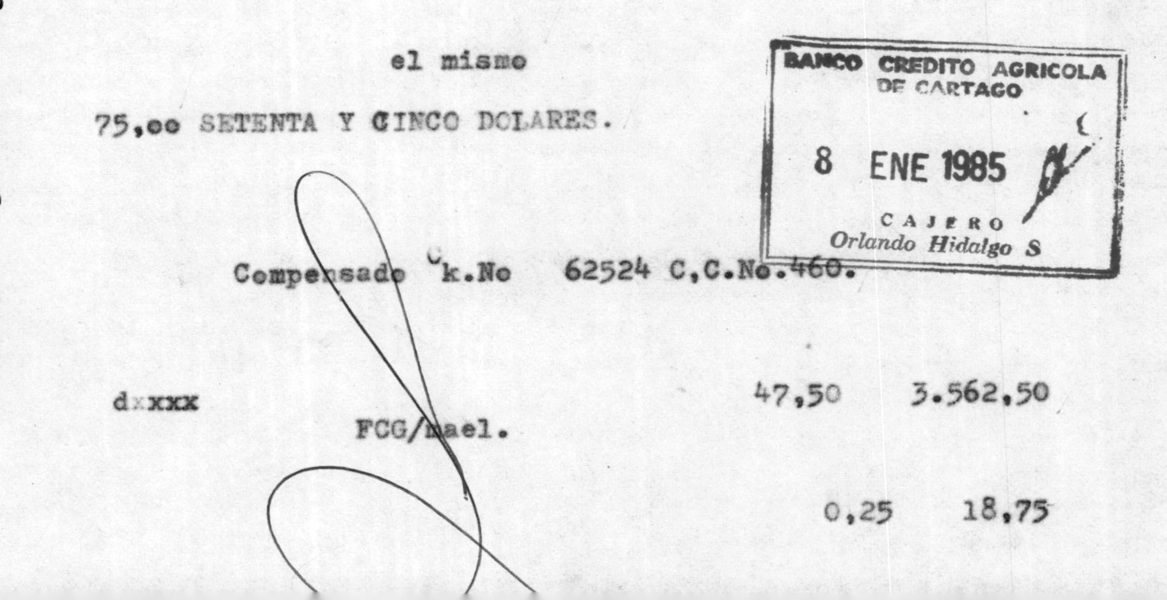

dxxxx 47,50 3.562,50

FCG/mael.

0,25 18,75

TABLE II *Continued*

Scientific name	Common name
Ornamental and weeds	
Viola tricolor	Pansy
Aster	Aster
Phlox drummondii	Drummond phlox
Campanula medium	Canterbury bell
Trees and shrubs	
Nyssa sylvatica var. *biflora*	Swamp tupelo
Nyssa aquatica	Water tupelo
Taxodium distichum	Bald cypress
Liquidambar styraciflua	Sweetgum
Fraxinus pennsylvanica	Green ash
Platanus occidentalis	Sycamore
Fraxinus caroliniana	Water ash
Good germination under water	
Crop plants	
Phleum pratense	Timothy
Axonopus compressus	Carpet grass
Poa compressa	Canadian bluegrass
Lactuca sativa	Lettuce
Apium graveolens	Celery
Cynodon dactylon	Bermuda grass
Oryza sativa	Rice
Ornamental and weeds	
Antirrhinum	Snapdragon
Oenothera	Evening primrose
Dianthus	Carnation
Petunia	Petunia
Xanthium pensylvanicum	Cocklebur
Lowland plants (herbaceous)	
Peltandra virginica	Arrow arum
Alisma plantago	Water plantain
Typha latifolia	Cattail
Cynodon dactylon	Bermuda grass
Nelumbo nucifera	Lotus
Euryale ferox	Golden euryale
Trapa natans	Water chestnut
Trees and shrubs	
Cephalanthus occidentalis	Buttonbush
Ulmus americana	American elm
Salix nigra	Black willow
Populus deltoides	Cottonwood

[a]The reader is referred to Morinaga (1926), Edwards (1933), DeBell and Naylor (1972), DuBarry (1963), Crocker and Davis (1914) for more complete listings.

tichum var. *nutans*), water tupelo (*Nyssa aquatica*), and swamp tupelo (*N. sylvatica* var. *biflora*)] do not germinate under water (Shunk, 1939; DeBell and Naylor, 1972). Natural regeneration in swamps is usually limited to drought periods when the soil surface is exposed; this occurs in deep swamps only during prolonged droughts. Swamp tree species typically grow very rapidly in height the first year or two. The rapid juvenile growth is believed to be an inherent trait of such species in that it enables them to extend their foliage as high as possible before the normal flood level returns (DeBell and Naylor, 1972; Hook *et al.*, 1983). A number of bottomland hardwood species may germinate under water (DuBarry, 1963). The flooding regime on such sites is usually shorter and shallower than in swamps, hence germination under water may have some ecological significance.

Caution is recommended in ecologically interpreting seed-germination characteristics under water because factors affecting seed dormancy such as germination stimulation and germination inhibition are involved with some species. For example, stratification requirements of cocklebur (*Xanthium pensylvanicum*), apple (*Pyrus malus*), and subterranean clover (*Trifolium subterraneum*) seeds may be partially replaced by anaerobic or reduced O_2 tension treatments (Esashi *et al.*, 1978) that are not related to adaptation to flooding. Also the seed coats of some species that germinate under water are very hard (e.g., *Alisma plantago, Typha latifolia, Peltandra virginica,* etc.), and mechanical breaking of the seed coat (pericarp) will partially release dormancy. The germination traits of seeds under water are listed for several species in Table II.

In summary, it appears that some competitive advantages can be achieved by germination under water if the flood regime is shallow or short lived. Under deep flooding, the capaicity of seeds to remain dormant under water tends to prevent losses of entire seed crops during prolonged flooding. However, the majority of land plant seeds lose their viability if submerged for prolonged periods.

III. GROWTH AND DORMANCY

It has been recognized for some time that dormant plants are not very susceptible to flooding damage (Broadfoot and Williston, 1973). In the Mississippi Delta of the United States, some tree species such as overcup oak (*Quercus lyrata*), green ash (*Fraxinus pennsylvanica*), and water hickory (*Carya aquatica*) do not leaf out until almost a month later than the other hardwoods in the surrounding upland areas and thereby avoid growth during prolonged flooding. These late emergers typically occur in areas of backwater flooding (areas subjected to floodwater that is backed into an area by a rising river and that may be retained in the area for an extended period).

Some marsh plants such as *Typha* sp. may exhibit a response opposite to that of tree species. They tend to emerge very early in the spring and complete most of their growth before the soil temperature rises and the soil becomes highly reduced.

IV. MORPHOLOGICAL CHARACTERISTICS

A. Stem Hypertrophy

Hypertrophy is observed on some herbaceous species [tomato (*Lycopersicon esculentum*) and sunflower (*Helianthus annuus*)] and on several tree species [bald cypress, pond cypress, swamp tupelo, water tupelo, *Eucalyptus globulus, E. camaldulensis,* and red maple (*Acer rubrum*), green ash (*Fraxinus pennsylvanica*), and white ash (*F. americana*) to lesser degrees], but it has not been clearly substantiated as an adaptation. However, it appears that stem hypertrophy in conjunction with knees and exposed roots is important in facilitating internal aeration of the roots of swamp tree species. The wood of swollen buttresses does not have aerenchyma, but it does have lower density, larger cells, and more parenchyma than noninundated wood of the trunk (Penfound, 1934). The changes in anatomy probably increase intercellular spaces and make them more pervious to gas. In sunflower and most herbaceous species, stem hypertrophy is associated with enlargement of the cortical layer (Kawase, 1981). Enlargement of the lower stem appears to increase the surface area of pervious tissue near the flood line and therefore increases the surface area over which O_2 may readily diffuse into the stem.

Stem hypertrophy that occurs in association with flooding appears to be caused by an interaction of ethylene and auxins that is more or less controlled by the flooding or "water effect" (Wample and Reid, 1979). Although ethylene was transported from the root to shoot in tomato plants, only a small quantity of the amount artificially applied was transported. In addition, ethylene rapidly escapes from the stem through lenticels (Chirkova and Gutman, 1972; Zeroni *et al.*, 1977). Bradford and Yang (1981) hypothesized that insufficient ethylene would be transported to the shoot to account for epinasty in flooded plants. They concluded that anaerobiosis inhibits ethylene production (a well-recognized fact), hence little ethylene is produced in flooded roots. On the other hand, they found that production of the ethylene precursor, 1-aminocyclopropane-1-carboxylic acid (ACC), was stimulated under anaerobiosis and was readily transported in the xylem sap. The precursor, ACC, in the presence of an enzyme that is constitutive in many plant tissues, is converted to ethylene in the presence of O_2. Therefore, they proposed that under flooding conditions and anaerobiosis in roots, ACC is produced and transported upward in the xylem sap. On exposure to O_2, ACC is converted to ethylene in quantities sufficient to cause epinasty and other physiological effects. Their theory helps explain stem hypertrophy. It is known that flooding causes indoleacetic acid (IAA) to accumulate in flooded tissue (Phillips, 1964; Wample and Reid, 1979) in sunflower and that IAA and ethylene in conjunction stimulate hypertrophy (Wample and Reid, 1979). Hence it appears that the signal that Wample and Reid (1978, 1979) suggest is transported from the roots probably is ACC. When ACC reaches the aerated portion of the stem, ethylene is produced and causes IAA to accumulate. The two hormones working in conjunction could contribute to development of stem hypertrophy. Unfortunately, data for this process in woody species are lacking.

B. Knees and Pneumatophores

Knees occur primarily on bald cypress and pond cypress and are the result of meristematic growth on the arched upper (adaxial) side of the root. Because knees occur primarily on sites where the water table fluctuates (not in lakes or on dry land), it appears the tissue is stimulated to grow by repeated cycles of anaerobiosis and aeration (Penfound, 1934; Whitford, 1956). Although no data exist to verify it, such growth may be a direct response to ACC accumulating on the adaxial side during anaerobiosis and conversion of ACC to ethylene during the aerobic phase.

The role that knees play on trees in flooded habitats is not clear. They do not occur on cypress trees growing in lakes, where the need would be greatest for aeration. Kramer *et al.* (1952) concluded that knees did not aid in aerating the remainder of the tree; on the other hand, it is known that gas exchange occurs freely through knees (Kramer, 1969; Cowles, 1975). It seems probable that knees are beneficial if they occur but are not necessary for survival.

Pneumatophores on black mangrove (*Avicennia nitida*) provide O_2 to submerged roots in relation to the tide cycle (Scholander *et al.*, 1955); hence they are beneficial in aerating the root system of the tree in its natural environment, primarily in the intertidal zone.

C. Roots

Several distinct forms of roots regenerate on flooded plants. Failure to distinguish between these has led to a misunderstanding of the function of different root types in relation to flooding. The most common type of root regeneration is the development of adventitious roots on the stem above the soil usually within the flood zone. They are initiated directly on the stem, and are designated *adventitious water roots*.

The second type of root regeneration occurs in the soil. This type of development is less common, or less frequently reported, for only a few researchers have reported their findings such that these distinctions can be made and they usually have dealt with rice or woody species (Hosner and Boyce, 1962; Yu *et al.*, 1969; Hook *et al.*, 1971; Coutts and Philipson, 1978a,b; Keeley, 1979; Alva *et al.*, 1980). The typical sequence observed is that on flooding, the original roots die back to major secondary roots or the primary root and new roots are initiated from these points (Hook *et al.*, 1971; Coutts and Philipson, 1978a,b). These are designated *soil water roots.*

A third type of root produced in response to flooding differs morphologically from the original roots. For example, in sweetgum (*Liquidambar styraciflua*), the new roots are more succulent and almost clear in appearance as compared to the original roots (Hook and Brown, 1973). The new roots are designated *altered soil roots.*

Still another response to flooding involves death of all or most of the root system without production of new roots. Obviously, plants exhibiting such response are intolerant to waterlogging and die if flooded conditions persist. Most agronomic plants exhibit this response as do some tree species, namely, yellow poplar (*Liriodendron tulipifera;* Hook and Brown, 1973) and, to a lesser extent, Sitka spruce (*Picea sitchensis;* Coutts and Philipson, 1978a,b).

The rooting traits of heathland species are so complex, specialized, and dependent on habitat that no attempt is made to include them here. The reader is referred to Malajczuk and Lamont (1981) and Lamont (1981) for coverage of this topic. With the distinctions just given it is easier to understand and to evaluate beneficial effects of each type of root.

Adventitious water roots have been observed on many species (Table III). Tomato, sunflower, and maize (*Zea mays*), which develop adventitious water roots, tolerate flooding or recover from flooding damage more quickly and completely than when such roots are experimentally removed. Epinasty was less pronounced and shoot growth was retarded less where adventitious water roots were left intact than when they were removed (Jackson, 1955; Drew *et al.*, 1979; Jackson *et al.*, 1981; Wenkert *et al.*, 1981). However, Wample and Reid (1978) found no apparent contribution of adventitious water roots to survival of flooded sunflower plants. Similarly, experiments on woody species have led to conflicting results. Gill (1975) found that removing adventitious water roots from flooded *Alnus glutinosa* had no significant effect on shoot growth, lateral shoot number, leaf dry weight, and bud number, but the treatment appeared to reduce leaf number. However, Sena Gomes and Kozlowski (1980) found that *Fraxinus pennsylvanica* seedlings with adventitious water roots had higher water-absorbing efficiency than seedlings without adventitious water roots, and there was a correlation between adventitious water root development and stomatal reopening.

For highly flood-tolerant species, the soil water roots probably become the primary root system and enable the flooded trees to flourish (Hook *et al.*, 1971; Keeley, 1979; Alva *et al.*, 1980). Soil water roots on rice penetrated 30–40 cm into a flooded soil (Yu *et al.*, 1969). On lodgepole pine (*Pinus contorta*) they penetrated up to 16.7 cm into a waterlogged peat soil at 10°C, but at 20°C penetration was greatly reduced (Coutts and Philipson, 1978a,b). It appears that the tendency of wetland species to develop soil water roots may account for their

TABLE III

Species that Produce Adventitious Water Roots in Response to Flooding

Scientific name	Common name
Herbaceous species	
Cassava sp.	Cassava
Helianthus annuus	Sunflower
Lycopersicon esculentum	Tomato
Molinia caerulea	Moorgrass
Oryza sativa	Rice
Zea mays	Maize or corn
Woody species[a]	
Acer rubrum	Red maple
Alnus glutinosa	European alder
Alnus rubra	Red alder
Amorpha fruticosa	Indiobush
Cephalanthus occidentalis	Buttonbush
Eucalyptus camaldulensis	Longbeak eucalyptus
Fraxinus americana	White ash
Fraxinus pennsylvanica	Green ash
Liriodendron tulipifera	Yellow poplar
Melaleuca quinquenervia	—
Nyssa aquatica	Water tupelo
Nyssa sylvatica var. *biflora*	Swamp tupelo
Picea sitchensis	Sitka spruce
Pinus contorta	Lodgepole pipe
Populus deltoides	Eastern cottonwood
Populus ×euramericana	Robusta, Heidemij, Regenerata, and Serotina cottonwoods
Populus nigra	Black poplar
Populus trichocarpa	Black cottonwood
Salix alba	White willow
Salix atrocinerea	—
Salix ×cinerea	Gray willow
Salix fragilis	Brittle Willow
Salix hookerana	Hooker willow
Salix lasiandra	Pacific willow
Salix nigra	Black willow
Salix repens	Creeping willow
Sequoia sempervirens	Redwood
Tamarix gallica	French tamarix
Taxodium distichum	Bald cypress
Thuja plicata	Western redcedar
Tsuga heterophylla	Western hemlock
Ulmus americana	American elm

[a]Modified from Gill (1975), with permission of the publisher.

capacity to withstand the alternating flooding and drying of the wetland site. Apparently, the roots harden off during dry periods and lose the capacity to oxidize their rhizosphere (i.e., the roots grown under well-aerated conditions do not oxidize their rhizospheres; Hook *et al.*, 1971). Presumably, on reflooding, initial roots die back and soil water roots develop that oxidize their rhizosphere. Keeley (1979) found that the soil water roots of swamp tupelo also tended to harden off within 1 year under flooding, and there was a net increase in O_2-diffusion rate from 1 month to 1 year of age. Responses of mature swamp tupelo and bald cypress trees in the field substantiated that soil water roots developed where the flood level was substantially increased (Harms *et al.*, 1980). *Pinus sylvestris* roots also oxidize their rhizosphere (Armstrong and Read, 1972). As shown in Table III, most woody species that develop soil water roots also develop adventitious water roots.

In a semicontrolled experiment, Hook *et al.* (1970b) found that many adventitious water roots formed on swamp and water tupelos that were flooded with moving water, but few or none developed after flooding with stagnant water. Also, in pond sites (stagnant water) in the Atlantic coastal plain, swamp tupelo

Fig. 1. Base of a mature swamp tupelo tree growing in a pond in the lower coastal plain of South Carolina. Base is devoid of adventitious water roots, and no hummock has developed.

Fig. 2. Hummock around the base of a mature swamp tupelo tree in a headwater swamp in the lower coastal plain of South Carolina. This hummock appears to have been formed from debris and soil collecting around the adventitious water roots near the base of the tree.

trees seldom develop adventitious water roots but do so prolifically in headwater and slough swamps, where the flood water is usually moving (Hook, 1968). Hummocks do not develop in ponds around the swamp tupelo trees (Fig. 1), but they do in headwater swamps (Fig. 2). Apparently, the adventitious water roots are not only stimulated by moving water but are functional in trapping debris and soil from the moving water and thereby over time build a hummock around the base of the tree. This process permits a portion of the roots to remain above the general flood level (Fig. 2).

In herbaceous and woody species, adventitious water roots may reduce flooding damage, but data substantiating such effects are somewhat conflicting. Circumstantial evidence from field observations indicates they help build hummocks around some swamp tree species and thereby many become permanent roots of the tree. On the other hand, soil water roots appear to be a major part of the root system in rice and some tree species that are extremely tolerant to flooding and therefore may be of the utmost importance to survival and growth of such species in wetland habitats. Their role on other species is not well known. In *Zea mays* the original roots died back and new roots did not develop in the soil, but adventitious water roots penetrated the flooded soil and the plants survived up to 13 days of flooding (Wenkert *et al.*, 1981).

The function or value of altered soil roots is not known. The one species that has been reported to exhibit this type of root response (sweetgum) will tolerate relatively long periods of flooding, but its growth is severely reduced.

Explanations of development of adventitious water roots are similar but differ slightly from that for stem hypertrophy in regard to which hormones are most important. Wample and Reid (1975, 1979) hypothesized that the higher ethylene concentrations in the stems of plants with flooded roots cause auxins to accumulate, in conjunction with other factors coming from the leaves (such as carbohydrates) that cause adventitious water roots to develop on sunflowers. They attributed the major rooting signal to auxin and other leaf factors. Jackson *et al.* (1981) and Drew *et al.* (1979) reported that ethylene stimulated adventitious water rooting of maize, but they could not exclude the possibility that ethylene interacted with other growth hormones to stimulate rooting.

V. ANATOMICAL CHARACTERISTICS

Considerable evidence shows that plants that tolerate prolonged soil flooding transport O_2 to the rhizosphere more readily than plants that are not tolerant to waterlogging (Van Raalte, 1940; Barber *et al.*, 1962; Armstrong, 1968, 1969, 1979; Hook *et al.*, 1971; Coutts and Armstrong, 1976; Keeley, 1979). For O_2 transport to occur, there must be continuity in intercellular space between the atmosphere and rhizosphere, and the intercellular space must be relatively nontortuous and sufficient in volume to account for the gas movement.

Although mass flow of gases apparently occurs in the floating water lily (*Nuphar luteum* ssp. *macrophylla* and *N. luteum* ssp. *variegatum*) (Dacey, 1981; Dacey and Klug, 1982), the majority of vascular plants are aerated by static-gas-phase diffusion (Armstrong, 1978, 1979).

The site of entry and pathway of O_2 diffusion varies considerably among species. In woody species O_2 appears to enter the plant primarily through the lenticels (Armstrong, 1968; Hook *et al.*, 1971), although Chirkova (1968) reported that leaves and lenticels are a source of O_2 entry into rooted cuttings of willow (*Salix alba*) and poplar (*Populus petrowskiana*). In nonwoody plants the point of entry is more obscure, but in general the leaves appear to be more important as a site of entry than is the case in woody plants (Sifton, 1945; Soldatenkov and Hsien-Tuan, 1961; Chirkova and Soldatenkov, 1965; Greenwood, 1967). The coleoptile of rice serves as the entry for O_2 to the germinating rice seed whether it germinates under water or in air (Kordan, 1976). The difference in size of woody and nonwoody plants may be a factor as to which entry point is most important.

A. Stomata

Stomata are the gas-exchange ports of the leaves and are an entry point for O_2 and other gases. Stomatal closure and reduced leaf growth in herbaceous species may be the most rapid external plant responses to flooding (Pereira and Kozlowski, 1977; Wenkert *et al.*, 1981), hence stomatal closure could be especially important in species that depend on internal aeration entirely through the leaf and possibly in young tree seedlings. Pereira and Kozlowski (1977) found considerable variation among woody species in how quickly stomata closed and reopened; similar variability may exist in nonwoody plants.

B. Lenticels

Lenticels become hypertrophied on stems within and just above the flood zone [Devaux, 1900; Templeton, 1926 (after Sifton, 1945); and Hook *et al.*, 1970a]. The hypertrophied lenticel is composed primarily of complementary cells having one or more closing layers at varying distances from the phellogen. The closing layers all have large breaks and provide little or no resistance to gas diffusion. Also, the phellogen appears to be pervious to gas diffusion, at least in swamp and water tupelo, for the active cells are spherical in shape and have columnar arrangement (Hook *et al.*, 1970a), which would provide the largest amount of intercellular space per unit cell size (Kawase, 1981). These changes in the anatomical structure of lenticels in relation to flooding increase the size of the lenticel appreciably and the amount of intercellular space; therefore, they would appear to enhance internal aeration.

Armstrong (1968) and Hook *et al.* (1971) showed that lenticels are primary entry points for O_2 diffusion to the rhizosphere of several woody species. Ethyl alcohol, acetaldehyde, and ethylene diffused out of lenticels on rooted cuttings of willow, and O_2 diffused into the lenticels of willow and poplar cuttings (Chirkova and Gutman, 1972).

C. Intercellular Space

Internal aeration arises through lysigenous and/or schizogenous processes (McPherson, 1939; Sifton, 1945; deWit, 1978; Kawase, 1981). Generally, intercellular space is more evident in cortical than in xylem tissue, but it does occur in the xylem tissue of some woody plants. Development of intercellular space in the cortex of roots of barley (*Hordeum vulgare;* Bryant, 1934), corn (*Zea mays;* McPherson, 1939), and *Pinus sylvestris* (Armstrong and Read, 1972) is assumed

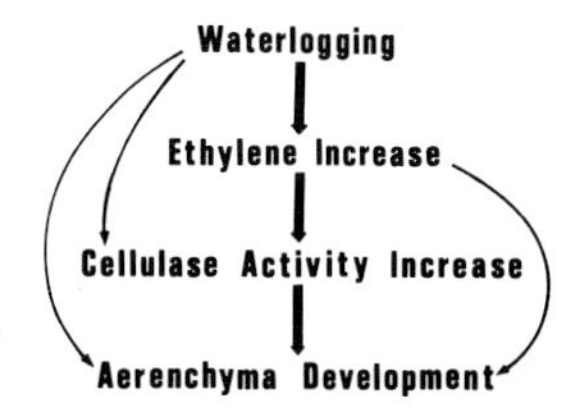

Fig. 3. Steps in aerenchyma (intercellular space) development. After Kawase (1981).

to provide pathways for O_2 diffusion in these species, but development of large intercellular spaces in the stele of roots of lodgepole pine (Coutts and Philipson, 1978b) and loblolly pine (*Pinus taeda;* (M. R. McKevlin and D. D. Hook, unpublished) apparently provides for limited gas diffusion in these species. Numerous herbaceous crop plants (McPherson, 1939; Greenwood, 1967), marsh plants (Bergman, 1920; Coult and Vallance, 1958; Armstrong and Boatman, 1967), and woody plants (Armstrong, 1969; Hook *et al.*, 1971; Chirkova and Gutman, 1972; Armstrong and Read, 1972; Philipson and Coutts, 1980) transport O_2 to their roots. In addition, flooding causes root porosity to increase in many species (McPherson, 1939; Schramm, 1960; Yu *et al.*, 1969) as well as lenticel formation and stem hypertrophy, as discussed previously. Based on evidence of O_2 diffusion from shoot to root and increased porosity of cortical and some stelar tissues, a large number of species of vascular plants appear to have a well-developed internal aerating system.

The stimulus for intercellular space development appears to be associated with increased ethylene concentrations and increased cellulase activity in the tissue. Kawase (1981) presented a partial explanation of how flooding stimulates aerenchyma (intercellular space) to develop (Fig. 3). Cellulase may play a role in cell softening but not in cell extension (Kawase, 1981). It seems probable that because IAA also tends to accumulate in tissue under anaerobiosis, it may be involved with ethylene and cellulase activity in intercellular space development. However, Kawase (1981) noted that there is a different response among cortical cells to cellulase activity, and as yet there is no suitable explanation why under this scheme some cells become stronger and increase in size while others become weaker and die.

VI. OXIDATION OF RHIZOSPHERE

For the wetland plant, oxidation of the rhizosphere is important even if not crucial to its survival (Armstrong, 1978, 1979; Coutts and Philipson, 1978b; Hook and Scholtens, 1978). Plants of wetlands live with their roots in highly reduced soils that contain numerous gaseous, organic, and inorganic phytotoxic compounds. Hence, in large part, their survival depends on their capacity to transform soil-borne toxins to less harmful products (see Ponnamperuma, Chapters 2, and Stolzy and Sojka, Chapter 7). Such transformations are usually oxidative and thereby related to internal aeration and oxidative capacity of roots (Armstrong, 1979).

Permeability of roots of wetland species to O_2 decreases rapidly with increasing distance from the root apex (Armstrong, 1964), and O_2 diffusion may cease 2 cm distal from the root apex (Armstrong, 1979). Such decreasing permeability distally from the root apex is important, otherwise aerobic populations of organisms could build up over time along the oxidized zone and exceed the capacity of roots to supply O_2 radially from the root. As a consequence, the radially oxidized zone could collapse and permit an influx of phytotoxins. Because only the growing tip oxidizes its surroundings, microbial populations do not have the opportunity to overpower it as it moves through the soil (Armstrong, 1979).

Rhizosphere oxidation occurs only when internal respiration demands are exceeded and root tissues are permeable to radial diffusion of O_2. This process is mitigated by increased porosity of roots, which not only reduces the amount of respiring tissue per unit area but also decreases resistance to longitudinal and radial diffusion. The radius of the oxidation zone around the root also depends on soil reduction rate and O_2 demand (see Ponnamperuma, Chapter 2).

Root respiration rate is greatest at the root tip and decreases distally to the apex until it stabilizes about 3–5 cm distal to the apex (Armstrong and Gaynard, 1976). Hence the internal O_2 loss to respiration is minimized along most of the O_2-diffusion pathway. This tends to provide for diffusion of a higher O_2 concentration to the root apical region.

Using equations developed from Fick's laws of diffusion, Armstrong (1979) illustrated the effect of porosity of plant tissues and O_2 consumption on the length of the O_2-diffusion pathway in plants. With an effective porosity of 60% and plant O_2 consumption of 1 ng cm^{-3} of tissue per second, the maximum diffusion path length would be ~250 cm. Such extremely high porosity and very low O_2 consumption probably would occur only in the roots of hydrophytic species such as rice, *Eriophorum angustifolium,* and swamp tree species. The stem or trunk of wood species would likely have a much lower porosity. Coutts and Armstrong (1976) proposed that gas-filled vessels and tracheids may be important in longitudinal diffusion in woody species, and Coutts and Philipson (1978a,b) demonstrated with excised mature roots of lodgepole pine that O_2 would diffuse through the xylem tissue of the root but would diffuse farther if the bark were kept on the roots. Hook and Brown (1972) showed that air could be pulled through the xylem elements of species nontolerant of waterlogging at lower tensions than through the lenticels. This indicated there was less resistance to longitudinal flow in the xylem of the nontolerant species than across the cambium. However, oxidation of the rhizosphere of *Nyssa aquatica* (a flood-tolerant species) was less pronounced and occurred more slowly when the stem was girdled, forcing O_2 diffusion through only the cambium–xylem pathway. In nongirdled stems, O_2 was free to diffuse through the cortex and cambium–xylem pathways (Hook and Brown, 1972).

Oxygen diffusion from the shoot through the root appears to account for root penetration of rice up to 35 cm in flooded soils (Valoras and Letey, 1966) and, theoretically, up to 250 cm in woody species (Armstrong, 1979). Yet even these distances are inadequate to aerate the lateral roots of mature trees and, probably, some herbaceous species under all waterlogged conditions.

For adequate lateral root aeration, entry of O_2 from sources other than the stem would appear to be beneficial. It is in this context that the porous knees of bald cypress and the exposed roots of tupelos (Penfound, 1934) appear to perform a beneficial function. One can theorize that their removal would not kill the tree, as O_2 diffusion through the stem would be adequate for survival. The presence of knees at varying distances from the trunk would greatly aid in aerating lateral roots that may be 5–10 m or more from the trunk. However, even these structures would fail to provide adequate aeration under deep and prolonged flooding (1–2 m deep), which occurs fairly regularly in some swamps.

Hypoxia or anoxia in the root environment appears to set off a chain reaction of enzymatic and hormonal responses in the roots in which some compounds may be quickly transported to the shoot. These signals from the root may quickly reduce growth and transpiration. If the species' genetic capabilities are lacking, these compounds seem to be generated in quantities or qualities that spell immediate death to the roots and, eventually, to the entire plant. If the genetic capabilities are present, compounds such as ethylene or its precursor ACC, abscisic acid, cellulase, IAA, and carbohydrates seem to work in various combinations to stimulate immediate development of adventitious water roots and/or soil water roots, aerenchyma or intercellular space, lenticels, and stem hypertrophy, which thereby increase the capacity of the plant to transport O_2 from the shoot to the root. In the process, anaerobic respiration is usually accelerated or stabilized at some metabolic level that provides the root with an energy source to carry on nutrient uptake (Section VII). Because less tissue is synthesized under flooded conditions, less energy is required. However, if flooding persists, the internal aeration must develop such that some aerobic-respiration needs of the root are met and sufficient excess O_2 diffuses into the surrounding reduced soil to oxidize reduced nutrients and transform toxins to less harmful products.

VII. METABOLIC ADAPTATIONS

A. Metabolism

The presence of well-developed aeration systems such as those in tupelos, rice, and *Spartina alterniflora* are not sufficient to avoid anoxia in roots under all field conditions. When stomata close and/or the soil becomes highly reduced [about −200 mV redox potential (E_h)], anaerobic metabolism of roots increases substantially (John and Greenway, 1976; Mendelssohn *et al.*, 1981). In the intertidal salt-marsh species *S. alterniflora,* alcohol dehydrogenase (ADH) activity was much higher in plants growing in soils with an E_h of about −200 mV than in those closer to tidal streams, where the E_h was 200–300 mV. Energy charge (Section VII,B) was also closely correlated with soil E_h and ADH activity of roots (Mendelssohn *et al.*, 1981). Root anaerobiosis is probably even more pronounced in swamp tree species, as mentioned previously, when deep seasonal flooding of 1–2 m covers the lower trunk and all knees and exposed roots for a few days to a few weeks.

The root apices of many species contain end products of anaerobic respiration even when grown in well-aerated conditions, apparently the result of the compact structure of cells in meristematic regions (Armstrong, 1978; Crawford, 1978; Smith and ap Rees, 1979). Vartapetian (1978) pointed out aptly that aerobic respiration succeeded anaerobic respiration in the process of evolution but did not supplant it. Aerobic respiration is built as a superstructure on the anaerobic foundation and, in the absence of O_2 switched off. If an organism has retained the appropriate enzymes, the anaerobic process (glycolysis) will continue to function, albeit at a lower efficiency per unit of carbohydrate metabolized.

Considering anaerobic respiration a beneficial mechanism for adapting to a flooded habitat has been criticized because of the inefficiency of anaerobic metabolism (yields only ~2 ATP per glucose molecule versus ~38 ATP under aerobic conditions) and because some products of anaerobic respiration are presumed to be phytotoxic. The chemistry of metabolic adaptation has been reviewed in depth (Crawford, 1978; Davies, 1980); therefore, this section concentrates primarily on the general mechanisms of metabolic adaptations.

Several theories have been advanced to explain the mode of metabolic adaptation of plants to flooding (Crawford, 1966, 1967; Crawford and McManmon, 1968; Crawford and Tyler, 1969) but have been modified by Crawford (1978) into two basic explanations. The response observed depends on the degree of flooding stress the plant is capable of sustaining. The first type of adaptation consists of plants that tolerate limited seasonal flooding and adapt metabolically to hypoxia or short-term anoxia by accumulating malate as the major end product of anaerobic respiration and by controlling the rate of glycolysis and ethanol production. In this manner, the problem of ethanol toxicity is reduced, but because malate production via glycolysis does not result in a net yield of ATP, energy yield is reduced per unit of carbohydrate metabolized also. Hence this type of adaptation is only valuable for short-term periods of flooding. In the second type of adaptation, it is proposed that plants living in habitats flooded for prolonged periods, particularly during the growing season, adapt by stimulating glycolysis, ethanol production, and ATP synthesis. Toxicity is avoided by loss of ethanol through the root, and malate is not accumulated to a significant degree.

This scheme seems to fit the known responses of plants better than previous ones, but it still has some flaws. Swamp tupelo, a highly flood-tolerant tree species, loses about 55% of the ethanol from its roots, whereas loblolly pine (moderately tolerant) loses about 85% of the ethanol. Also, both species accumulate ethanol and malate, although ethanol appears to be the primary end product (Hook *et al.*, 1983) as it is in rice (Avadhani *et al.*, 1978; Smith and ap Rees, 1979; Mocquot *et al.*, 1981).

It is also argued that malate production is beneficial because it is less toxic than ethanol and it may function in transferring the O_2 deficit from roots to shoots by its upward transport in the xylem sap (Crawford, 1978). Avadhani *et al.* (1978) reported that the malate present in rice seedlings was not rapidly consumed when they were exposed to air, which raises some doubt as to the efficiency of this process. In contrast, ethanol is rapidly metabolized on exposure to aerated conditions (Rowe, 1966; Effer and Ranson, 1967). Keeley (1979) suggested that malate may accumulate in response to an ionic imbalance resulting from flooding, but Avadhani *et al.* (1978) found that the maximum amount of malate produced in rice could not account for more than 40% of the electrical balance of K^+ in rice coleoptiles. But in support of Crawford's strategy, Avadhani *et al.*, (1978) found that rice seedlings accumulated more malate at low O_2 tensions than in air or nitrogen.

Amino acid compositions of roots and plant sap are significantly altered by anaerobiosis in the roots. In pumpkins (*Cucurbita* sp. cv. Mozoleevskaya), tomato (*Lycopersicon* sp. cv. Bison), and willow (*Salix cinera*), root anaerobiosis resulted in increases of glutamic acid, γ-aminobutyric acid, and alanine and decreases in glutamine relative to the amount in aerated roots (Dubinina, 1961). D. S. DeBell (unpublished) found similar responses in the xylem sap of loblolly pine grown in seasonally and continuously flooded soils. Decreased glutamine synthesis and increased alanine synthesis under hypoxia have been reported in other species and were attributed to a low energy charge and high pyruvate concentrations (Kohl *et al.*, 1978). Also, increased CO_2 concentration, which accompanies ethanol fermentation, favors synthesis of γ-aminobutyric acid via α-ketoglutaric acid and glutamate (Zemlianukhin and Ivanov, 1978). D. S. DeBell (unpublished) suggest that alterations in the amino acid composition in the roots and its subsequent transport to the shoot via the xylem sap may provide another means of transferring the O_2 debit of roots to shoots.

The persistent concern over the toxicity of ethanol to plants seems to be largely unfounded. Because ethanol is not ionized, it is readily eliminated from cells (Davies, 1980; and as substantiated in the previous discussion). Furthermore, anaerobiosis appears to increase membrane permeability, thereby facilitating the elimination of metabolites (Crawford, 1978; Smith and ap Rees, 1979; Hook *et al.*, 1983). In addition, the tissue of species tolerant to waterlogging appears to withstand high concentrations of ethanol with little adverse effect (Chirkova,

1978). The mechanism for this increased tolerance, if it exists, is not well understood but may be related to interactions between enzymes that regulate the cell pH (Pradet and Bomsel, 1978; Zemlianukhin and Ivanov, 1978) and/or induction of quasi-dormancy in tissue (Rowe, 1966). Drew and Lynch (1980) concluded that, as yet, ethanol has not been identified as a factor in waterlogging injury to roots or shoots.

Accumulation of organic acids occurs under anaerobiosis and potentially could induce toxicity. However, this is apparently prevented by a pH-stat in the cells. For instance, if the malic enzyme and phosphoenolpyruvate carboxylase are present with malate dehydrogenase, they interact to control pH. If the pH rises, the carboxylating system is stimulated and decarboxylation is reduced, and the reverse occurs if the pH falls (Davies, 1980).

B. Energy Charge and Regulating Mechanisms

Pradet and Bomsel (1978) pointed out that adaptation by plants or other organisms to changing environmental conditions involves both coarse and fine control mechanisms. The coarse control occurs at the gene level and requires a long time to become effective. But the fine control mechanisms, which involve reactions between enzyme and substrate already present in the cell, respond almost instantaneously.

The energy charge of a cell seems to function as a fine-control mechanism and appears to have some implications in measuring adaptation to environmental changes. However, there is some controversy among researchers as to whether concentrations or ratios and mole fractions of the adenine nucleotides is more important (Davies, 1980). The energy charge is defined as the concentration of the various adenine nucleotides:

$$\frac{[(\mathrm{ATP}) + \frac{1}{2}(\mathrm{ADP})]}{[(\mathrm{ATP}) + (\mathrm{ADP}) + (\mathrm{AMP})]}$$

A normal metabolizing cell usually has an energy charge between 0.8 and 0.95 (Pradet and Bomsel, 1978). Plants intolerant to flooding generally show a reduction in energy charge from 0.2 to 0.6 under anaerobic conditions. Rice, however, shows a small drop in energy charge on exposure of anoxia and then slowly recovers and remains stable at an energy charge of ~0.8 (Pradet and Bomsel, 1978). More recently, Mocquot *et al.* (1981) showed that rice seedlings kept under anoxia maintained an energy charge of ~0.8 and some proteins were produced. They interpreted this to mean that metabolic adaptations accounted for the survival of rice seedlings for several days in the absence of O_2.

Spartina alterniflora showed reduced growth and metabolic adaptations along a transect perpendicular to a tide stream in a Louisiana salt marsh, apparently in response to increased hypoxia or anoxia (Mendelssohn *et al.*, 1981). Growth was best near the stream where soil E_h was ~300 mV, root energy charge high (~0.8), and ADH activity and malate concentration in roots low. About 8 m from the stream, growth was slightly reduced, E_h was about 100 mV, malate concentration was higher, and energy charge was near zero. At 14 m, growth was about half of that at streamside, E_h was around −200 mV, ADH activity and energy charge increased (~0.7), and malate concentration was near zero. At 18–24 m dieback occurred in the marsh grass, E_h was around −200 mV, ADH activity and energy charge (~0.6) were lower, but malate concentration increased. Ethanol concentration in roots was relatively constant across the transect, indicating that it readily leaked from the roots. These data are the most comprehensive available on any species that relate soil-waterlogging conditions to root physiology as it occurs in the soil. They indicate strong correlations among soil E_h, root metabolism, and growth of waterlogged plants. Hence, it seems probable that metabolic adaptations are as essential to survival and growth as are internal aeration of plants and oxidation of the rhizosphere.

VIII. MINERAL RELATIONS AND MYCORRHIZAE

Nutrient uptake by flooded plants is related to (1) degree of tolerance of the plant to soil waterlogging, (2) soil reduction and level of toxins generated during the flooding periods, and (3) type of soil (see also Ponnamperuma, Chapter 2, and Kozlowski and Pallardy, Chapter 5).

A. Flood-Tolerant Species

Flood-tolerant species apparently absorb nutrients from waterlogged soils with no ill effects (Valoras and Letey, 1966; Hook *et al.*, 1983) until soil reduction overpowers the capacities of the roots to oxidize reduced compounds and transform toxins to less harmful products (Howeler, 1973; Armstrong, 1979; and see Ponnamperuma, Chapter 2, and Stolzy and Sojka, Chapter 7). Under highly reduced conditions, species such as rice, *Spartina alterniflora,* heathland species, and swamp tree species show reduced growth and numerous injury symptoms (Table I) and become susceptible to diseases associated with inorganic and organic toxins [International Rice Research Institute (IRRI)], Hook *et al.*, 1970b; Howeler, 1973; Rao and Mikkelsen, 1977; Armstrong, 1979; Drew and Lynch, 1980; Mendelssohn *et al.*, 1981).

The bronze and orange diseases of rice are caused by direct and indirect iron toxicity in soils. These responses occur after flooding of soils that are very acid and high in iron. Too much iron (300–500 ppm) in leaves of rice plants results in the bronze disease, and iron oxide coatings on roots interfere with nutrient uptake, particularly P and Mg, and cause deficiencies in these nutrients in plants, resulting in the orange disease (Howeler, 1973). Several other diseases of rice are related to mineral toxicities or imbalances brought about by highly reduced conditions in the soil, low soil pH, high organic matter and low silica content, and poor drainage (IRRI, 1964).

Lowland rice generally does not respond to P fertilization under flooded conditions (Alva *et al.*, 1980), even though native P becomes more available under such conditions (Ponnamperuma, 1972; Gambrell and Patrick, 1978). Lack of response to P is attributed to release of O_2 by rice roots, which oxidizes ferrous iron to ferric oxide. In this process, available P is immobilized in the soil because of coprecipitation with ferric oxide (Alva *et al.*, 1980). Swamp tupelo is insensitive to P fertilization in flooded soils, and the lack of response in this flood-tolerant species has been attributed to oxidation of the rhizosphere also (Hook *et al.*, 1983).

Sedberry *et al.* (1971) found that rice grown in the coastal plain of Louisiana was very responsive to zinc application under flooded conditions but showed no response under well-drained conditions on the same soils. This is somewhat surprising because zinc solubility usually increases with soil reduction (Ponnamperuma, 1972). A possible explanation of this may be the formation of bicarbonates or other ion pairs that may be soluble in solution but unavailable to plants (McKee, 1980). This could be a factor in the toxic effect of high CO_2 partial pressure reported in the following discussion (Rao and Mikkelsen, 1977).

Keeley (1979) found that flooding of upland blackgum (*Nyssa sylvatica*) caused large increases in 10 nutrients in roots, particularly Mg, Fe, Mn, B, and Na. In contrast, the flood-tolerant swamp tupelo showed increases in only 3 nutrients. In a pot experiment, swamp tupelo had lower concentrations of N, K, and Mn under flooded than under well-drained conditions (Hook *et al.*, 1983).

Rice grown in nutrient solutions gassed with 99.9% pure CO_2, CH_4, N_2, and air showed dry-matter reductions only with CO_2, but N_2 and CO_2 reduced nitrogen and phosphorus uptake in the shoots. Nitrogen concentration in the roots was increased 30% by CO_2, and phosphorus in roots was increased by CH_4 and N_2 and decreased by CO_2 (Rao and Mikkelsen, 1977). Rao and Mikkelsen concluded that P was probably immobilized in the roots by CO_2, CH_4, and N_2. In comparison, Hook *et al.* (1971) found that swamp tupelo tolerated 2 and 10% CO_2 with 1% O_2 with no ill effects, but 31% CO_2 with 1% O_2 reduced root development, height growth, transpiration rate, and aerobic respiration rate. However, sweetgum was killed by 10 and 31% CO_2 within 15 and 10 days, respectively. The initial roots of swamp tupelo that survived the 10 and 31% CO_2 were coated with white, flaky compounds, probably inorganic carbonates and bicarbonates of Fe, Mn, and Mg. The soil water roots were not coated with these compounds (Hook *et al.*, 1971).

B. Moderately Tolerant and Nontolerant Species

The degree of disturbance of nutrient relations by flooding depends on the relative tolerance even within this group. Generally, a lack of O_2 in the root environment causes an immediate decrease in nutrient uptake and redistribution of nitrogen from older to younger leaves (Greenwood, 1967; Epstein, 1972; Drew and Lynch, 1980). Also, early work substantiated that intact roots take up nutrients better under anoxia than excised roots. This suggests there is a metabolic link between ion absorption in the shoot and root that is dependent on O_2 transport from shoots to roots (Larkum and Loughman, 1969). Because a number of crop plants exhibit varying degrees of O_2 transport from shoots to roots (Greenwood, 1967), such relationships may be important even in relatively non-flood-tolerant species.

In 4-month-old loblolly pine seedlings, flooding reduced total concentrations of N, P, K, Ca, Mg, Zn, and Mn as compared to those grown in well-drained conditions, but Fe concentration was much higher in flooded plants. Excessive uptake of Fe was attributed to increased permeability of the roots and possible loss of discriminatory uptake and transfer of nutrients to xylem tissue, which probably resulted from a low energy status in the roots (Hook *et al.*, 1983).

In 2-year-old loblolly pine seedlings, flooding decreased N and P concentration in the foliage but increased it in the root. Flooding effects on the other nutrients were minimal, but there was a trend of slightly higher levels of Mg, Na, and Fe in the foliage of flooded seedlings (W. H. McKee, Jr., unpublished).

Loblolly pine growing on P-deficient wet soils show large responses to P fertilizers (Pritchett, 1979). Langdon and McKee (1981) and others proposed that addition of P fertilizer may substitute for drainage of loblolly pine on wet sites in some instances. Results in semicontrolled experiments and in a growth-chamber pot study indicate that the added P increases the tolerance of loblolly pine to soil waterlogging (Hook *et al.*, 1983; D. S. DeBell, unpublished; W. H. McKee, Jr., unpublished).

In 2-year-old seedlings, P stimulated CO_2, ethanol, and malate accumulation in excised roots of flooded plants. Four-month-old seedlings grown in a growth chamber did not show the same metabolic responses, but growth was stimulated by P application, and uptake of other nutrients appeared to be stabilized by the addition of P (D. S. DeBell, unpublished).

Anaerobic conditions are known to inhibit development of mycorrhizae. Therefore, mycorrhizae have received very little attention in plants growing in flooded soils. Some data suggest that this may be a very fertile area for research in the future. For instance, in heathland vegetation waterlogging reduced but did not prevent mycorrhizal development (Malajczuk and Lamont, 1981). In peat bogs mycorrhizae are restricted to the upper soil surface (Mikola, 1973), which is typically oxidized even in a flooded soil (Patrick and DeLaune, 1972). D. D. Hook (unpublished) found mycorrhizae on roots of flooded 4-month-old loblolly pine seedlings in the upper 1–2 cm of the sand–peat mixture. Read and Armstrong (1972) noted that mycorrhizae formed on the unsuberized roots of lodgepole pine and Sitka spruce in anaerobic media and demonstrated that establishment and growth of the fungi were restricted to the portions of the roots that released O_2 into the root environment. In a glasshouse pot study using forest soils, D. D. Hook and W. H. McKee, Jr. (unpublished) found mycorrhizae growing on loblolly pine roots at all depths in the flooded pots throughout a 32-week period. The E_h of these soils averaged 0.0 mV. The latter response may be related to O_2 release from the pine roots, strains of mycorrhizae in the soil that can withstand highly reduced soils, or a combination of the two.

ACKNOWLEDGMENTS

I am indebted to Drs. William R. Harms and William H. McKee, Jr., of the Southeastern Forest Experiment Station, Charleston, South Carolina, and Dr. Robert Teskey, University of Georgia, Athens, for reviewing the manuscript and providing helpful hints, and to Myra Warren, for typing the manuscript and for editorial assistance.

REFERENCES

Alva, A. K., Larsen, S., and Bille, S. W. (1980). The influence of rhizosphere in rice crop on resin-extractable phosphate in flooded soils at various levels of phosphate application. *Plant Soil* **56,** 17–33.

Applequist, M. B. (1959). A study of soil and site factors affecting the growth and development of swamp blackgum and tupelo gum stands in southeast Georgia. D.F. Thesis, School of Forestry, Duke Univ., Durham, North Carolina.

Armstrong, W. (1964). Oxygen diffusion from the roots of some British bog plants. *Nature (London)* **204,** 801.

Armstrong, W. (1968). Oxygen diffusion from the roots of woody species. *Physiol. Plant.* **21,** 539–543.

Armstrong, W. (1969). Rhizosphere oxidation in rice: an analysis of intervarietal differences in oxygen flux from the roots. *Physiol. Plant.* **22,** 296–303.

Armstrong, W. (1978). Root aeration in the wetland condition. *In* "Plant Life in Anaerobic Environments" (D. D. Hook and R. M. M. Crawford, eds.), pp. 269–297. Ann Arbor Sci. Publ., Ann Arbor, Michigan.

Armstrong, W. (1979). Aeration in higher plants. *Adv. Bot. Res.* **7,** 225–332.

Armstrong, W., and Boatman, D. J. (1967). Some field observations relating the growth of bog plants to conditions of soil aeration. *J. Ecol.* **55,** 101–110.

Armstrong, W., and Gaynard, T. J. (1976). The critical oxygen pressure for respiration in intact plants. *Physiol. Plant.* **37,** 200–206.

Armstrong, W., and Read, D. J. (1972). Some observations on oxygen transport in conifer seedlings. *New Phytol.* **71,** 55–62.

Avadhani, P. N., Greenway, H., Lefroy, R., and Prior, L. (1978). Alcoholic fermentation and malate metabolism in rice germinating at low oxygen concentrations. *Aust. J. Plant Physiol.* **5,** 15–25.

Barber, D. A., Ebert, J., and Evans, N. T. S. (1962). The movement of ^{15}O through barley and rice plants. *J. Exp. Bot.* **13,** 397–403.

Bergman, H. F. (1920). The relation of aeration to the growth and activity of roots and its influence on the ecesis of plants in swamps. *Ann. Bot. (London)* **34,** 13–33.

Bergman, H. F. (1959). Oxygen deficiency as a cause of diseases in plants. *Bot. Rev.* **25,** 418–485.

Bradford, K. J., and Yang, S. F. (1981). Physiological responses of plants to waterlogging. *HortScience* Special Insert 16, 25–30.

Broadfoot, W. M., and Williston, H. L. (1973). Flooding effects on southern forests. *J. For.* **71,** 584–587.

Bryant, A. E. (1934). Comparison of anatomical and histological differences between roots of barley grown in aerated and non-aerated culture solutions. *Plant Physiol.* **9,** 389–391.

Chirkova, T. V. (1968). Oxygen supply to roots of certain woody plants kept under anaerobic conditions. *Fiziol. Rast. (Moscow)* **15,** 565–568.

Chirkova, T. V. (1978). Some regulatory mechanisms of plant adaptation to temporal anaerobiosis. *In* "Plant Life in Anaerobic Environments" (D. D. Hook and R. M. M. Crawford, eds), pp. 137–154. Ann Arbor, Sci. Publ., Ann Arbor, Michigan.

Chirkova, T. V., and Gutman, T. S. (1972). Physiological role of branch lenticels in willow and poplar under conditions of root anaerobiosis. *Sov. Plant Physiol. (Engl. Transl.)* **19,** 289–295.

Chirkova, T. V., and Soldatenkov, S. V. (1965). Paths of movement of oxygen from leaves to roots kept under anaerobic conditions. *Fiziol. Rast. (Moscow)* **12,** 216–225.

Cochran, P. H. (1972). Tolerance of lodgepole and ponderosa pine seedlings to high water tables. *Northwest Sci.* **46,** 322–331.

Coult, D. A., and Vallance, K. B. (1958). Observations on the gaseous exchanges which take place between *Menyanthes trifoliata* L. and its environment. *J. Exp. Bot.* **9,** 384–402.

Coutts, M. P., and Armstrong, W. (1976). Role of oxygen transport in the tolerance of trees to waterlogging. *In* "Tree Physiology and Yield Improvement" (M. G. R. Cannell and F. T. Last, eds.), pp. 361–385. Academic Press, New York.

Coutts, M. P., and Philipson, J. J. (1978a). Tolerance of tree roots to waterlogging. I. Survival of Sitka spruce and lodgepole pine. *New Phytol.* **80,** 63–69.

Coutts, M. P., and Philipson, J. J. (1978b). Tolerance of tree roots to waterlogging. II. Adaptation of Sitka spruce and lodgepole pine to waterlogged soil. *New Phytol.* **80,** 71–77.

Cowles, S. W., III (1975). Metabolism measurements in a cypress dome. M.S. Thesis, Univ. of Florida, Gainesville.

Crawford, R. M. M. (1966). The control of anaerobic respiration as a determining factor in the distribution of the genus *Senecio*. *J. Ecol.* **54,** 403–413.

Crawford, R. M. M. (1967). Alcohol dehydrogenase activity in relation to flooding tolerance in roots. *J. Exp. Bot.* **18,** 458–464.

Crawford, R. M. M. (1978). Metabolic adaptation to anoxia. *In* "Plant Life in Anaerobic Environments" (D. D. Hook and R. M. M. Crawford, eds.), pp. 119–154. Ann Arbor Sci. Publ., Ann Arbor, Michigan.

Crawford, R. M. M., and McManmon, M. (1968). Inductive responses of alcohol and malic dehydrogenases in relation to flooding tolerance in roots. *J. Exp. Bot.* **19,** 435–441.

Crawford, R. M. M., and Tyler, P. D. (1969). Organic acid metabolism in relation to flooding tolerance in roots. *J. Ecol.* **57,** 235–244.

Crocker, W., and Davis, W. E. (1914). Delayed germination in seed of *Alisma plantago*. *Bot. Gaz. (Chicago)* **58,** 285–321.

Dacey, J. W. H. (1981). Pressure ventilation in the yellow water lily. *Ecology* **62,** 1137–1147.

Dacey, J. W. H., and Klug, M. J. (1982). Ventilation by floating leaves in *Nuphar*. *Am. J. Bot.* **69,** 999–1003.

Davies, D. D. (1980). Anaerobic metabolism and the production of organic acids. *In* "The Biochemistry of Plants" (D. D. Davies, ed.), Vol. 2, pp. 581–611. Academic Press, New York.

DeBell, D. S., and Naylor, A. W. (1972). Some factors affecting germination of swamp tupelo seeds. *Ecology* **53,** 504–506.

Devaux, H. (1900). Recherches sur les lenticelles. *Ann. Sci. Nat. VI. Bot.* **12,** 1–240.

deWit, M. C. J. (1978). Morphology and function of roots and shoot growth of crop plants under oxygen deficiency. *In* "Plant Life in Anaerobic Environments" (D. D. Hook and R. M. M. Crawford, eds.), pp. 333–350. Ann Arbor Sci Publ., Ann Arbor, Michigan.

Dickson, R. E., Hosner, J. F., and Hosley, N. W. (1965). The effects of four water regimes upon the growth of four bottomland tree species. *For. Sci.* **11,** 299–305

Drew, M. C., and Lynch, J. M. (1980). Soil anaerobiosis, micro-organisms, and root function. *Annu. Rev. Phytopathol.* **18,** 37–66.

Drew, M. C., Jackson, M. B., and Giffard, S. C. (1979). Ethylene-promoted adventitious rooting and development of cortical air spaces (aerenchyma) in roots may be adaptive responses to flooding in *Zea mays* L. *Planta* **147,** 83–88.

DuBarry, A. P. (1963). Germination of bottomland tree seeds while immersed in water. *J. For.* **61,** 225–226.

Dubinina, I. M. (1961). Metabolism of roots under various levels of aeration. *Fiziol. Rast. (Moscow)* **8,** 314–322.

Edwards, T. I. (1933). The germination and growth of *Peltandra virginica* in absence of oxygen. *Bull. Torrey Bot. Club* **60,** 573–581.

Effer, W. R., and Ranson, S. L. (1967). Respiratory metabolism in buckwheat seedlings. *Plant Physiol.* **42,** 1042–1052.

Epstein, E. (1972). "Mineral Nutrition of Plants: Principles and Perspectives." Wiley, New York.

Esashi, Y., Tsukada, Y., and Ohhara, Y. (1978). Interrelation between low temperature and anaerobiosis in the induction of germination of cocklebur seed. *Aust. J. Plant Physiol.* **5,** 337–345.

Gambrell, R. P., and Patrick, W. H., Jr. (1978). Chemical and microbiological properties of anaerobic soils and sediments. *In* "Plant Life in Anaerobic Environments" (D. D. Hook and R. M. M. Crawford, eds.), pp. 375–423. Ann Arbor Sci. Publ. Ann Arbor, Michigan.

Gill, C. J. (1975). The ecological significance of adventitious rooting as a response to flooding in woody species with special reference to *Alnus glutinosa* (L). Gaertn. *Flora* **164,** 85–97.

Greenwood, D. J. (1967). Studies of the transport of oxygen through the stems and roots of vegetable seedlings. *New Phytol.* **66,** 337–347.

Haas, A. R. C. (1936). Growth and water relations of the avocado fruit. *Plant Physiol.* **11,** 383–400.

Harms, W. R. (1973). Some effects of soil type and water regime on growth of tupelo seedlings. *Ecology* **54,** 188–193.

Harms, W. R., Schreuder, H. T., Hook, D. D., Brown, C. L., and Shropshire, F. W. (1980). The effects of flooding on the swamp forest in Lake Ocklawaha, Florida. *Ecology* 61, 1412–1421.

Heinicke, A. J. (1932). The effect of submerging the roots of apple trees at different seasons of the year. *J. Am. Soc. Hortic. Sci.* **29,** 205–207.

Heinicke, A. J., Boynton, D., and Reither, W. (1940). Cork experimentally produced in Northern Spy apples. *Proc. Am. Soc. Hortic. Sci.* **37,** 47–52.

Hook, D. D. (1968). Growth and development of swamp tupelo [*Nyssa sylvatica* var. *biflora* (Walt.) Sarg.] under different root environments. Ph.D. Dissertation, Univ. of Georgia, Athens.

Hook, D. D., and Brown, C. L. (1972). Permeability of the cambium to air in trees adapted to wet habitats. *Bot. Gaz. (Chicago)* **133,** 304–310.

Hook, D. D., and Brown, C. L. (1973). Root adaptations and relative flood tolerance of five hardwood species. *For. Sci.* **19,** 225–229.

Hook, D. D., and Scholtens, J. R. (1978). Adaptations and flood tolerance of tree species. *In* "Plant Life in Anaerobic Environments" (D. D. Hook and R. M. M. Crawford, eds.), pp. 299–331. Ann Arbor Sci. Publ., Ann Arbor, Michigan.

Hook, D. D., Brown, C. L., and Kormanik, P. P. (1970a). Lenticel and water root development of swamp tupelo under various flooding conditions. *Bot. Gaz. (Chicago)* **131,** 217–224.

Hook, D. D., Langdon, O. G., Stubbs, J., and Brown, C. L. (1970b). Effect of water regimes on the survival, growth, and morphology of tupelo seedlings. *For. Sci.* **16,** 304–311.

Hook, D. D., Brown, C. L., and Kormanik, P. P. (1971). Inductive flood tolerance in swamp tupelo [*Nyssa sylvatica* var. *biflora* (Walt.) Sarg.]. *J. Exp. Bot.* **22,** 78–89.

Hook, D. D., DeBell, D. S., McKee, W. H., Jr., and Askew, J. L. (1983). Responses of loblolly pine (Mesophyte) and swamp tupelo (Hydrophyte) seedlings to soil flooding and phosphorus. *Plant Soil* **71,** 383–394.

Hosner, J. F. (1960). Relative tolerance to complete inundation of fourteen bottomland tree species. *For. Sci.* **6,** 246–251.

Hosner, J. F., and Boyce, S. G. (1962). Tolerance to water saturated soil of various bottomland hardwoods. *For. Sci.* **8,** 180–186.

Howeler, R. H. (1973). Iron-induced Orange disease of rice in relation to physico-chemical changes in a flooded oxisol. *Soil Sci. Soc. Am. Proc.* **37,** 898–903.

International Rice Research Institute (IRRI) (1964). Annual report. Los Baños, Laguna, Philippines.

Jackson, M. B., and Campbell, D. J. (1975). Ethylene and waterlogging effects on tomato. *Ann. Appl. Biol.* **81,** 102–105.

Jackson, M. B., Drew, M. C., and Giffard, S. C. (1981). Effects of applying ethylene to the root system of *Zea mays* on growth and nutrient concentration in relation to flooding tolerance. *Physiol. Plant.* **52,** 23–28.

Jackson, W. T. (1955). The role of adventitious roots in recovery of shoots following flooding of the original root systems. *Am. J. Bot.* **42,** 816–819.

John, C. D., and Greenway, H. (1976). Alcoholic fermentation and activity of some enzymes in rice roots under anaerobiosis. *Aust. J. Plant. Physiol.* **3,** 325–326.

Jones, H. E., and Etherington, J. R. (1970). Comparative studies of plant growth and plant distribution in relation to waterlogging. VI. The survival of *Erica cinerea* L. and *E. tetralix* L. and its apparent relationship to iron and manganese uptake in waterlogged soil. *J. Ecol.* **58,** 487–496.

Kawase, M. (1981). Anatomical and morphological adaptation of plants to waterlogging. *Hort-Science* Special Insert 16, 30–34.

Keeley, J. E. (1979). Population differentiation along a flood frequency gradient: physiological adaptations to flooding in *Nyssa sylvatica*. *Ecol. Monon.* **49,** 89–108.

Kohl, J. G., Baierova, J., Radke, G., and Ramshorn, K. (1978). Regulative interaction between anaerobic catabolism and nitrogen assimilation as related to oxygen deficiency in maize roots. *In* "Plant Life in Anaerobic Environments" (D. D. Hook and R. M. M. Crawford, eds.), pp. 473–496. Ann Arbor Sci. Publ., Ann Arbor, Michigan.

Kordan, H. A. (1974). Patterns of shoot and root growth in rice seedlings germinating under water. *J. Appl. Ecol.* **11,** 685–690.

Kordan, H. A. (1976). Adventitious root initiation and growth in relation to oxygen supply in germinating rice seedlings. *New Phytol.* **76,** 81–86.

Kozlowski, T. T. (1982). Water supply and tree growth. II. Flooding. *For. Abstr.* **43,** 145–161.

Kramer, P. J. (1951). Causes of injuries to plants resulting from flooding of soil. *Plant Physiol.* **26,** 722–736.

Kramer, P. J. (1969). "Plant and Soil Water Relationships: A Modern Synthesis." McGraw-Hill, New York.

Kramer, P. J., and Jackson, W. F. (1954). Causes of injury to flooded tobacco plants. *Plant Physiol.* **29,** 241–245.

Kramer, P. J., and Kozlowski, T. T. (1979). ''Physiology of Woody Plants.'' Academic Press, New York.

Kramer, P. J., Riley, W. S., and Bannister, T. T. (1952). Gas exchange of cypress knees. *J. Ecol.* **33,** 117–121.

Lamont, B. B. (1981). Specialized roots of non-symbiotic origin in heathlands. *In* ''Ecosystems of the World'' 9B, Heathlands and Related Shrublands. Analytical Studies (R. L. Specht, ed.), pp. 183–195. Elsevier, Amsterdam.

Langdon, O. G., and McKee, W. H., Jr. (1981). Can fertilization of loblolly pine on wet sites reduce the need for drainage? *U.S. For. Serv. Southeast. For. Exp. Stn. Gen. Tech. Rep. No. SE–34, pp. 212–218.*

Langdon, O. G., DeBell, D. S., and Hook, D. D. (1978). Diameter growth of swamp tupelos: seasonal pattern and relation to water table level. Proc. N. Am. For. Biol. Workshop 5th, pp. 326–333.

Larkum, A. W. D., and Loughman, B. C. (1969). Anaerobic phosphate uptake by barley plants. *J. Exp. Bot.* **20,** 12–24.

Lindsey, A. A., Petty, R. O., Sterling, D. K., and van Asdall, W. (1961). Vegetation and environment along the Wabash and Tippecanoe Rivers. *Ecol. Monogr.* **31,** 105–154.

Malajczuk, N., and Lamont, B. B. (1981). Specialized roots of symbiotic origin in heathlands. *In* ''Ecosystems of the World 9B Heathlands and Related Shrublands. Analytical Studies'' (R. L. Specht, ed.), pp. 165–182. Elsevier, Amsterdam.

McDermott, R. E. (1954). Effect of saturated soil on seedling growth of some bottomland hardwood species. *Ecology* **35,** 36–41.

McKee, W. H., Jr. (1980). Changes in solution of forest soil with waterlogging and three levels of base saturation. *Soil Sci. Soc. Am. J.* **44,** 388–391.

McPherson, D. C. (1939). Cortical air spaces in the roots of *Zea mays* L. *New Phytol.* **35,** 64–73.

Mazé, M. P. (1900). Recherches sur le rôle de l'oxygen dans la germination. *Ann. Inst. Pasteur* **14,** 350–368.

Mendelssohn, I. A., McKee, K. L., and Patrick, W. H., Jr. (1981). Oxygen deficiency in *Spartina alterniflora* roots: metabolic adaptations to anoxia. *Science (Washington, D.C.)* **214,** 439–441.

Mikola, P. (1973). Application of mycorrhiza symbiosis in forestry practice. *In* ''Ectomycorrhizae'' (G. C. Marks and T. T. Kozlowski, eds.), pp. 383–411. Academic Press, New York.

Mocquot, B., Prat, C., Mouches, C., and Pradet, A. (1981). Effect of anoxia on energy charge and protein synthesis in rice embryo. *Plant Physiol.* **68,** 636–640.

Morinaga, T. (1926). The favorable effect of reduced oxygen supply upon the germination of certain seed. *Am. J. Bot.* **13,** 159–166.

Nagai, I. (1916). Some studies on the germination of the seed of *Oryza sativa. J. Agr. Coll. Tokyo Imp. Univ.* **3,** 109–156.

Oskamp, J., and Batjer, L. (1932). Soils in relation to fruit growing in New York. II. Size, production, and rooting habit of apple trees on different soil types in the Hilton and Morton area, Monroe County. Cornell Univ. Agric. Exp. Stn. Bull. No. 550.

Parker, J. (1949). The effects of flooding on the transpiration and survival of some southeastern forest tree species. *Plant Physiol.* **25,** 453–460.

Patrick, W. H., Jr., and DeLaune, R. D. (1972). Characterization of the oxidized and reduced zones in flooded soil. *Soil Sci. Soc. Am. Proc.* **36,** 573–576.

Penfound, W. T. (1934). Comparative structure of the wood in the ''knees,'' swollen bases, and normal trunks of the tupelo gum (*Nyssa aquatica* L.). *Am. J. Bot.* **21,** 623–631.

Pereira, J. S., and Kozlowski, T. T. (1977). Variations among woody angiosperms in response to flooding. *Physiol. Plant.* **41,** 184–192.

Philipson, J. J., and M. P. Coutts (1980). The tolerance of free roots to waterlogging. IV. Oxygen transport in woody roots of Sitka spruce and lodgepole pine. *New Phytol.* **85,** 489–494.

Phillips, I. D. J. (1964). Root-shoot hormone relations. II. Changes in endogeneous auxin concentration produced by flooding the root system in *Helianthus annus*. *Ann. Bot. (London)* **23,** 37–45.

Ponnamperuma, F. N. (1972). The chemistry of submerged soils. *Adv. Agron.* **24,** 29–95.

Pradet, A., and Bomsel, J. L. (1978). Energy metabolism in plants under hypoxia and anoxia. *In* "Plant Life in Anaerobic Environments" (D. D. Hook and R. M. M. Crawford, eds.), pp. 89–118. Ann Arbor Sci. Publ., Ann Arbor, Michigan.

Pritchett, W. L. (1979). "Properties and Management of Forest Soils." Wiley, New York.

Rao, D. N., and Mikkelsen, D. S. (1977). Effects of CO_2, CH_4, and N_2 on growth and nutrition of rice seedlings. *Plant Soil* **47,** 313–322.

Read, D. J., and Armstrong, W. (1972). A relationship between oxygen transport and the formation of the ecotrophic mycorrhizal sheath in conifer seedlings. *New Phytol.* **71,** 49–53.

Regehr, D. L., Bazzaz, F. A., and Boggess, W. R. (1975). Photosynthesis, transpiration, and leaf conductance of *Populus deltoides* in relation to flooding and drought. *Photosynthetica* **9,** 52–61.

Rowe, R. N. (1966). Anaerobic metabolism and cyanogenic glycoside hydrolysis in differential sensitivity of peach, plum, and pear roots in water-saturated conditions. Ph.D. Thesis, Univ. of California, Davis.

Sasaki, T. (1927). On an abnormal type of germination of rice seed under reduced air supply. *J. Sci. Agr. Soc.* **228,** 101–102.

Scholander, P. F., Van Dam, L., and Scholander, S. I. (1955). Gas exchange in roots of mangroves. *Am. J. Bot.* **42,** 92–98.

Schramm, R. J., Jr. (1960). Anatomical and physiological development of root in relation to aeration of the substrate. *Diss. Abstr.* **21,** 2089.

Sedberry, J. E., Peterson, F. J., Wilson, E., Nugent, A. J., Engler, R. M., and Brupbacher, R. H. (1971). Effect of zinc and other elements on the yield of rice and nutrient content of rice plants. *La. Agric. Exp. Stn. Bull.* No. 653.

Sena Gomes, A. R., and Kozlowski, T. T. (1980). Growth responses and adaptations of *Fraxinus pennsylvanica* seedlings to flooding. *Plant Physiol.* **66,** 267–271.

Shadan, M. (1980). Fixation, translocation, and root exudation of $^{14}CO_2$ by *Phaeseolus vulgaris* L. subjected to root anoxia. M.S. Thesis, Michigan State Univ., East Lansing.

Shunk, I. V. (1939). Oxygen requirements for germination of *Nyssa aquatica*—tupelo gum. *Science (Washington, D.C.)* **90,** 565–566.

Sifton, H. B. (1945). Air-space tissue in plants. *Bot. Rev. (Chicago)* **11,** 108–143.

Smith, A. M., and ap Rees, T. (1979). Effects of anaerobiosis on carbohydrate oxidation by roots of *Pisum sativum. Phytochemistry* **18,** 1453–1458.

Snow, L. M. (1904). The effects of external agents on the production of root hairs. *Bot. Gaz. (Chicago)* **37,** 143–145.

Soldatenkoz, S. V., and Hsien-Tuan, C. (1961). The role of bean and corn leaves in respiration of oxygen-deprived roots. *Fiziol. Rast. (Moscow)* **8,** 385–394.

Tang, Z. C., and Kozlowski, T. T. (1982). Some physiological and morphological responses of *Quercus macrocarpa* seedlings to flooding. *Can. J. For. Res.* **12,** 196–202.

Templeton, J. (1926). Hypotrophied lenticels on the roots of cotton plants. Ministry Agr., Egypt. *Bull.—Tech. Sci. Serv.* **59.**

Valoras, N., and Letey, J. (1966). Soil oxygen and water relationships to rice growth. *Soil Sci.* **101,** 210–215.

van Raalte, M. H. (1940). On the oxygen supply of rice roots. *Ann. Jard. Bot. Buitenzorg.* **50,** 99–113.

Vartapetian, B. B. (1978). Life without oxygen. *In* "Plant Life in Anaerobic Environments" (D. D. Hook and R. M. M. Crawford, eds.), pp. 1–11. Ann Arbor Sci. Publ., Ann Arbor, Michigan.

Vartapetian, B. B., Andreeva, I. N., and Nuritdinov, N. (1978). Plant cells under oxygen stress. *In*

"Plant Life in Anaerobic Environments" (D. D. Hook and R. M. M. Crawford, eds.), pp. 13–88. Ann Arbor Sci. Publ., Ann Arbor, Michigan.

Wample, R. L., and Reid, D. M. (1975). Effect of aeration on the flood-induced formation of adventitious roots and other changes in sunflower (*Helianthus annuus*). *Planta* **127,** 263–270.

Wample, R. L., and Reid, D. M. (1978). Control of adventitious root production and hypocotyl hypertrophy of sunflower (*Helianthus annuus*) in response to flooding. *Physiol. Plant.* **44,** 351–358.

Wample, R. L., and Reid, D. M. (1979). The role of endogenous auxins and ethylene in the formation of adventitious roots and hypocotyl hypertrophy in flooded sunflower plants (*Helianthus annuus*). *Physiol. Plant,* **45,** 219–226.

Weaver, J. E., and Himmell, W. J. (1930). Relations of increased water content and decreased aeration to root development in hydrophytes. *Plant Physiol.* **5,** 69–92.

Wenkert, W., Fausey, N. R., and Watters, H. D. (1981). Flooding responses in *Zea mays* L. *Plant Soil* **62,** 351–366.

Whitford, L. A. (1956). A theory on the formation of cypress knees. *J. Elisha Mitchell Sci. Soc.* **72,** 80–83.

Whitlow, T. H., and Harris, R. W. (1979). Flood tolerance in plants: a state of the art review. *U.S. Army Corps Eng. Waterways Exp. Stn. Environ. Lab. Tech. Rep.* E–79–2.

Williamson, R. E., and Splinter, W. E. (1968). Effect of gaseous composition of root environment upon root development and growth of *Nicotiana tobaccum* L. *Agron. J.* **60,** 365–368.

Yelenosky, G. (1964). The tolerance of trees to poor soil aeration. *Diss. Abstr.* **25,** 734–735 (FA 26#3432).

Yu, P. T., Stolzy, L. H., and Letey, J. (1969). Survival of plants under prolonged flooded conditions. *Agron. J.* **61,** 844–847.

Zemlianukhin, A. A., and Ivanov, B. F. (1978). Metabolism of organic acids of plants in the conditions of hypoxia. *In* "Plant Life in Anaerobic Environments" (D. D. Hook and R. M. M. Crawford, eds.), pp. 169–202. Ann Arbor Sci. Publ., Ann Arbor, Michigan.

Zeroni, M., Jerie, P. H., and Hall, M. A. (1977). Studies on the movement and distribution of ethylene in *Vicia faba* L. *Planta* **134,** 119–125.

CHAPTER 9

Adaptations of Plants to Flooding with Salt Water

S. J. WAINWRIGHT
Department of Botany and Microbiology
University College, Swansea
Swansea, Wales

I. INTRODUCTION

Soils may be saline for entirely natural reasons. For example, coastal soils may be regularly flooded with saline water, or inland soils may become saline when they occur adjacent to lakes that have net water loss resulting from evaporation. Soils may become secondarily saline as a result of human activities. Soils may become flooded with saline water because of brine pumping, and roadside soils may suffer from saline runoff water from highways treated with deicing salt. Irrigation can, for a variety of reasons, result in increased soil salinity.

Although the causes of soil salinity are diverse, and not all of the causes are related to the inundation of the soil by saline water, many of the problems experienced by plants as a result of soil salinity are shared by them, whatever the immediate cause of the salinity. Consequently, the adaptations that plants have

ISBN 0-12-424120-4

evolved to tolerate salinity, whatever its cause, are likely to be relevant to plants which are able to tolerate flooding by saline water. Hence, this chapter does not concentrate exclusively on salinity resulting from flooding with saline water, but also considers other aspects of salinity and salt tolerance. The mechanisms present in various plants that enable them to tolerate saline conditions and some of the ways in which it may be possible for humans to increase the salt tolerance of selected species are examined.

II. SOIL SALINITY AND ADAPTATION AT THE LEVEL OF THE POPULATION

A. True Halophytes, Salt-Tolerant Glycophytes, and Salt-Susceptible Glycophytes

Degree of Adaptation to the Saline Environment and Variation in Salt Tolerance in Populations of Glycophytes

Soil salinity can be present for a variety of reasons. There are areas of land that are naturally saline. The most important of these habitats are saline because of their maritime location and the fact that they are regularly inundated with seawater or saline estuarine water. These naturally saline habitats sustain communities of plants such as salt-marsh or mangrove vegetation, depending on the latitude in which they are situated. These communities are in equilibrium with the salinity of the environment. Although the equilibrium is dynamic, there are successions of plant communities that invade or retreat from these areas as the pedological processes of substrate accretion or erosion bring about changes in the surface level and, hence, the frequency and duration of inundation.

As these saline habitats have always existed at the interface of terrestrial and marine environments, it is to be expected that species will have evolved that are exclusive to salt marsh, spray zone, strand line, and mangrove communities. The high proportion of halophytes in families such as the Chenopodiaceae may be indicative of the ancient origin of halophytic angiosperms (Fitter and Hay, 1981).

That species exist that are distributed exclusively in saline habitats does not necessarily mean they reflect a physiological requirement for a saline substrate. It is common for physiological and ecological optima to be different with respect to specific environmental factors (Rorison, 1969).

The ecological response of a species toward gradients in environmental factors such as salinity is moderated by its competitive relationship with other species in the community. In mesic environments, the success of a species is largely determined by its capacity to compete with (or avoid competition with) other members of the community. In extreme environments such as saline habitats, a species must not only be able to tolerate various environmental stresses, but must also be able to compete with other successful members of the community.

There has been discussion in the literature concerning the classification of species as *halophytes* or *glycophytes*. Stoker (1928) defined halophytic conditions as soils containing at least 0.5% NaCl on a dry-weight basis. Chapman (1960) suggested that a more realistic criterion based on the salt actually available to plants would be soil content of at least 0.5% NaCl in the soil solution. Richards (1954) suggested that a salinity of 100 mol m^{-3} NaCl of saturated soil extracts is the concentration limiting growth of crop species. Adriani (1956) and Waisel (1972) discussed various schemes for classifying plants as halophytes or glycophytes. A system of classification based simply on the salinity of soil in which plants grow does not account for interactions between salinity and other factors. Gale *et al.* (1970) showed that growth of *Atriplex halimus* in saline culture was affected by relative humidity (RH). Hoffman and Jobes (1978) showed that high RH increases the salt tolerance of barley. Stelzer and Läuchli (1977) showed that the effect of salinity on growth of *Puccinellia peisonis* is affected by oxygen supply to the roots. They also concluded (Stelzer and Läuchli, 1978) that the double endodermis and aerenchyma present in *Puccinellia* roots appear to be involved in salt and flood tolerance. There is strong interaction between temperature and salinity for seed germination, seedling emergence, and seedling dry weight in *Beta vulgaris* (Mahmoud and Hill, 1980). For example, salinity had little effect on seedling emergence at 10–15°C but was increasingly inhibitory between 25 and 35°C. In the absence of salt, germination was maximal at 25°C.

It has been suggested that growth stimulation by salt be incorporated in the definition of a halophyte [Tsopa, 1939 (quoted by Chapman, 1960); Waisel, 1972]. However, although this would include species such as *Suaeda maritima* (Yeo and Flowers, 1980), which shows stimulation of both fresh and dry weight by salinity with an optimum NaCl concentration of 170 mol m^{-3}, it would exlude species such as *Atriplex hastata* (Black, 1956). Moreover, such a scheme would include even the salt-susceptible ecotype of *Agrostis stolonifera* studied by Hodson *et al.* (1982), which showed a 60% increase in dry-weight increment by 50 mol m^{-3} NaCl in culture solution but whose growth was completely inhibited by 200 mol m^{-3} NaCl. This contrasted with the response of the salt-tolerant ecotype, which showed maximum growth (a 30% stimulation) at 100 mol m^{-3} NaCl, but only a 36% inhibition at 200 mol m^{-3}. Yeo and Flowers (1980) suggested that descriptions of growth stimulation by salt in halophytes are somewhat arbitrary. They pointed out that increased growth with increased salt concentration results from the relationship between turgor pressure and extension growth. They suggested that salt tolerance in halophytes is a phenomenon separate from the growth response. They found that large increases in fresh and dry weight of *S. maritima* represented only small increases in organic dry weight, which is the measure of true growth. Much of the increase in weight was the result of the salt taken up (and increased succulence). They found that although growth increase was progressive with increased salinity, it did not increase proportionately and a plateau of organic dry weight was reached. Plants grown in saline conditions had larger cells rather than more cells. Neales and Sharkey (1981) showed that in the halophyte *Disphyma australe* grown at 500 mol m^{-3} NaCl, the inorganic component represented 55% of total leaf dry weight. Flowers (1975) pointed out that as salt tolerance in plants forms a continuous distribution, it may not be meaningful to attempt to draw a line in this distribution in order to place plants into one of two categories. Wyn Jones (1981) did not believe that an exclusive classification of plants as halophytes or glycophytes could be sustained.

Notwithstanding these difficulties, the concepts of halophytes and glycophytes are useful. In general, halophytes can be regarded as plants that complete their life cycles in saline environments, and glycophytes as plants that cannot. Flowers (1975) was more specific about the degree of salinity of the environments inhabited by halophytes and defined halophytes as plants completing their life cycle in the presence of a salt content of 0.5 to 6% on a weight-per-volume basis in the aqueous phase. So far as individual species are concerned, it is probably more meaningful to describe the stresses to which they are tolerant, for example, saline soil, saline soil and poor aeration, periodic total submersion in saline water, high magnesium concentration, salinity, etc. However, care must be taken to relate tolerance to specific factors that are *present* in the plant's natural environment. For example, a salt-marsh ecotype of *Agrostis stolonifera* was more tolerant to chlorides of all the alkali metals than was an inland ecotype, even though only Na^+ was present in high concentrations in the salt marsh (Hodson *et al.*, 1982). It seems likely that evolution of tolerance to sodium increased tolerance to the other alkali metals, possibly because of similarities in their chemistry. There is, however, no apparent ecological significance to the increased cotolerance toward the other alkali metals. Halophytes mainly do not have an obligate requirement for high salinity, although *Halogeton glomeratus* will not survive in nonsaline culture solution (Williams, 1960), and both *Salicornia europaea* and *S. bigelowii* do not grow well if NaCl is absent from their culture medium (Webb, 1966). Rather than having an obligate requirement for high salinity, it is likely that these plants have a higher than normal requirement for NaCl. This would be expected if these plants possess very efficient sequestering mechanisms to remove NaCl from sites of metabolism, thus creating an artificial deficiency within the cytoplasm. Analogous situations have been reported for elevated zinc requirements in zinc-tolerant plants (Antonovics *et al.*, 1971). However, Romney and Wallace (1980), in studying the ecotonal distribution of salt-tolerant shrubs in the northern Mojave Desert, in Nevada, found species ranging in response to salt from facultative halophytes to those requiring high soil salinity for normal growth and development.

Thus, in the primary, naturally saline environments of the world, over millions of years, species have evolved that are well adapted to life in the presence of high salt concentrations. They are able to complete their life cycles in saline conditions. These are the true halophytes.

Although the naturally saline coastal soils produce problems for the plants inhabiting them, they do not, in general, create difficulties for humans in exploitation of vegetation. The halophytes of these saline soils are well adapted to prevailing conditions, and human exploitation is largely limited to use of the *natural* vegetation, often as a source of grazing for animals, as exemplified by the north Gower salt marshes of Wales. Secondary salinization is a far more serious problem, as it usually results in loss of once fertile agricultural land, and unlike natural coastal saline land, secondary saline land is continuously increasing in area. It is difficult to obtain precise information concerning the areas of land involved (Flowers *et al.*, 1977), but it is possible to make realistic estimates that put soil salinity into perspective. Of the ~13,300 million ha of the earth's land surface, ~1406 million ha are under cultivation and about 16% of the cultivated land is subject to irrigation. Irrigation involves diversion of large quantities of water, either to provide water in areas of uncertain rainfall in arid or semiarid regions or to supplement an existing substantial rainfall to make up slight water deficits (as done in southeast England during the summer months). Of all irrigated land, 64.8% is in Asia, 10% in the United States, 8.3% in the U.S.S.R., and 5.9% in Europe. The remaining 11% is distributed around the rest of the world (Cantor, 1967).

Irrigation thus brings about changes in the hydrological cycle of vast areas of land. About one-third of the irrigated land is affected by salinity, and, for example, in the Punjab in India in the 1960's ~40,000 ha of land per year were going out of use because of increased salinity, primarily a result of irrigation (Raheja, 1966).

Irrigation schemes obtain and apply water in various ways, and these can variously lead to soil salinization. Surface water can be diverted, or water can be extracted from the ground by free-flowing wells or by pumping. The extraction of water from underground can lead to a lowering of the water table. In coastal regions, if the water table falls below sea level, incursion of saline water can render ground supplies unsuitable for further irrigation. Good irrigation water contains salts at concentrations of ~17 mol m^{-3}, and marginal water contains between 17 and 51 mol m^{-3} of dissolved salts. As irrigation accounts for about 80% of the world's water use during the farming season (Flowers *et al.*, 1977), appreciable quantities of salt are obviously added to the soil by irrigation.

Many irrigation schemes involve application of flowing water directly to the soil surface. This surface water causes a sorting of the surface particles; the finer particles remain in suspension for longer periods of time. When they eventually sediment out, the fine particles cause clogging of the soil pores. This in turn reduces water-infiltration rates, which increases evaporative losses and causes salts to concentrate in the surface layers. Moreover, application of surface water, particularly overwatering, can quickly raise the water table, and in extreme cases this can bring about waterlogging, which impedes drainage and downward leaching of salts. In the first 25 years of irrigation of the Snake Pass plain (Idaho), the water table was raised by up to 100 m (Tivy, 1975). In regions of high insolation and low rainfall, irrigation, its associated effects of soil-water regimes, and evapotranspiration led to salinization of large areas of agricultural land and reduction in crop yields or wholesale abandonment of land (Carter, 1975).

Agricultural land is lost for a variety of reasons, such as salinization, alkalinization, erosion, acidification, pulverization, compaction, and petrifaction. It has been estimated that the total area of degraded land that was once biologically productive is $\sim 2.0 \times 10^9$ ha, an area more than the entire cultivated arable area at present (King, 1980). In a world with an increasing population and diminishing agricultural land, losses from salinity on its present scale represent a serious and unacceptable strain on the food-producing capacity of the earth. Considerations such as these led Casey (1972) to question the wisdom of long-term irrigation of arid lands. Although it may not be technologically intractable to alleviate some of the soil salinity arising from irrigation, for example, by pumping off excess groundwater to lower the water table and thus facilitate downward drainage and leaching of salts (Tivy, 1975), energy constraints are likely to inhibit an early technological solution to the problem. A biological approach to the salinity problem may provide solutions in the medium term as means of prolonging exploitation of existing saline land and, possibly, partially desalinizing saline soils.

Most agricultural plant species are not halophytes, although some of them (e.g., sugar beet) have halophytic ancestors. Consequently, the agricultural exploitation of saline land must involve attempts to grow species normally regarded as glycophytes under saline conditions or to adapt true halophytes for agricultural use.

Salt tolerance is not an all-or-none phenomenon but rather a continuum of response to saline conditions. True halophytes are at one end of the continuum of response, and very salt-susceptible glycophytes are at the other extreme. Very salt-tolerant halophytes such as *Suaeda maritima* can survive and grow in salinities in excess of 500 mol m^{-3} NaCl (Greenway and Munns, 1980). In contrast, growth of very salt-susceptible glycophytes is reduced by very low levels of salt. For example, *Vitis vinifera* is severely affected by 4 mol m^{-3} Cl^- (Edelbauer, 1978), growth of *Glycine max* cv. Jackson is severely reduced by 10 mol m^{-3} Cl^- (Läuchli and Wieneke, 1979), and growth of avocado (*Persea americana*) is severely inhibited in the presence of 20 mol m^{-3} NaCl (downton,

1978). In addition to growth inhibition, salinity also induces symptoms of injury in salt-sensitive species. Chlorophyll bleaching and leaf necrosis occur in peach trees (Hayward *et al.*, 1946) as well as in beans and peas (Strogonov, 1964, 1970). In the middle of the continuum there is some degree of overlap in the salt tolerance of species normally regarded as halophytic (e.g., *Puccinellia peisonis*) and those normally considered glycophytic (e.g., barley) (Greenway and Munns, 1980).

Not only is there considerable variation in salt tolerance and susceptibility among distantly related species, but there is also appreciable difference in the salt tolerance of closely related species. The mangrove *Rhizophora mangle* can inhabit the American Pacific coast, which, however, is not a suitable habitat for the less salt tolerant *R. racemosa* (Breteler, 1977). *Plantago media* is more salt susceptible than *P. coronopus,* which in turn is more salt susceptible than *P. maritima* (Laszlo and Kuiper, 1979). *Atriplex halimus* is more salt tolerant than *A. calatheca,* which in turn is more salt tolerant than *A. nitens* (Priebe and Jäger, 1978). The cultivated tomato (*Lycopersicon esculentum*) is more salt susceptible than *L. cheesmanii,* a coastal species from the Galapagos (Rush and Epstein, 1976); these two species can interbreed (Norlyn, 1980).

There is, moreover, considerable variation in salt tolerance within species. *Glycine max* is a very salt sensitive species, but it has a cultivar (Lee) that is moderately tolerant compared with the very susceptible cultivar Jackson (Wieneke and Läuchli, 1979). El-Sharawi and Salama (1977) grew cultivars of wheat and barley in cultures simulating conditions in areas that have become saline through irrigation with underground water. The introduced Mexican wheat cultivar Super-X was more tolerant than the local variety Giza-155. The former maintained higher relative water content in its leaves when under high stress and under conditions favoring high transpiration rates. Similarly, the barley cultivar Giza-177 was more salt tolerant than Borg El-Arab. Bhatti *et al.* (1976) demonstrated differences in salt tolerance among varieties of barley. Rathore *et al.* (1977) also found differences in salt tolerance among 22 cultivars of *Hordeum vulgare*. Variations in salt tolerance have been demonstrated among cultivars of musk melon (Shannon and Francois, 1978), *Triticum aestivum* (Maliwal *et al.*, 1976), and sorghum (Taylor *et al.*, 1975). Field experiments involving irrigation of soil with saline water have demonstrated differences in salt tolerance with respect to yield and quality of sweet pepper (*Capsicum annuum;* Fernandez *et al.*, 1977).

Variation in salt tolerance is even observed within cultivars. Compared with subspecies or even ecotypes, cultivars have considerable genetic uniformity. This is because during their development cultivars have been selected for a number of phenotypic characteristics such as morphology, growth, minimum temperature at which growth starts, digestibility, palatability, disease resistance, yield of fruit, etc. Moreover, commercially available cultivars are derived from

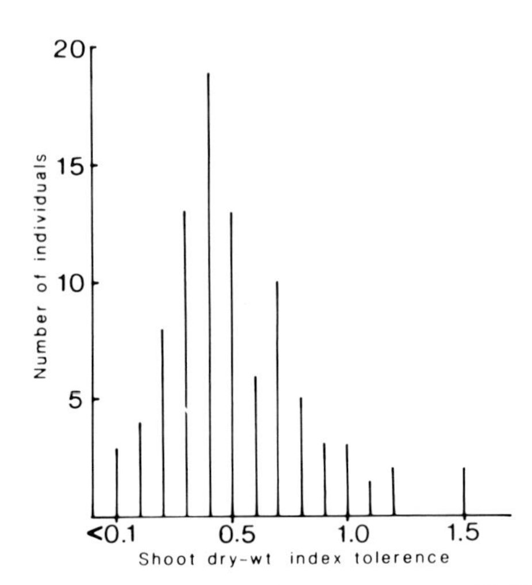

Fig. 1. Variation in rooting index of tolerance of *Brassica napus* cv. Norli grown at 80 mol m^{-3} NaCl. Rooting index of tolerance is root length expressed as a proportion of the mean root length in a nonsaline culture solution. From Mary Clark (unpublished).

relatively small genetic pools. Nevertheless, it has been possible to select for salt tolerance within the lettuce cultivar Empire (Shannon, 1980). There is great variability in sodium uptake and salinity tolerance within cultivars of rice (*Oryza sativa*) that are not homozygous lines (Flowers and Yeo, 1981). Figure 1 shows variability in the rooting index of salt tolerance in the Norli cultivar of oil seed rape (M. F. Clark, unpublished).

The great variability in salt tolerance that can be observed at all levels of classification within the plant kingdom is evidence of the ready availability of selectable genetic material leading to the evolution of halophytic plants.

B. Selection Pressures in Saline Environments and the Evolution of Salt-Tolerant Ecotypes

Salt-tolerant glycophytes fall into two categories. First, there are many species of nonsaline habitats that when subjected to salinity, either by introduction of salt into their natural habitat or by dispersal from their normal, nonsaline habitat into an adjacent saline one, evolve halophytic physiological races or ecotypes. These species include opportunistic species such as *Agrostis stolonifera* and *Festuca rubra* (Aston and Bradshaw, 1966; Hannon and Bradshaw, 1968; Tiku and Snaydon, 1971; Ahmad and Wainwright, 1977; Venables and Wilkins, 1978; Wu, 1981), which also evolve edaphic ecotypes tolerant of heavy metals and other stresses (Antonovics *et al.*, 1971). Hannon and Bradshaw (1968) showed the presence of salt-tolerant ecotypes of *A. stolonifera* and *F. rubra* in the lower salt marshes as well. *Agrostis* was able to achieve the same level of rooting tolerance as *Festuca,* but the former was restricted to the upper salt marshes. This implies that factors other than salinity may be involved in preventing *Agrostis* from invading the lower marshes. These other factors could be physicochemical, such as soil aeration. This serves to highlight the easily overlooked fact that evolution of a strain capable of living in a particular saline soil is likely to involve evolution of associated characteristics (e.g., tolerance of poor aeration) and not only tolerance of NaCl. Indeed, Ahmad and Wainwright (1977) showed that a salt-tolerant clone of *A. stolonifera* was also more tolerant of low O_2 tension in the root environment than was an inland clone. However, the factors limiting *Agrostis* to the upper salt marshes could be biotic. It was discussed previously that if a plant is to be successful in a saline environment, it must not only be able to tolerate high concentrations of salt but must also be able to grow successfully in the presence of the other plants in the habitat, either by competing successfully with them or by avoiding competition with them.

It seems likely that competitive relationships are involved in restriction of salt-tolerant ecotypes to saline environments, even in the face of considerable gene flow from the salt marsh toward nonsaline adjacent inland sites. Hannon and Bradshaw (1968) and Wu (1981) demonstrated that populations of *Agrostis stolonifera* on nonsaline land adjacent to saline seashore sites were considerably less salt tolerant than those of salt-marsh and seashore habitats. Moreover, the salt-marsh populations of both *A. stolonifera* and *Festuca rubra* were significantly less vigorous than the inland populations when grown in nonsaline culture (Hannon and Bradshaw, 1968; Tiku and Snaydon, 1971).

Whereas *Agrostis stolonifera* has the capacity to rapidly evolve fully salt-tolerant plants from seed collected from nontolerant populations in a single generation, *A. tenuis* does not seem to have the capacity to evolve salt-tolerant races. It is notable that even inland populations of *A. stolonifera* are relatively salt tolerant compared to many other species and that this species thus has a predisposition toward salt tolerance. This may be associated with its capacity to evolve necessary associated characteristics. Wu (1981) found that $MgCl_2$ completely inhibited root growth of *A. tenuis* but that *A. stolonifera* produced and grew roots in the presence of $MgCl_2$. As Mg^{2+} concentrations are high in seashore sites (Wu, 1981; Hodson *et al.*, 1982), it is possible that lack of seashore populations of *A. tenuis* is related to intolerance of $MgCl_2$. It would be interesting to screen seed of Magnesian limestone ecotypes of *A. tenuis* (Woolhouse, 1970) for tolerance to seashore or salt-marsh soils.

Simushina and Morozova (1980) have demonstrated salt-tolerant ecotypes of *Kochia prostrata* from various arid regions in the southern U.S.S.R. These habitats contrast markedly with coastal saline habitats.

The use of salt as a deicing agent on highways produces saline roadside soils as a result of runoff from the road and direct dumping of salt by the roadside.

Shaw and Hodson (1981) have shown the toxic effect of salt dumping on roadside trees. Salt present in the roadside environment presents a selection pressure sufficient to produce salt-tolerant roadside populations of various herbaceous species. For example, Pitelka and Kellog (1979) demonstrated salt-tolerant populations of *Agropyron trachycalum* and *Solidago juncea* in roadside soils in the northern United States. Briggs (1978) demonstrated salt-tolerant *Senecio vulgaris* from saline roadside soils.

Venables and Wilkins (1978) studied a site at Upton Warren near Droitwich (United Kingdom), where brine pumping over a 30-year period led to subsidence and flooding of a permanent pasture with saline water. Many of the original species of plants had been eliminated, and others such as *Puccinellia distans* and *Spergularia marina* had invaded the site. Figure 2 shows the effects of increasing NaCl concentration on root growth of six species of pasture grass from five soils of different salinities. *Festuca rubra* showed optimum root growth at ~100 mol m^{-3} NaCl in plants from all soil types. Optimum root growth of *Agrostis*

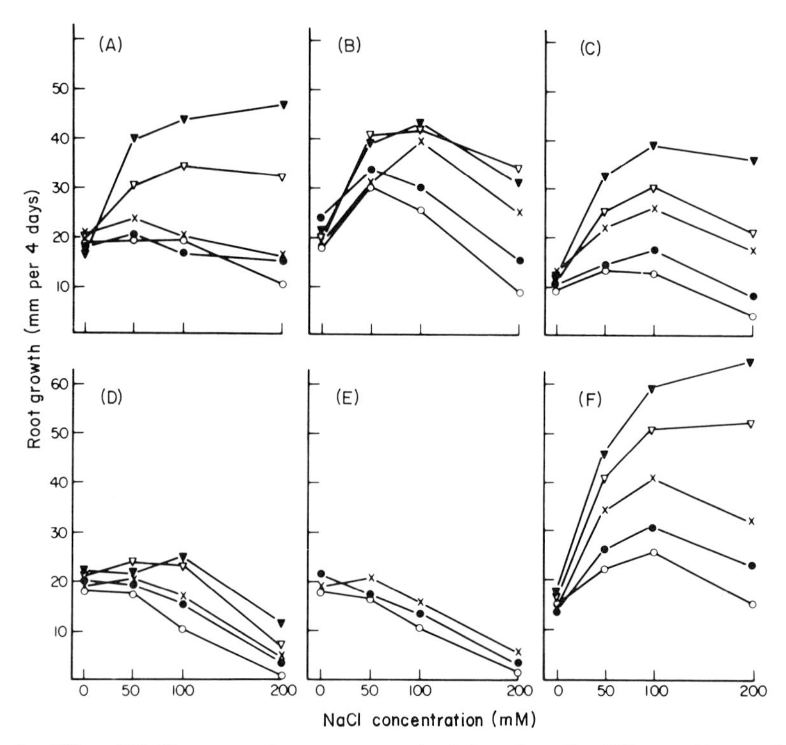

Fig. 2. Effect of NaCl concentration on root growth of plants from five different soil types, from highest to lowest salinity: ▲, △, x, ●, ○. (A) *Agropyron repens*. (B) *Agrostis stolonifera*. (C) *Festuca rubra*. (D) *Hordeum secalinum*. (E) *Lolium perenne*. (F) *Puccinellia distans*. From Venables and Wilkins (1978).

stolonifera occurred at 100 mol m^{-3} NaCl in plants from the most saline soils and at 50 mol m^{-3} in plants from the less-saline soils. *Agropyron repens* and *P. distans* showed increased root growth at concentrations up to 200 mol m^{-3} in plants from the most-saline sites. Root growth was inhibited by salt in *Hordeum secalinum* and *Lolium perenne;* with the exceptions of these two species, all grasses showed strong adaptation to salinity of soils in which they were growing. Breeding experiments with *F. rubra* showed that salt tolerance was highly heritable. The presence of *P. distans* in this inland site can probably be explained by importation by seabirds. It also follows that the tolerant strains of the other grasses could have been imported by birds also, but it is more likely that they arose by selection from existing populations. There was a high correlation between root growth in salt and salinity of the soil in which the plants were growing. This suggests very strong selection pressures in the face of gene flow, as the experimental plants were collected at distances of only 10–100 m from each other. Venables and Wilkins's (1978) interpretation of the situation was that the meadow was once a flat and homogeneous site, with little variation within species. As a result of subsidence and subsequent flooding by saline water, salt gradients were imposed and acted as selection pressures on the plants *in situ*. This might explain the fact that the plants did not have greater tolerance than they "needed" to on their own soil type.

C. Breeding of Salt-Tolerant Plant Cultivars

The second category of salt-tolerant glycophytes comprises those species, already discussed, that have various degrees of salt tolerance or that have strains variously sensitive to salinity but that may not have evolved salt tolerance under the selection pressures of a saline environment. Moreover, they may not have been selected particularly for their salt tolerance. It follows that the degree of salt tolerance present in these species is fortuitous in the sense that these plants do not need to be salt tolerant in order to survive in their normal habitats. It may be that selection of tolerance toward some other stress such as drought has imposed a degree of salt tolerance. This could happen if part of the physiological mechanism complex of tolerance to, say, drought, were common to the mechanism complex to salinity. Although this is discussed in more detail in the context of investigating mechanisms involved in salt tolerance, it is evident that accumulation of organic solutes such as proline or glycine betaine is associated with salt tolerance and occurs when the plants are placed under salt stress. It is known, however, that when plants are placed under water stress, for example, by growing them in the presence of polyethylene glycol 6000, they respond by accumulating organic solutes (Hanson and Nelson, 1978; Jefferies *et al.*, 1979). Proline accumulation is associated with drought resistance in barley (Singh *et al.*, 1972) and in *Carex stenophylla* (Hubac and Guerrier, 1972). It is likely, then, that a drought-tolerant plant may to some extent be more salt tolerant than one intolerant of drought, in the sense that salinity will impose a degree of water stress. However, it is unlikely that a drought-tolerant plant will be fully salt tolerant, because although it may be able to cope with the water stress imposed by certain levels of salinity, it will not possess the means to combat other aspects of salt toxicity. Thus, if a fully salt-tolerant halophytic plant can be regarded as possessing the full complex of physiological adaptations to a saline environment, the partially salt-tolerant glycophyte that has not evolved tolerance in response to the selection pressures of a saline environment may be regarded as possessing just a part of the complex of physiological adaptations necessary to adapt to a saline environment. Moreover, if the degree of salt tolerance possessed by a species is not related to the evolution of tolerance toward another stress such as drought, then the salt tolerance it does have will depend on the chance possession of characteristics associated with tolerance.

There is nevertheless, as previously discussed, considerable variation in salt tolerance of species normally regarded as being glycophytes, and many of these are crop species. This variation constitutes the raw material for selection in plant breeding programs that attempt to produce crop varieties capable of surviving and producing economic yields on the annually increasing areas of saline land.

Norlyn (1980) reported on a project to determine whether the salt tolerance of a barley crop could be improved by selection and breeding. It was shown that different lines of barley have different salt tolerances and that these are under genetic control. Figure 3 shows the difference in emergence of four cultivars of barley at different salinities. Arivat was the most sensitive cultivar, and California Mariout the most tolerant. Wyn Jones and Storey (1978) also showed California Mariout to be a salt-tolerant cultivar. The cultivars Briggs and Numar were derived from backcrossing, with Briggs having California Mariout twice in its pedigree, and Numar having it four times. It was also shown that the different

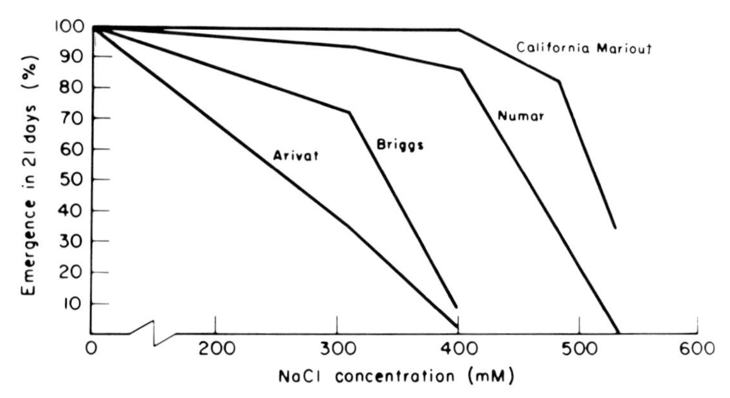

Fig. 3. Effect of salinity on emergence of four cultivars of barley. From Norlyn (1980).

cultivars have different degrees of salt tolerance at various stages in their life cycles. Ideally, a salt-tolerant crop should be tolerant at all stages in its life cycle, because it would probably be necessary to sow the seed in saline soil.

Tolerance at all stages of the life cycle is not always evolved by perennial halophytes, however. Germination of *Limonium vulgare* seed is inhibited by increasing concentrations of seawater. However, increased salinity causes an increase in subsequent germination in fresh water (Boorman, 1967). Germination in fresh water was about 25% but was reduced to about 5% in 100% seawater. However, on subsequent transfer from 100% seawater to fresh water, the germination rate increased to about 55%. The transfer from seawater to fresh water could signal the onset of temporarily favorable conditions for seedling growth, possibly during periods of high rainfall. Norlyn (1980) described a two-stage screening experiment by R. Kingsburg in which 5000 lines of wheat were screened and compared for emergence against a tolerant check line. Seeds recovered from the selected plants in the germination- and emergence-screening experiments were germinated at lower salt concentrations (e.g., 50% seawater), and the survivors were then grown in water culture at the same concentration. Field experiments with barley cultivars grown at a sand-dune site and irrigated 24 times during the growing season with undiluted seawater produced low yields, but it was significant that a grain yield was produced at all. Norlyn also reported on work done by D. W. Rush on tomatoes. Various commercial varieties of tomatoes were similarly susceptible to salt. However, a closely related species (*Lycopersicon cheesmanii*), which grows on the seashore in the Galapagos Islands, could survive with its roots in undiluted seawater and could fruit when grown in culture solution containing 50% seawater. *Lycopersicon cheesmanii* does not itself bear commercially valuable tomatoes, as its fruit are small and bitter. However, it hybridizes freely with *L. esculentum,* and it was found that plants produced by crossing *L. cheesmanii* with the large-fruiting cultivar Walters and then backcrossing twice with screening for salt tolerance, produced small but acceptable tomatoes with good flavor when they are irrigated with up to 70% seawater. The F_2 plants survived at high salinities even better than the Galapagos plants. The relative success of this experiment was probably the result of introduction of genes for salt tolerance from the Galapagos plants and, then, selection of the plants with the highest salt tolerance in the screening process. It is likely that the gene pool of the Galapagos plants contained variation allowing for improvement in salt tolerance by selection but that the wild population was well adapted to the levels of salinity present in its habitat.

Ramage (1980) discussed genetic methods of breeding salt tolerance in plants. Salt tolerance is a complex phenomenon involving a number of physiological and biochemical systems. Ramage suggested that selection for a suitable background genotype is at least as important as selection for salt tolerance itself. This is likely to be true for two reasons. First, it would be of little value to select for salt tolerance in a crop if, in the process of selection, the high-yield characteristics of the crop were lost. Second, by selection of a suitable background genotype one would wish to select a genome more suitable for expression of the genes associated with salt tolerance. The recurrent-selection technique of Sprage and Brumhall (1950) has been used to select simultaneously for salt tolerance and for a suitable background genotype. The technique is designed to accumulate desirable genes in a population without significantly reducing genetic variation in the population. From the baseline population, plants with the highest salt tolerance are selected and selfed. The seeds from the selfed plants are grown and all possible combinations of crosses made between the progenies of selfings. Equal numbers of seeds from each of the crosses are mixed together and then grown for one more generation to allow recombination to occur. The process is then repeated by selecting the most salt tolerant plants and selfing them as before. The cycle is repeated until no more improvement in salt tolerance is obtained. In addition to improvements in salt tolerance of economic species through breeding and selection programs, techniques deriving from biotechnological research are available that are likely to play an increasing role in this field. They are discussed briefly later.

III. ADAPTATION AT THE LEVEL OF THE INDIVIDUAL

A. The Nature of Physiological Stress in Saline Environments

1. Water Stress

The growth and metabolic efficiency of plant cells are optimal when the cells are fully turgid. However, conditions that bring about stomatal opening and that favor absorption of CO_2 by leaves also favor transpirational water loss, leading to some loss of turgor and a reduction of leaf water potential ψ. A ψ gradient becomes established in the plant down which water is drawn from the soil to the leaf. In normal nonsaline soils, the solute concentrations are usually very low, and as a result, the matric forces are the major contributors to the ψ of the soil pores and cause retention of water:

$$\psi_{\text{soil}} = \psi_{\text{matric}} + \psi_{\text{solute}}$$

The matric potential in megapascals is given by

$$\psi_{\text{matric}}\ (\text{MPa}) = -0.3/d$$

where d is the pore diameter in micrometers (Russell, 1973). Only pores less than 60 μm in diameter can retain water in well-drained situations as gravity exerts a force of 5×10^{-3} MPa, and this brings soils to field capacity (FC) by spontaneous drainage (Webster and Beckett, 1972). Water is held in soil pores down to diameters below 0.1 μm, giving ψ values of −3 MPa. However, the permanent wilting point of soils is achieved when water has been removed from pores down to ~0.2 μm diameter, where the ψ is −1.5 MPa. The permanent wilting point is achieved in various soils at various soil-water percentages (Brady, 1974). In saline soils, however, the ψ is largely determined by the solute potential, and this is particularly true in waterlogged or partly waterlogged conditions, such as those in salt-marsh soils. On 30 July 1981, the writer took soil samples at 5-cm depth from the upper salt marsh at Llanrhidian, North Gower, Wales. Evacuated porcelain porous-cup soil-moisture samplers were placed in auger holes at 5-cm depth, and soil-moisture samples were extracted. The ψ values of the soil samples and the soil-solution samples were measured by dew-point hygrometry using a Wescor hygrometer C52 sample chamber. The ψ of both the soil and the soil solution was −1.4 MPa, and analysis of the soil-moisture samples showed that the solute potential was virtually entirely because of NaCl. This solute potential was equivalent to ~0.6 seawater concentration. It is clear that the wet salt-marsh soil had a ψ approximately equivalent to the permanent wilting point of a well-drained, nonsaline soil. If a soil is inundated with seawater, the ψ will be about −2.4 MPa at 20°C. It would seem that on purely thermodynamic grounds, a plant would be under severe water stress when growing in a very saline soil even if the soil were waterlogged.

High concentrations of salt in the growth medium affect the water status of plants. The relative water content of leaves is reduced in tomato (Hayward and Long, 1943) and bean (O'Leary, 1969). Gates (1972) showed that the relative water content of *Acacia harpophylla* and *Atriplex numularia* increased when grown in saline conditions. The percent water content of salt-susceptible *Agrostis stolonifera* leaves is severely reduced when the plants are grown in 200 mol m^{-3} NaCl in culture solution, particularly in the older leaves. Table I shows the effect of NaCl on water absorption of *Agrostis stolonifera* roots of salt-tolerant and salt-

TABLE I

Changes in Root Water Absorption of *Agrostis stolonifera* in Response to Salinity[a]

Control NaCl (mol m^{-3})	Salt-marsh ecotype[b]	Inland ecotype[b]
	382	357
100	371	326
150	340	247
200	326	96
300	322	8

[a]Modified from Ahmad *et al.* (1981).

[b]Values of experimental plants in grams of water per gram root dry weight per day.

susceptible ecotypes (Ahmad *et al.*, 1981). The salt-marsh ecotype was still absorbing water at 84% of the control rate, even when grown in 300 mol m^{-3} NaCl in culture solution. Absorption by the inland ecotype, however, was reduced to 27% of the control rate by 200 mol m^{-3} NaCl, and to 2.24% of the control rate by 300 mol m^{-3} NaCl in culture solution.

Water stress can cause disruption in metabolic processes within the cell. Immersing intact chloroplasts isolated from *Spinacea oleracea* in increasing

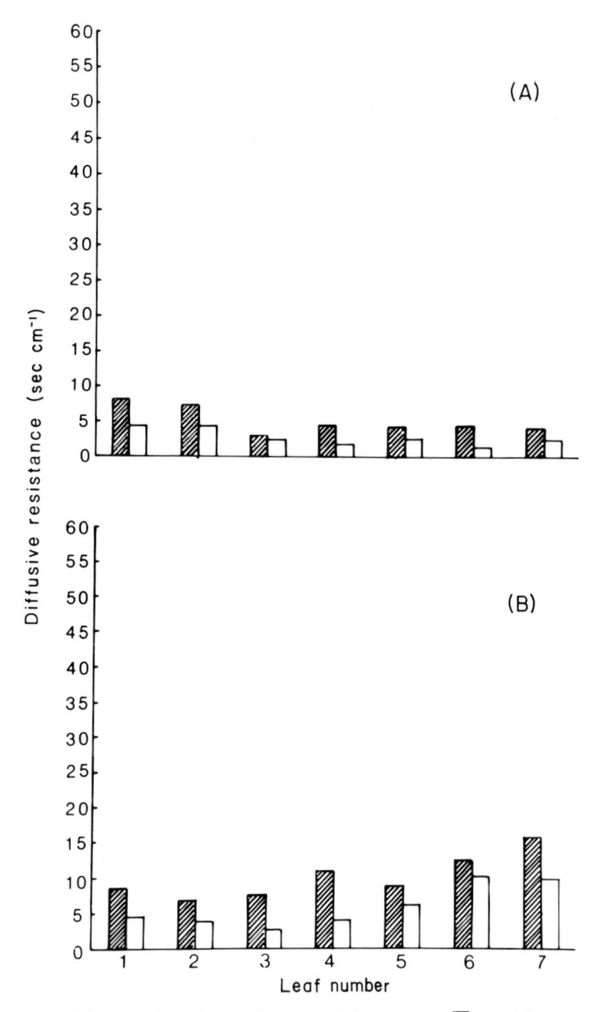

Fig. 4. Effect of salinity on diffusive resistance of the upper (▨) and lower (□) surfaces of *Chenopodium album* leaves. NaCl concentration (mol m^{-3}): A, 0; B, 100; C, 200; D, 300. From P. Kempster and S. J. Wainwright (unpublished).

concentrations of sorbitol induced fast and nonspecific efflux of metabolites. The dark reactions of photosynthesis by intact chloroplasts are inhibited at low ψ's and after chloroplasts are ruptured the activity of stromal enzymes are decreased by high solute concentrations (Kaiser and Heber, 1981). Salt stress has also been reported to reduce the protein content of plants (Strogonov, 1970; Waisel, 1972; Ahmad, 1978). Eaton (1941) and Hayward and Spurr (1944) showed that water uptake by plant roots decreased as the osmotic potential of the growth medium

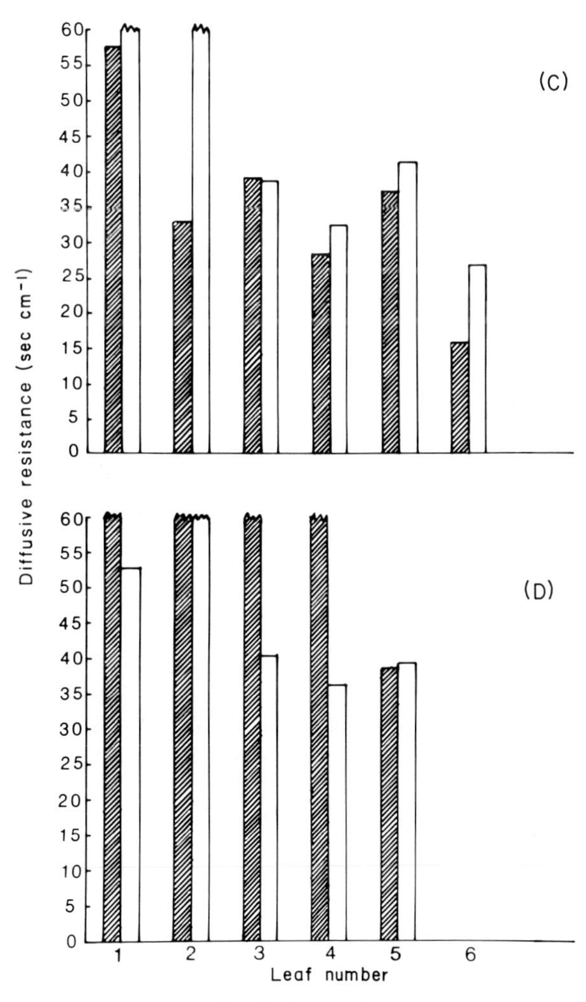

Fig. 4. *Continued*

was lowered. At first sight it might appear that the inhibition of water uptake by a low external ψ might be the result of a ψ imbalance between the plant and the medium and that the susceptible plant is unable to reduce its ψ below that of the medium in the same way that it is unable to cope with the low ψ of a normal soil at the permanent wilting point. However, Walter (1955) objected to this interpretation because the increase in concentration of the external medium is normally accompanied by ion uptake and a concomitant reduction in the solute potential of the leaf cell sap. Bernstein (1961, 1963) showed that osmotic adjustment takes place in roots as well as shoots. Slatyer (1961) showed that when osmotic adjustment occurred in salt-stressed tomato plants, water uptake increased, thereby restoring the water balance of the plants. M. J. Hodson has shown (reported in Wainright, 1980) that there is more than adequate uptake of salt into roots and shoots of both salt-tolerant and salt-susceptible ecotypes of *Agrostis stolonifera* to effect an osmotic adjustment of salt-treated plants, and yet the salt-tolerant ecotype maintains a normal water balance whereas the salt-susceptible ecotype undergoes dehydration, and then in the leaves with the highest salt content.

The water balance of a plant is determined by the resultant water uptake via the roots and water loss by transpiration. It seems unlikely that salt stress would cause an increase in water loss via normal transpiration, as the effect of salt stress on plants is to inhibit transpiration by closure of stomata (Meyer, 1931; Gale *et al.*, 1967, 1970; Meiri and Poljakoff-Mayber, 1970; Kaplan and Gale, 1972; Ahmad, 1978). Figure 4 shows the effect of increasing salinity on diffusive resistance of the upper and lower surfaces of *Chenopodium album* leaves. In the absence of salt in the culture solution, the diffusive resistances to water loss were very low, mainly below 5 s cm^{-1} (Fig. 4A). Increasing salinity brought about considerable increase in diffusive resistances, with the greatest effects in the oldest leaves (Fig. 4,B–D). Oertli and Richardson (1968) theorized that whereas osmotic adjustment may occur, the hydraulic conductivity of roots could be impaired by saline conditions. O'Leary (1969) showed this to be the case in salt-stressed *Phaseolus vulgaris* plants that, despite osmotic adjustment, underwent leaf dehydration. Moreover, permeability of the roots to water was reduced severely in salt-stressed plants even when the water was put under pressure. Kaplan and Gale (1972) demonstrated a similar effect in roots of the halophyte *Atriplex halimus*. The reduction in water uptake may be the result of osmotic rather than specific ion effects, because Eaton (1941) showed that similar reductions in water uptake were induced by isoosmotic concentrations of a variety of salts. Ahmad *et al.* (1981) showed that when a salt-marsh ecotype of *Agrostis stolonifera* was grown in the presence of 200 mol m^{-3} mannitol, its leaf water content was reduced by 6.5%, whereas the leaf water content of an inland ecotype grown in the same manner was reduced by 24.4%. Similarly, when roots were immersed in 15% polyethylene glycol 6000, the water content of leaves of the salt-marsh ecotype was reduced by 1.7%, whereas that of the inland ecotype was reduced by 26%.

In root-growth experiments in the presence of polyethylene glycol 6000 in culture solution, Ahmad and Wainwright (1977) showed that the salt-marsh ecotype was more tolerant to the imposed osmotic stress than was the inland ecotype of *Agrostis stolonifera*. This suggests that osmotic stress may form part of the selection pressure imposed by the saline environment that brings about ecotypic differentiation in salt tolerance in *A. stolonifera* and possibly other species. However, whether the osmotic stress involved in the selection pressure is related to ψ imbalance between the plant and soil is open to question and is discussed later in the context of the mechanism of tolerance. Greenway and Munns (1980) suggested that the most convincing evidence that adverse water relations occur in the presence of high concentrations of salt is given by the facts that growth inhibition is smaller at high relative humidity than at low relative humidity, and that the reduction may be greater for salt-sensitive than for salt-tolerant species. A plant has less difficulty in maintaining a favorable water balance in a humid atmosphere than a dry one because it will have a low transpiration rate. However, as ion transport within the plant is related to transpiration rates, and as ion concentration in leaves tend to be higher when transpiration rates are high (Nieman and Poulson, 1967; Greenway and Munns, 1980), it is not possible with any certainty to ascribe growth inhibition to adverse water relations rather than ion excess.

2. *Ion Toxicity*

a. Na^+ and Cl^- Toxicity. Strogonov (1964) favored a major role for chloride toxicity in plants undergoing salt stress. Leaf burning caused by the effects of salinity on peach trees was attributed to accumulation of Cl^- in leaves. These conclusions may have been based on the study of plants that take more Cl^- than Na^+ into the leaves. In fact, many of the more salt-susceptible species, such as maize, cress, sunflower, pepper, and bean, have higher Cl^- than Na^+ concentrations in the leaf (Lessani and Marschner, 1978). This is also true of *Acer saccharum* (Holmes and Baker, 1966), *Glycine* (Wilson *et al.*, 1970), *Panicum repens* (Ramati *et al.*, 1979), *Spinacea oleracea* (Cloughlan and Wyn Jones, 1980), and *Fagus sylvatica* (Shaw and Hodson, 1981). On the other hand, some of the most salt tolerant species have higher concentrations of Na^+ than of Cl^- in the leaves, for example, *Spartina* ×*townsendii* (Storey and Wyn Jones, 1978b), *Plantago maritima* (Ahmad and Wyn Jones, 1979), *Suaeda monoica* and *Atriplex spongiosa* (Storey and Wyn Jones, 1979), and *Suaeda maritima* and *Puccinellia maritima* (Stewart *et al.*, 1979). Figure 5A,B shows Na^+ and Cl^- concentrations in leaves of *Chenopodium album* and *Atriplex glabriuscula* grown in water culture at various salinities (A. Majeed, unpublished). In both species,

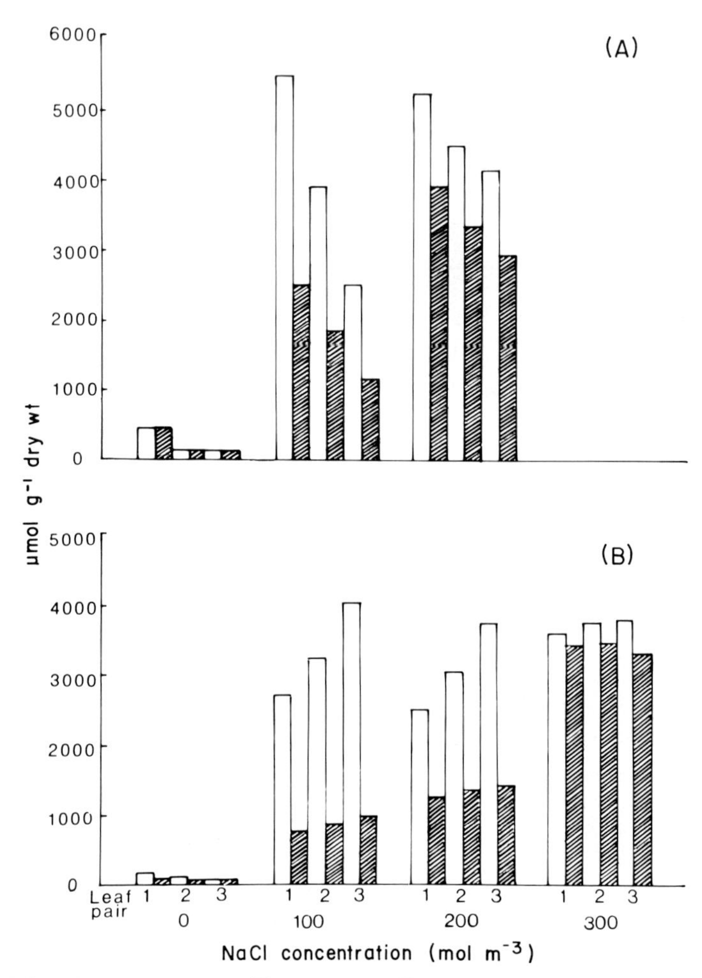

Fig. 5. Effect of salinity on sodium (□) and chloride (▨) contents of leaves of *Chenopodium album* (A) and *Atriplex glabriuscula* (B) when grown for 1 week in saline culture. From A. Majeed (unpublished).

Na^+ concentrations were higher than Cl^- concentrations in the leaves. Moreover, over a range of salinities, Na^+ concentrations were far less dependent on salt concentration in the medium than were Cl^- concentrations. High Na^+ concentrations were achieved in the leaves at moderate salinities and did not increase much as salinity of the culture solution increased. Chloride concentration in the leaves, on the other hand, was more strongly correlated with salinity of the culture solution.

The response of *Agrostis stolonifera* is midway between these two extremes. There is a stoichiometric relationship between Na^+ and Cl^- concentrations of

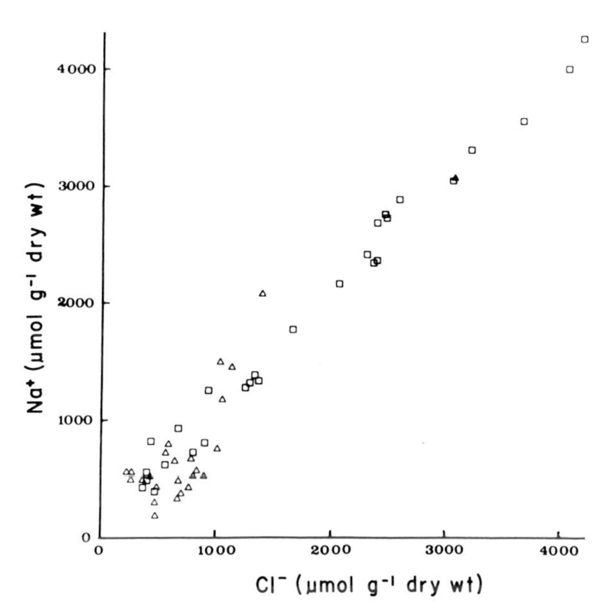

Fig. 6. Relationship between leaf Na^+ and Cl^- concentrations in *Agrostis stolonifera* grown at various salinities. □, Inland clone; △, salt-marsh clone. From M. J. Hodson (unpublished).

the leaves of both salt-tolerant and salt-susceptible clones grown over a range of salinities. Figure 6 shows that leaves of both clones grown over a range of salinities contained the same concentrations of Na^+ and Cl^-, even though the inland clone took up far more salt than did the salt-marsh clone (M. J. Hodson, unpublished). There are both glycophytes and halophytes that show this kind of relationship between Na^+ and Cl^- concentrations, for example, barley (Storey and Wyn Jones, 1978a) and *Atriplex littoralis, Spartina anglica,* and *Limonium vulgare* (Stewart *et al.*, 1979). Yeo *et al.* (1977) showed that the relationship between the distribution of Na^+ and Cl^- in *Zea mays varied between different tissues.*

b. Toxicity of Other Ions. In studies of heavy-metal toxicity there is no ambiguity as to whether toxicity and tolerance are due to the cations or anions involved. Typically, heavy metals at very low concentrations (e.g., 10^{-3} mol m^{-3} Cu) inhibit plant growth (Wainwright and Woolhouse, 1977). At these concentrations, the associated anions such as nitrate or chloride do not have toxic effects. This can be concluded because, for example, sodium chloride or sodium nitrate at a concentration of 10^{-3} mol m^{-3} has no inhibitory effects at all. Toxicity resulting from salinity, however, is not so clear-cut, because the concentrations of NaCl required to inhibit plant growth are much higher, for example, 100 mol m^{-3} (Ahmad and Wainwright, 1977). At this order of concentration both cations and anions are potentially toxic. Moreover, the potential

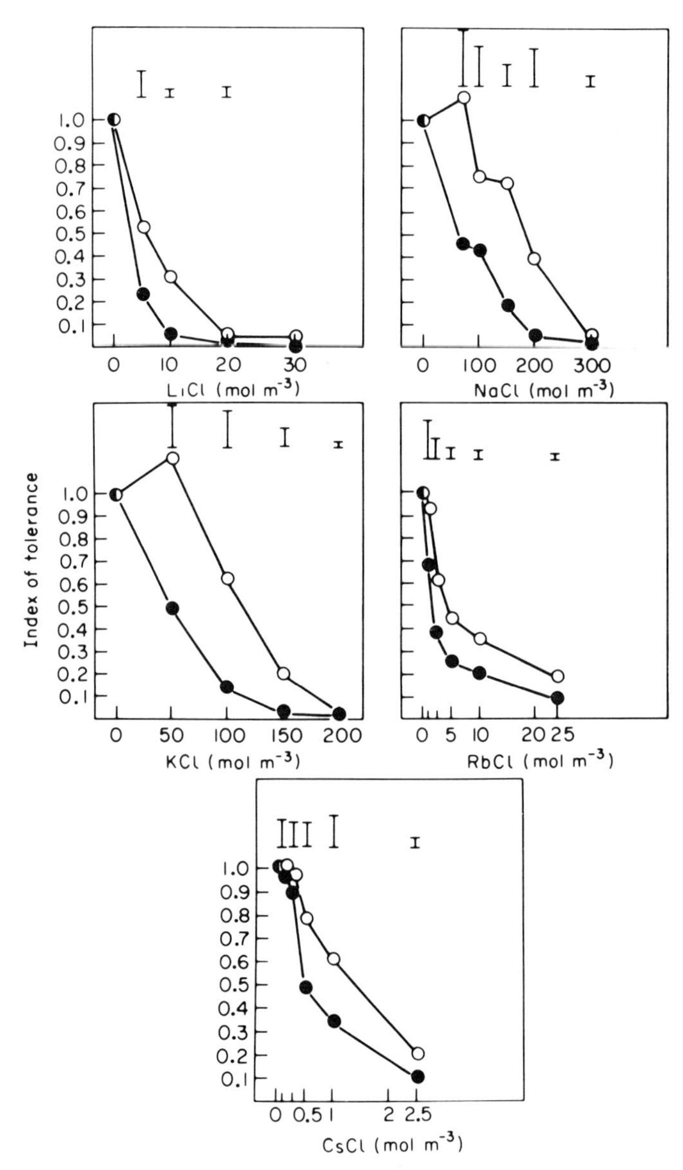

Fig. 7. Effect of the alkali metal chlorides on the rooting index of tolerance of a salt-marsh clone (○) and an inland clone (●) of *Agrostis stolonifera*. Vertical bars are 5% LSD. From Hodson *et al.* (1982).

osmotic effects become significant. It must be accepted as a limitation to the investigation of the problem that if the ions concerned are potentially toxic only at concentrations at which they significantly affect the activity of water, then it is in principle impossible to directly separate the effects of the ions themselves from those resulting from the changed osmotic potential. Indirect inferences can, however, be made. Cation toxicity is suggested by such observations as those by Hodson *et al.* (1982) that, over a range of concentrations, KCl was more inhibitory than NaCl to root growth of *Agrostis stolonifera.*

Hodson *et al.* (1982) found potentially toxic concentrations of Mg^{2+} in salt-marsh soil-solution samples and demonstrated that a salt-marsh clone of *Agrostis stolonifera* was considerably more tolerant to Mg^{2+} than was an inland clone. In the same soil-moisture samples, Na^+ was the only other cation present in potentially inhibitory concentration, and yet the salt-marsh ecotype of *A. stolonifera,* which had evolved in this environment, was also more tolerant to Li^+, K^+, Rb^+, and Cs^+ (Fig. 7). It was also more tolerant to the other alkaline earth Ca^{2+} than was the inland ecotype. It appeared that part of the selection pressure of the salt-marsh environment was the result of the presence of high concentrations of Na^+ and, possibly, Mg^{2+}. Physiological adaptations leading to increased tolerance toward Na^+ and possible Mg^{2+} had also produced cotolerance toward the other alkali metals and the alkaline earth Mg^{2+}. Increased tolerance to Na^+ did not produce a broad spectrum of increased tolerance, however, as the salt-marsh plants were not also tolerant to the heavy metals Cu^{2+} and Zn^{2+} (M. J. Hodson, unpublished). It is likely that the chemical similarities between the alkali metals render the tolerance mechanism partially effective for all of them. The same may be true for the alkaline earths.

c. Effects on Enzymes and Membranes. Halophilic bacteria have a positive requirement for salt in order to achieve normal activity of many of their enzymes. In *Halobacterium salinarium* at low salinity, a variety of enzymes show little or no activity but have high activity when the cells are grown in saline culture (Larsen, 1967). The possibility that halophytes have salt-tolerant or even salt-requiring enzyme systems has been investigated, and frequently the conclusions reached were that the enzymes from halophytes exhibited salt sensitivities similar to those of enzymes from glycophytes, and that in both groups of plants, salt inhibited enzyme activity (Greenway and Osmond, 1972; Osmond and Greenway, 1972; Flowers, 1972a,b; Hall and Flowers, 1973; Stewart and Lee, 1974; Greenway and Sims, 1974; Ahmad *et al.*, 1979).

Enzymes might be affected directly by high solute concentrations or by reduction in hydration of proteins. Dark reactions of spinach catalyzed by soluble enzymes of the chloroplast stroma are more sensitive to water stress than is electron transport from water to methyl viologen, which is catalyzed by the electron-transport chain of chloroplasts. Except for plastocyanin and proteins, such as ferredoxin and NADP reductase (which are not involved in methyl viologen reduction), the catalysts of electron transport are proteins embedded in the lipid phase of thylakoid membranes. These proteins are hydrophobic and are less accessible to the effects of hydrophilic solutes than are soluble proteins (Kaiser and Heber, 1981).

Plaut (1974) provided evidence for reversible dehydration of nitrate reductase enzyme molecules in wheat seedlings exposed to and recovering from water stress and salinity. Enzyme activities in different parts of the plant are variously affected by salinity. In a number of halophytes, an increase in external salinity reduces the levels of glutamine synthetase and glutamate dehydrogenase in roots, but increases the level of glutamine synthetase in shoots. This may reflect a tendency of shoots to play a greater role in nitrogen assimilation under saline conditions and could be of significance in the fact, discussed later, that shoots of some species accumulate high levels of amino acids and quaternary ammonium compounds under salt stress (Stewart and Rhodes, 1978). Sometimes enzyme activities are increased when plants are grown under saline conditions. Phosphoenolpyruvate carboxylase (PEPC) activity of *Cakile maritima* was stimulated when NaCl was added to the reaction mixture in concentrations up to 200 mol m^{-3}. Plants grown in the presence of 100 mol m^{-3} NaCl showed a 30% higher PEPC activity than plants grown without salt. This stimulatory effect was restricted to chlorides, as sulfates and nitrates were inhibitory (Beer *et al.*, 1975). It was shown by von Willert (1974) that malate dehydrogenase (MDH) from *Aster tripolium* had maximum activity in the presence of 150 mol m^{-3} NaCl.

There is other evidence for adaptation of certain enzymes to salt. Austenfeld (1976) showed that glycolate oxidase of *Pisum* was somewhat more sensitive to salt than that from *Salicornia*, although the difference was not great. Gettys *et al.* (1980) studied the effect of salt on *in vitro* activities of leucine aminopeptidase, peroxidase, and MDH in the highly salt tolerant *Spartina alternifolia* and the less salt tolerant *S. patens*. Both malate dehydrogenase and peroxidase were more salt tolerant in *Spartina alternifolia* than in *S. patens*. Cavalieri and Huang (1977) studied the effect of NaCl on *in vitro* activity of MDH in *Spartina alternifolia, S. patens, Distichlis spicata, Juncus roemerianus, Salicornia virginica,* and *Borrichia frutescens*. The effect of NaCl on MDH activity of the first five species was similar to the effect on glycophytes, although the root MDH activity was more tolerant than the shoot MDH activity. In shoots, NaCl concentrations of 50 mol m^{-3} were optimal for all species except *B. frutescens,* which had optimal activity at 250 mol m^{-3} NaCl. In roots, NaCl concentrations of 100 mol m^{-3} were optimal for all species except *B. frutescens,* which showed optimal activity at 500 mol m^{-3}, equivalent to environmental salt concentrations. *Borrichia frutescens* does not have salt glands or much succulence, so it is possible that the selection pressure may be greater on its enzyme systems than in other species that possess these characteristics. Moreover, the cytosol MDH is

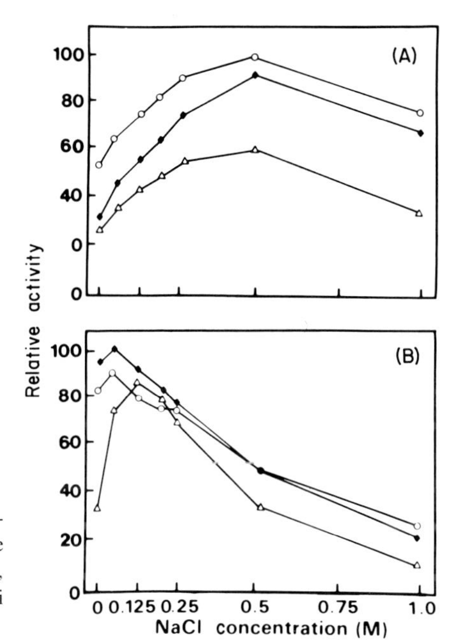

Fig. 8. Effect of salinity on the malate dehydrogenase activity of the cytosol (A) and the mitochondria (B) of *Borrichia frutescens*. ○, pH 6.5; ◆, pH 7.5; △, pH 8.5. From Cavalieri and Huang (1977).

more tolerant than that of the mitochondrial enzyme (Fig. 8), and this may reflect the fact that the mitochondria have their own uptake system, which may be able to control salt uptake from the cytosol. Priebe and Jaeger (1978) showed that glutamate dehydrogenase from salt-sensitive *Vicia faba* was inhibited by NaCl, whereas this enzyme was stimulated by salt in salt-tolerant *Atriplex* species. Thus, although the possession of salt-tolerant enzymes is not a universal component of the mechanism of salt tolerance, it does seem that in some cases there is adaptation to salt at the level of the enzyme.

It would be surprising if membranes did not play an important role in the mechanism of tolerance, because whether they are involved in transfer of salt from one site to another within the cell or whether they are involved in exclusion of salt from cells or organelles, they are present at the interface between regions of high salt concentrations and regions potentially susceptible to salt.

Laszlo *et al.* (1980) studied the effects of salt on lipid composition of roots and shoots of salt-tolerant and salt-sensitive species of *Plantago*. In roots of the salt-sensitive *P. media,* salinity caused a strong decrease in the levels of phospho-, galacto-, and sulfolipids, indicating decreased control of permeability of root-cell membranes. In roots of the salt-tolerant species of *Plantago,* the levels of most lipid classes were maintained at concentrations up to 75 mol m^{-3} NaCl, and decreases were only observed at higher salinities.

Sodium chloride can cause potassium leakage from plant roots (Nassery, 1975, 1979; Bates, 1976). Wyn Jones and Storey (1978) suggested that one of the reasons why barley is less adapted than *Spartina* to salinity is the inability of the former to exploit the osmotic effects of Na^+ and Cl^-, possibly because it undergoes Na^+-induced membrane damage and subsequent solute leakage. Nassery (1975) showed that salt-induced potassium leakage from barley and bean roots is not the result of osmotic effects. M. M. Smith, reported by Wainwright (1980), showed salt-induced potassium leakage from roots of *Agrostis stolonifera* and demonstrated that there was less salt-induced leakage from roots of a salt-tolerant clone than from roots of a salt-susceptible clone. Roy and Hendrix (1980) showed that the electrical cell membrane potential of *Salicornia bigelovii* was insensitive to salinity of the bathing solution. Evidence for involvement of membranes within the cell in manifestations of toxicity is given by the rapid, nonspecific efflux of metabolites from intact isolated chloroplasts of *Spinacea oleracea* subjected to increasing sorbitol concentrations. More direct evidence of salt involvement was given by M. M. Smith, reported by Wainwright (1980), on the effects of salt treatment on ultrastructure of root tips of *A. stolonifera*. The mitochondria of root-tip cells from roots grown in nonsaline culture solution had a normal structure with abundant cristae in evidence. However, in root tips of a salt-susceptible strain treated with salt, membrane structure was disrupted and cristae virtually absent. In the salt-tolerant strain, however, the salt treatment had no effect on mitochondrial structure.

3. The Unbalanced Nutrient Status of Saline Soils

Saline soils may or may not be deficient in certain nutrients. However, the excess of salt makes them unbalanced nutrient media, which can lead to deficiencies in plants of ions present in the soil only in low concentrations. Nitrate is the most important form of nitrogen available in salt marshes. Plant species from the strand line and the lower marsh have higher nitrate reductase levels than those from the upper marsh. Competition for nitrate is more important in the upper than in the lower marsh (Stewart *et al.*, 1973). *Suaeda maritima* is nitrogen deficient on the upper marsh and has nitrate reductase levels only one-fiftieth of those on the lower marsh, where nitrate is more available (Stewart *et al.*, 1972). The unbalanced nutrient status of the salt-marsh soil results from the high concentrations of Na^+ and Cl^- relative to other ions. Tissues high in Cl^- show greatly reduced nitrate uptake rates (Smith, 1973; Cram, 1973). Salt-treated *Agrostis stolonifera, Chenopodium album, Atriplex glabriuscula,* and *Brassica napus* all show reduced nitrate reductase levels (A. Majeed, unpublished). Whether this is the result of inhibition of the enzyme system directly, or indirectly through reduced nitrate uptake is not known.

Unless an uptake system is very selective, competition with Na^+ or Cl^- could

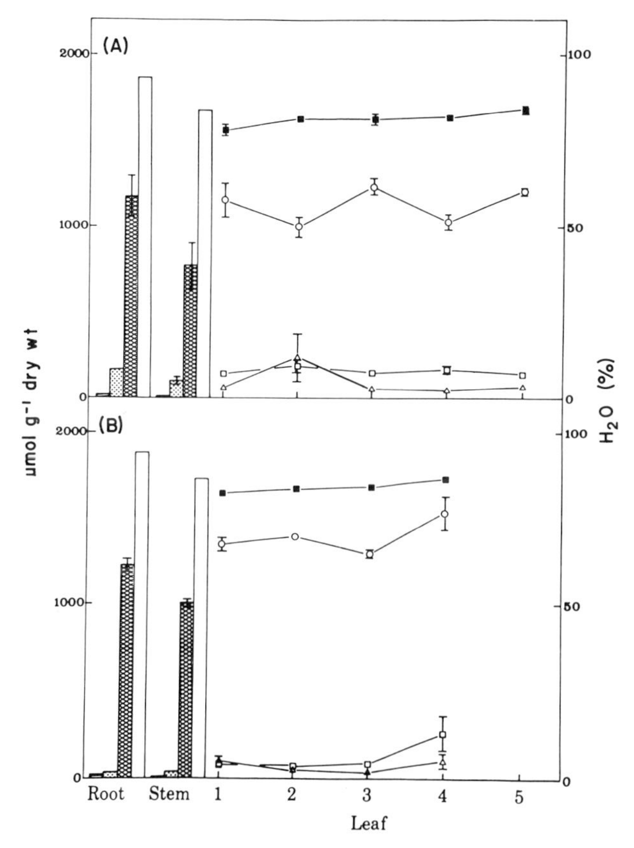

Fig. 9. Plants grown for 4 weeks in control solution. (A) 4W SM0. (B) 4W INL0. Ionic concentration and water percentage determinations for individual leaves, stems, and roots. Water contents are expressed as a percentage of fresh weight. Ionic concentrations are expressed on a dry weight basis. Results are shown as means ±SE. Key for stem and root histograms: ▤, sodium concentration; ▨, chloride concentration; ▩, potassium concentration; □, percent water content. Key for leaf graphs: △, sodium concentration; □, chloride concentration, ○, potassium concentration; ■, percent water content. SM, Salt-marsh plants; INL, inland plants. See also Figs. 10 and 11. From M. J. Hodson (unpublished).

be a serious problem in a saline environment. The cell membrane must not only be able to resist disruption, but must also have a highly selective mechanism for absorption of K^+ in the presence of a high concentration of Na^+. An inhibition of K^+ uptake by Na^+ was shown by Epstein and Hagen (1952). In the relatively salt-tolerant glycophyte barley, K^+ uptake showed saturation kinetics at low K^+ concentrations, but K^+ uptake increased again at higher concentrations, indicating two K^+ uptake systems (Epstein, 1961; Rains, 1972). The mechanism

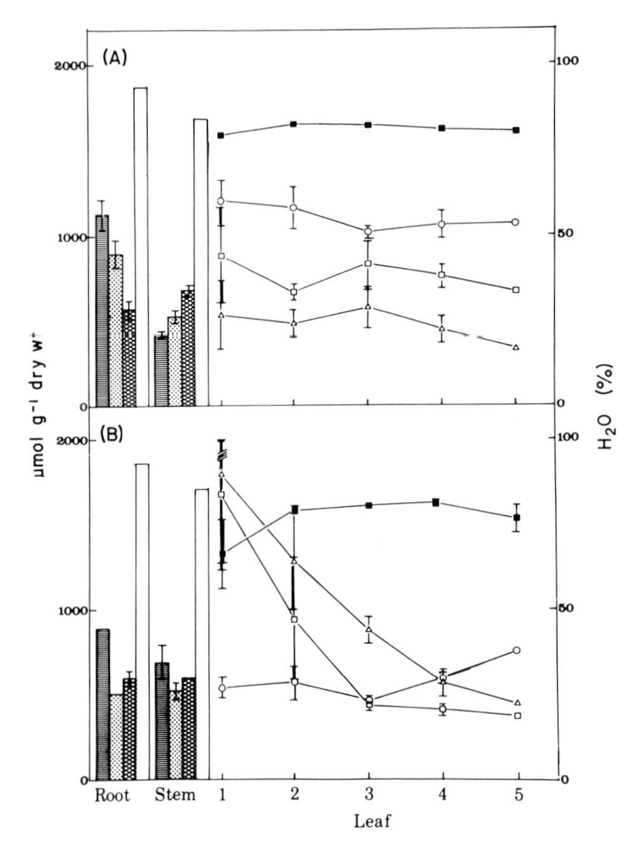

Fig. 10. Plants grown for 2 weeks in control solution followed by 2 weeks in solution containing 100 mol m^{-3} NaCl. (A) 4W SM100. (B) 4W INL10. See legend to Fig. 9 for details. From M. J. Hodson (unpublished).

operating at low K^+ concentration had high affinity for K^+ and high selectivity for K^+ in the presence of Na^+ (Rains and Epstein, 1967a). In the mangrove *Avicennia marina,* leaf tissue selectively absorbed K^+ from a medium containing 10 mol m^{-3} and K^+ and up to 500 mol m^{-3} Na^+. The rate of K^+ uptake was not inhibited by any of the Na^+ concentrations used; indeed, uptake of K^+ was approximately 30% stimulated in the presence of 100 mol m^{-3} Na^+ (Rains and Epstein, 1967b). Storey and Wyn Jones (1979) observed that 500 mol m^{-3} NaCl in culture solution caused a substantial increase in the concentration of potassium in roots of *Suaeda monoica*. Figures 9, 10, and 11 show that a salt-tolerant clone of *Agrostis stolonifera* was able to maintain higher levels of potassium in roots, stem, and leaves than a salt-susceptible clone when grown in saline culture solution.

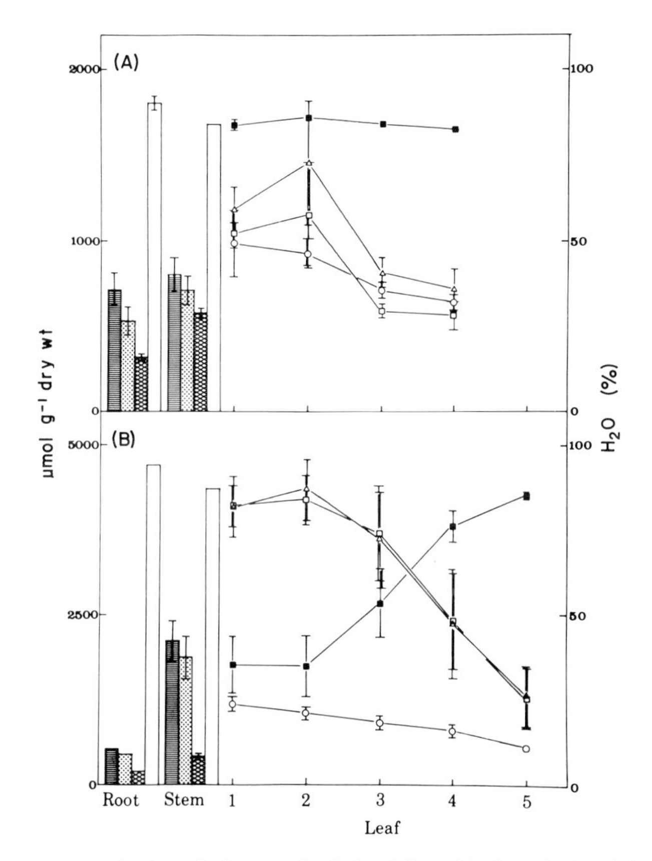

Fig. 11. Plants grown for 2 weeks in control solution followed by 2 weeks in solution containing 200 mol m^{-3} NaCl. (A) 4W SM200. (B) 4W INL200. See legend for Fig. 9 for details. From M. J. Hodson (unpublished).

Potassium is not the only cation that may be affected by excess Na^+. Calcium and Mg^{2+} concentrations of avocado leaves are decreased with an increase in exchangeable Na^+ in the soil (Martin and Bingham, 1954). A salt-induced reduction in the levels of nutrient ions could result in their deficiency and contribute to inhibition of growth. Greenway (1963) suggested that sensitivity of *Hordeum vulgare* to excess salt increased with a decrease in mineral-nutrient levels in the culture solution. Calcium is involved in maintaining the integrity of membranes and is essential for selective transport of alkali cations (Epstein, 1961; Läuchli and Epstein, 1970). Increasing the concentration of Ca^{2+} in a salinized culture solution dramatically reduced growth inhibition in the salt-sensitive species *Phaseolus vulgaris* (La Haye and Epstein, 1969). The rate of Na^+ uptake was maintained proportional to growth. Elzam and Epstein (1969) showed that salinity resulted in lower Ca^{2+} levels in roots of *Agropyron intermedium* and *A. elongatum;* Ahmad (1978) showed a similar effect in *Agrostis stolonifera.* Lessani and Marschner (1978) showed that for sugar beet, maize, cress, sunflower, safflower, pepper, and bean, salt treatment caused a smaller decrease in K^+ content than in Ca^{2+} content.

4. *Aeration Problems Resulting from Waterlogging, and Interactions between Stresses in Saline Environments*

The effects of waterlogging on soil aeration are dealt with in detail by Ponnamperuma (Chapter 2 this volume), but it is important to realize that the resulting problems for aeration are thus an important feature of soils flooded with saline water. As a result, plants growing in salt-marsh soils are exposed to the interaction between stresses as diverse as salinity and varying degrees of anaerobiosis. Various morphological adaptations (e.g., pneumatophores and aerenchyma) are present in plants characteristic of waterlogged saline soils that enable them to maintain adequate internal aeration. However, gross morphological adaptations are not always present in species from these environments. Cooper (1982) grew *Festuca rubra, Juncus gerardii, Armeria maritima, Aster trifolium, Triglochin maritima, Puccinellia maritima,* and *Salicornia europaea* in drained and waterlogged salt-marsh soils under saline and nonsaline conditions. Tolerance to waterlogging and salinity differed in a way consistent with the position of the species on the salt-marsh ecotone. Yield of *S. europaea* was greatest under the saline treatment, whereas *P. maritima* showed a preference for waterlogged soils. It was suggested that factors other than the greater availability of iron and manganese in anaerobic salt-marsh soils were involved in determining species distribution along the salt-marsh ecotone.

John *et al.* (1977) studied the interactive effects of anaerobiosis and salinity on barley and rice. At an early tillering stage, the plants were exposed simultaneously to anaerobiosis and high NaCl. Anaerobiosis only slightly aggravated the effects of salt on root and shoot growth in both species. Anaerobiosis increased permeability of the root system to Na^+ and Cl^-; however, under anaerobic conditions ion levels were decreased, possibly because of inhibition of active ion uptake. The most important interaction between anaerobiosis and salinity occurred in the shoot, where anaerobiosis increased Na^+ and decreased K^+ content. M. M. Smith, reported by Wainwright (1980), showed that partial anaerobiosis created by nitrogen "aeration" of the culture solution caused an increase in the levels of Na^+ and Cl^- in shoots of *Agrostis stolonifera* grown in saline culture and that the effect was greater in an inland ecotype than in a salt-marsh ecotype. Ahmad and Wainwright (1977) showed that a salt-marsh ecotype of *A. stolonifera* was more tolerant to partial anaerobiosis than were inland or spray-zone ecotypes. The salt-marsh plants were also more tolerant to the simultaneously applied stresses of salinity and partial anaerobiosis (Fig. 12).

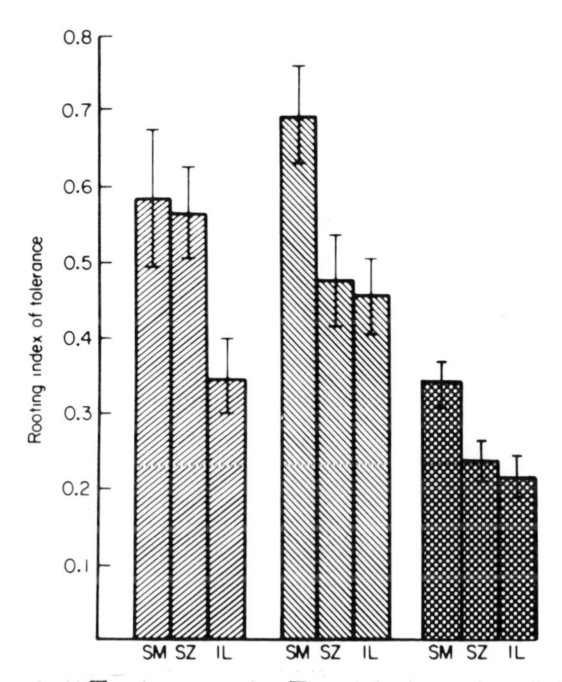

Fig. 12. Effects of salt (▨), nitrogen aeration (▧), and simultaneously applied salt and nitrogen aeration (▩) on the rooting index of tolerance. SM, Salt-marsh plants; SZ, spray-zone plants; IL, inland plants. Values plotted are means ±95% confidence limits. From Ahmad and Wainwright (1977).

TABLE II

Mean Dry Weight of *Brassica napus* cv. Brutor Plants[a]

NaCl in culture solution (mol m^{-3})	$CaSO_4$ in culture solution (mol m^{-3})	
	0	1
0	100 ± 18[b]	92 ± 11[b]
25	54 ± 10	64 ± 18
50	19 ± 0.7	52 ± 10
100	7 ± 1.4	13 ± 0.8

[a]From P. Kempster and S. J. Wainwright (unpublished).

[b]Values of experimental plants expressed as a percentage of mean dry weight of control plants ±SE.

As mentioned before, there is an interactive effect between Ca^{2+} and salinity on growth of various plants. There is a strong interaction between Ca^{2+} and NaCl on growth in *Brassica napus* (P. Kempster and S. J. Wainwright, unpublished; see Table II). Increased Ca^{2+} is antagonistic to the inhibitory effects of NaCl on growth. Nassery (1979) showed that Ca^{2+} could prevent salt-induced K^+ loss from plant roots but Mg^{2+} could not. Undoubtedly, interactions of higher orders exist for growth of plants in saline environments, for example, NaCl–Mg^{2+}–Mn^{2+}–anaerobiosis–nitrogen deficiency, etc.

B. Mechanisms of Tolerance or Avoidance of Salt Damage

Tolerance of saline conditions in a variety of plants is not the result of a single mechanism of salt tolerance, but rather possession of a subset of adaptations (physiological, morphological, etc.) from a larger set of possible adaptations. The adaptations possessed by a particular species determine the degree of salt tolerance that it can achieve, but in principle it is possible for the same degree of tolerance to be achieved by different subsets of adaptations.

1. Halophytes

High ion uptake is considered to be the major halophytic adaptation (Greenway and Munns, 1980). There is a constitutively high uptake of salt by halophytes that are able to produce very low solute potentials even in the absence of high salt concentrations. In order to maintain turgor and to ensure water uptake from the soil solution, ions are taken up and accumulated, thus lowering the osmotic potential of the cell sap. However, as ions are continuously taken up by plants growing in saline habitats, various ways of avoiding the effects of excess internal salt are incorporated by different species into their subsets of adaptations. Avoidance of the effects of excess salt can be achieved in various ways. In each of the possible ways of avoiding these effects, plants must still be able to absorb water from saline substrates.

Some halophytes (e.g., *Triplex hastata*) show reduced uptake of salt by their root systems (Black, 1956). Such species are not truly salt excluders because they absorb appreciable quantities of salt. However, if salt uptake is regulated to keep pace with growth, then excess ion concentrations will not be experienced by the plant. The interplay of growth and ion uptake can constitute a feedback system involving increased salt uptake, via turgor relations, and increased extension growth and succulence (Yeo and Flowers, 1980), which in turn can compensate for the increased ion burden. However, for this to be a continuing process, organic growth must take place, producing new cells capable of extension. Albert (1975) considered increases in growth and succulence as possible compensatory mechanisms for regulating salt concentration by dilution. Greenway

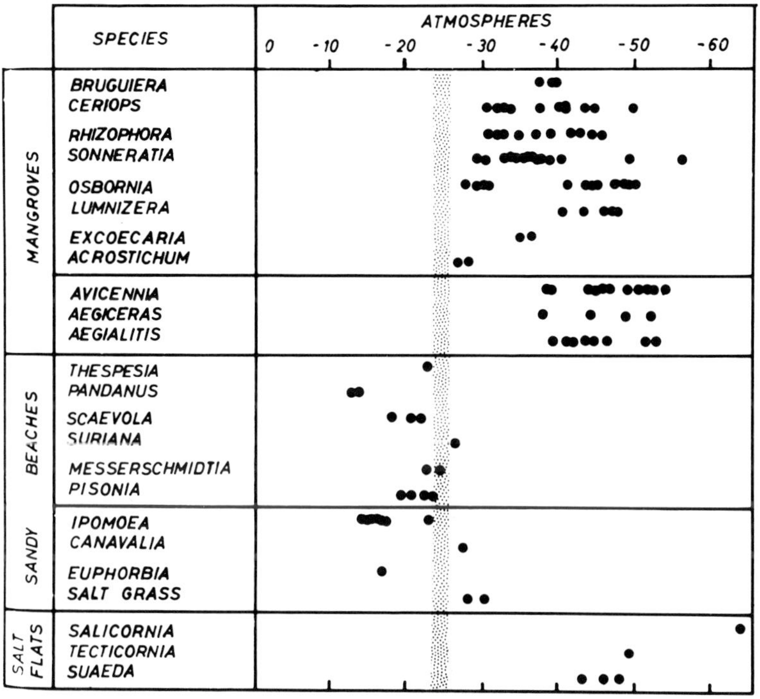

Fig. 13. Xylem sap tension in a number of mangroves and other halophytes. Stippled area is the negative solvent pressure of seawater. From Scholander (1968).

(1965) defined a simple mathematical model describing the maintenance of a constant ionic concentration in the shoot as a result of the diluting effects of growth via increases in volume. Scholander (1968) showed that mangroves had a high negative pressure in the xylem of between about 3 and 6 MPa, which is more than adequate to balance the osmotic potential of seawater (Fig. 13). When ~4 MPa of pressure was applied to the root system of a decapitated seedling of *Sonneratia* potted in seawater, a flow of virtually fresh water resulted and was maintained for several hours. Neither low temperature nor metabolic toxins inhibited this ultrafiltration process, but damage to membranes caused by chloroform destroyed the salt-preparation process (Fig. 14).

Lawton *et al.* (1981) compared the structure of roots of a salt-excluding mangrove, *Bruguiera gymnorrhiza,* and a salt-excreting mangrove, *Avicennia marina,* which does not exclude salt. A casparian strip is present in the cell walls of the endodermis of both species at a distance of 5 mm from the distal end of the root. In the salt excluder, the Casparian strip is immediately adjacent to the proximal end of the root cap. In the salt excreter, there is a 3-mm gap between them. Suberization of the outer layers of the roots and the development of corky

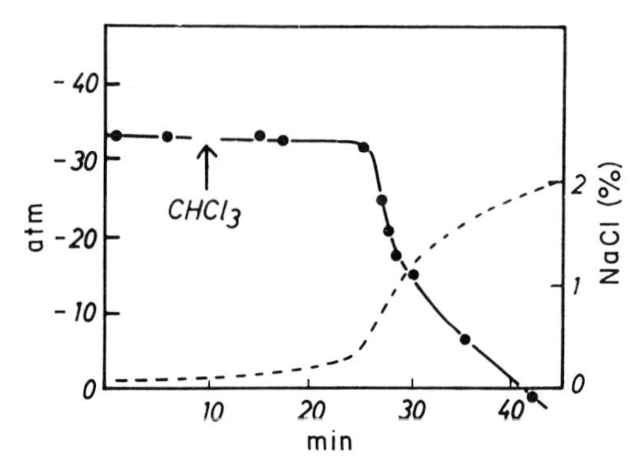

Fig. 14. Effect of chloroform on sap tension and concentration on the twig (*Sonneratia*) *in situ*. - - -, NaCl. From Scholander (1968).

layers occurred 3 mm above the proximal end of the root cap in both species. There was thus a permeable gap in the roots of *A. marina* that allowed free access of salt to the stele. In *B. gymnorrhiza,* this gap was absent. Drennan and Pammenter (1982) studied the physiology of salt excretion from the leaves of *A. marina.* There was a strong positive correlation between total salt excreted and total transpiration in the long term, but a negative correlation with the daily transpiration pattern. Excretion was maximal at night and minimal in the day. That salt arriving at the leaf during daytime transpiration is sequestered prior to secretion explains this apparent contradiction. Continued excretion after removal from salt supports the idea of a storage phenomenon, as does preliminary ultrasturctural evidence presented by Berjak (1978). Excess salt is sequestered in special hairs in several species of *Atriplex* (Osmond *et al.*, 1969; Mozafa and Goodin, 1970). In *Allenrolfia, Holoinemum,* and *Salicornia,* excess salt is removed by the physical loss of salt-saturated organs (Chapman, 1968).

Excess salt may be secreted by salt glands in some halophytes [e.g., *Spartina* ×*townsendii* (Skelding and Winterbotham, 1939) and *Limonium* (Ziegler and Lüttge, 1967)]. During salt stress, Na^+ and Cl^- ions are selectively secreted from the leaves of *Diplachne fusca,* whereas shoot K^+ content on a tissue water basis remains constant (Sandhu *et al.*, 1981). Ion pumps that enable salt uptake against an electrochemical potential gradient are involved in salt secretion in *Atriplex* (Osmond *et al.*, 1969); it is likely that they are involved in salt glands of *Avicennia marina,* which has two to four collecting cells, one stalk cell, and eight secretory cells. The salt content of the glands is lower than that of the mesophyll cells, so a gradient is established down which salt moves (Shimony *et al.*, 1973). In *Tamarix aphylla,* the apoplast serves as a pathway from the xylem to the salt glands under conditions of high salinity, and the subcuticular space acts as a collecting chamber. Here water moves down a ψ gradient into the collecting chamber, where hydrostatic pressure develops, eventually forcing open the cuticular pores and releasing the concentrated secretion (Campbell and Thompson, 1975). In *Atriplex,* salt is secreted to epidermal bladders, as mentioned previously. These bladders collapse and dry out, leaving a reflective salt crust on the leaves. By transporting salt to the epidermal bladder rather than to vacuoles of the mesophyll cells, salt stress of the photosynthesizing tissues may be avoided to some extent (Osmond *et al.*, 1969).

The vacuole is central to the current model of a mechanism of salt tolerance, which involves compartmentation of NaCl away from sensitive sites of metabolism. The model originated from the previously mentioned observations that various aspects of metabolism, both at the organelle level (Hall and Flowers, 1973; Flowers, 1974) and at the enzyme level (Flowers, 1972a; Greenway and Osmond, 1972; Stewart and Lee, 1974), are sensitive to amounts of salt present in halphytes and that, in most cases, the enzymes of halophytes are just as sensitive to salt as those of glycophytes. Adriani (1956) and Jennings (1968) suggested that halophytes have a capacity to accumulate large amounts of salt in their vacuoles. Flowers *et al.* (1977) reviewed the evidence for this hypothesis. Accumulation of Cl^- in vacuoles has been shown in *Salicornia pacifica* (Hess *et al.*, 1975), *Suaeda maritima* (Harvey *et al.*, 1976), and *Aegiceras corniculatum* (Van Steveninch *et al.*, 1976). Hall *et al.* (1974) demonstrated localization of rubidium in the vacuoles of cells of rubidium-treated *Suaeda maritima*. Campbell and Thomson (1976) showed that salt is accumulated in the apoplast and specialized vacuoles in *Franklinia* leaves and that chloroplast Cl^- levels are lower in cells in which the specialized vacuoles are present than in cells lacking them. Yeo (1981) showed that when *Suaeda maritima* was grown in a saline medium, the Na^+ concentration in the leaf was discontinuous across the tonoplast, with high Na^+ concentrations (600 mol m^{-3}) in the vacuole, and low concentrations (165 mol m^{-3}) in the cytoplasm.

If halophytes do have a capacity to sequester Na^+ and Cl^- in the vacuole away from the sensitive sites of metabolism, they would have to accumulate nontoxic compatible solutes in the cytoplasm to balance the osmotic potential of the vacuole and prevent the cytoplasm from underoing water stress. Moreover, if salt is also sequestered to some extent in the apoplast, the cell would need to maintain still lower osmotic potentials, and this in turn would require osmotic balancing of the cytoplasm. As the cytoplasm comprises only 5–10% of the volume of a vacuolated cell, it is possible to accumulate moderate absolute quantities of organic solutes and still achieve the cytoplasmic concentrations needed to bring about an osmotic balance between the cytoplasm and vacuole. Moreover, the cytoplasm develops a significant matric potential because of its colloidal contents, which will thus aid the balancing of potentials on each side of the tonoplast.

There are many observations of accumulation of organic solutes under water stress and salt stress. Frequently, these compatible organic solutes are nitrogenous compounds such as proline and quaternary ammonium compounds (Stewart and Lee, 1974; Treichel, 1975; Storey and Wyn Jones, 1975, 1977, 1978b, 1979; Boggess *et al.*, 1976; Chu *et al.*, 1976; Stewart *et al.*, 1977; Wyn Jones *et al.*, 1977a,b; Bar-Nun and Poljakoff-Mayber, 1977; Cavalieri and Huang, 1977; Huber and Schmidt, 1978; Flowers and Hall, 1978; Hanson and Nelson, 1978; Wyn Jones and Storey, 1978; Stewart and Rhodes, 1978; Wyn Jones and Ahmad, 1978; Tully *et al.*, 1979; Huang and Cavalieri, 1979; Ahmad and Wyn Jones, 1979; Leigh *et al.*, 1981; Neals and Sharkey, 1981; Gorham *et al.*, 1981; Briens and Larher, 1982; Larher *et al.*, 1982). There is evidence that glycine betaine is localized in the cytoplasm in beet (Wyn Jones *et al.*, 1977b) and *Suaeda maritima* (Hall *et al.*, 1978). Stewart *et al.* (1979) classified halophytic plant species into four groups, depending on the organic solutes they accumulate under stress. Group 1 plants mainly accumulate methylated quaternary ammonium compounds (e.g., *S. maritima*). Group 2 plants mainly accumulate proline (e.g., *Puccinellia maritima*). Group 3 plants accumulate both proline and methylated quaternary ammonium compounds (e.g., *Spartina anglica*). Group 4 plants accumulate sorbitol (e.g., *Plantago maritima*). Gorham *et al.* (1981) have shown that *Plantago coronopus* accumulates sorbitol under salt stress and that *Honkenya peploides* accumulates the cyclitol pinitol.

Briens and Larher (1982) have classified halphytic plants into three groups. Group 1 plants accumulate high sucrose and or polytols only (e.g. *Plantago maritima, Juncus maritimus, Phragmites communis,* and *Scirpus maritimus*). Group 2 plants accumulate carbohydrates and nitrogenous compounds (e.g., *Aster tripolium, Beta maritima, Festuca rubra,* and *Spartina townsendii*). Group 3 plants accumulate more nitrogenous than carbohydrate solutes (e.g., *Agropyron pungens, Atriplex hastata, Halimione portulacoides, Limonium vulgare, Salicornia europaea, Suaeda macrocarpa,* and *Triglochin maritima*).

Although the model of salt tolerance at its simplest assumes that all of the salt is sequestered in the vacuole and that all of the compatible organic solutes accumulate in the cytoplasm, halophytes can sometimes overproduce organic osmotica. Jefferies (1973) showed that the proline produced by *Triglochin maritima* ould induce an osmotic potential of − 5.6 MPa in the cytoplasm, whereas the measured osmotic potential of the cell sap was −2.9 MPa. Some of the proline must therefore be located in the vacuole as well as the cytoplasm if osmotic equlibrium is maintained. Leigh *et al.* (1981) showed that glycine betaine was present in beet vacuoles, but its concentration was always higher in the cytoplasm than in the vacuole. These vacuolar pools of organic osmotica could serve the dual purpose of a nitrogen store, which could be important to plants in a nitrogen-limiting environment and as an osmotic buffer system in which osmotica could be transported across the membrane in response to changing external osmotic conditions. An essential property of accumulated organic osmotica is that they do not inhibit enzyme activity when present in high concentrations

(Stewart and Lee, 1974; Pollard and Wynn Jones, 1979). Indeed, there is evidence that glycine betaine can, to some extent, protect enzyme activity from salt inhibition (Pollard and Wyn Jones, 1979). Schobert (1977) suggested that water-like substances of the general structure ROH could help prevent the hydrophobic groups of biopolymers from precipitating under water stress by participating in the hydration shell around hydrophobic sites. On the other hand, amphiphile molecules, such as proline or betaine, could associate with hydrophobic groups on biopolymers, effectively converting them to hydrophilic groups that would require less water to remain soluble.

2. *Salt-Tolerant Glycophytes*

Salt tolerance in nonhalophytes has been reviewed by Greenway and Munns (1980). A major difference between salt-tolerant glycophytes and many true halophytes is that the former achieve a certain degree of tolerance by excluding salt. Many halphytes, as mentioned, do not exclude salt but rather take up large quantities of salt, and many are capable of excreting excess salt. Figures 9–11 show that the salt-tolerant ecotype of *Agrostis stolonifera* takes up less salt than does the salt-susceptible ecotype. The salt-exclusion strategy itself may not impose an upper limit to salt tolerance because, in salt-tolerant varieties, the internal levels of salt achieved are sufficient to produce the required osmotic adjustment to the external medium. The salt-susceptible varieties, which take up even more salt, are in an even better position to osmotically adjust to the external medium and yet, as can be seen in Figs. 9–11, they undergo severe water stress. The salt-exclusion mechanism of salt-tolerant glycophytes is likely related to protection of some inherently intolerant structures within the plant. It seems likely that the membranes of glycophytes may constitute the physiological weak link. It may be the case that when certain internal levels of salt are attained, membranes are damaged and the cells are unable to maintain turgor despite their high salt content. Preliminary data obtained by A. Majeed (unpublished) indicate that when an inland ecotype of *A. stolonifera* was treated with 200 mol m^{-3} NaCl, the plants absorbed salt and were able to maintain turgor for 2 days. However, as salt uptake increased, turgor pressure was not maintained and the ψ declined to a value equal to the solute potential, and the leaf precipitously lost water and died.

Salt-tolerant glycophytes have enhanced capacities to accumulate compatible organic solutes in a way similar to that of halophytes (Wainwright, 1980; Ahmad *et al.*, 1981). It seems that the basic model of salt compartmentation in the vacuole and, possibly to some extent, the apoplast is similar for halophytes and salt-tolerant glycophytes. Smith *et al.* (1982) demonstrated that Cl^- was present in vacuoles and cell walls of salt-treated, salt-tolerant *Agrostis stolonifera*, with little Cl^- present in the cytoplasm. However, the major difference between halophytes and salt-tolerant glycophytes appears to be in capacity to take up salt. It is interesting however, that despite the fact that the salt-tolerant ecotype of *A. stolonifera* restricts the uptake of salt from highly saline media, it maintains significantly higher levels of Na^+ and Cl^- in its roots and shoots than does the salt-susceptible variety when grown in nonsaline media (A. Majeed, unpublished).

In an environment subject to flooding with saline water, another route for entry of salt into the plant is through the leaves. The problems encountered by leaves submerged in salt water are not osmotic. There is a far lower ψ gradient between a leaf submerged in salt water and the surrounding water than there is

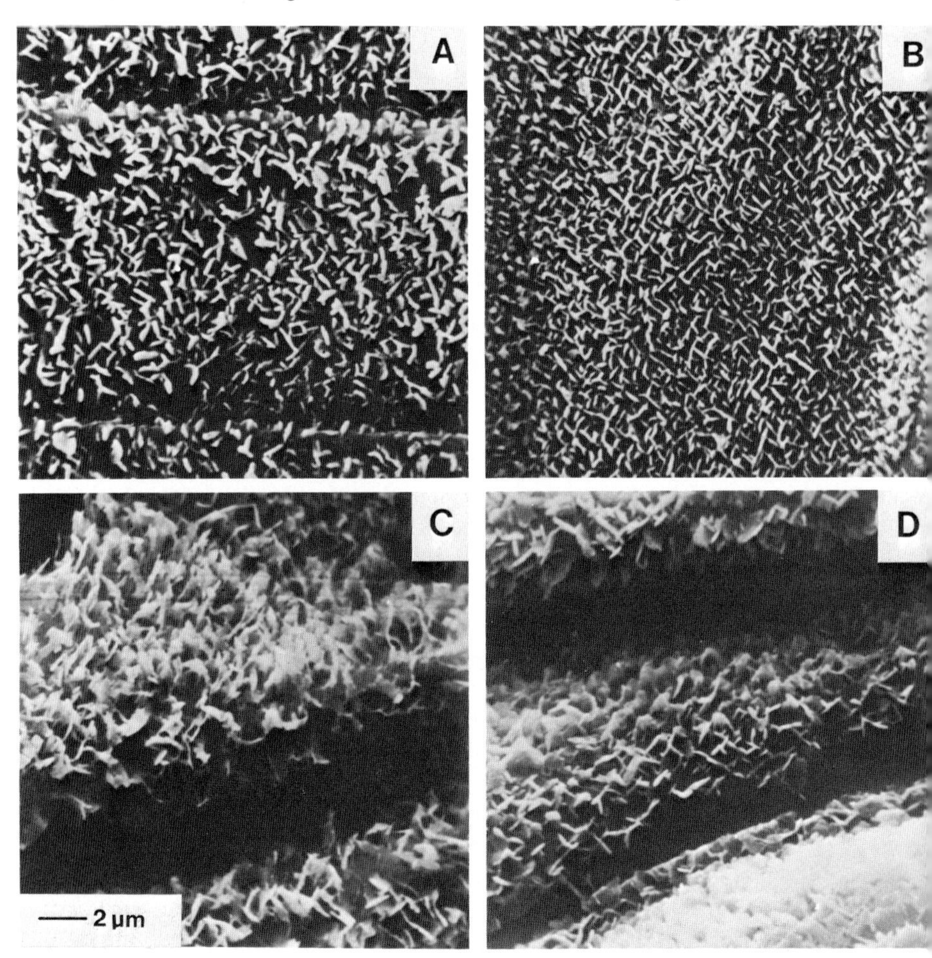

Fig. 15. Scanning electron micrographs of the leaf surfaces of *Agrostis stolonifera*. (A) Inland plant upper surface. (B) Inland plant lower surface. (C) Salt-marsh plant upper surface. (D) Salt-marsh plant lower surface. From Ahmad and Wainwright (1976).

between a leaf and an atmosphere of moderate humidity. For example, the ψ of an atmosphere at 50% relative humidity is -94.4 MPa. Therefore, adaptations that protect the leaf from the direct contact of seawater are likely to involve prevention of entry of Na^+ and Cl^- rather than the loss of water. Ahmad and Wainright (1976) showed that after the leaves of the inland ecotype of *Agrostis stolonifera* were immersed in salt water, they retained 18 times as much Na^+ as did the leaves of the salt-marsh ecotype. Moreover, the leaf surfaces of the salt-marsh plants were covered with large wax flakes, whereas those of the inland plants had small, sparse wax rodlets (Fig. 15A–D). These differences in leaf surfaces substantially reduced the wettability of the salt-marsh plants.

IV. ADAPTATIONS AT THE LEVEL OF THE CULTURED CELL

It was mentioned previously in relation to the breeding of salt-tolerant plants that biotechnological methods are potentially available to augment classical plant breeding programs. The relative salt tolerance of the rather tolerant species *Hordeum jubatum* and the more sensitive *H. vulgare* is preserved when they are grown as callus cultures (Orton, 1980), indicating that salt tolerance is expressed at the tissue or cell level. Warren and Gould (1982) gave evidence for salt tolerance at the level of the cell in suspension cultures of the halophytic grass *Distichlis spicata*.

Nabors *et al.* (1975) recognized that by a combination of plant tissue-culture techniques and techniques for isolating spontaneous or induced mutations, it might be possible rapidly to select salt-tolerant crop plants. By growing tobacco cells in suspension culture, treating them with the mutagen, ethyl methane sulfonate, and then culturing the mutagen-treated cells in a saline selection medium, it was possible to select for salt-tolerant cell lines that would grow in medium containing 89 mol m^{-3} NaCl, whereas normal tobacco suspensions are killed by concentrations of NaCl as low as 27.4 mol m^{-3}. Nabors *et al.* (1975) calculated that because spontaneous mutation occurs at a rate of ~1 per 10^5 gene copies per generation, it is likely that large volumes of suspension cultures would contain at least one mutant cell for each mutant phenotype requiring a dominant or codominant allele. However, if one wished to select for double-recessive mutant cells, the rate of spontaneous appearance would only be 1 in 10^{10}. Therefore, mutation induction would be required. Dix and Street (1975) selected salt-tolerant lines of cells of *Nicotiana sylvestris* and *Capsicum annuum*. Nabors *et al.* (1980) had managed to select lines of tobacco cells that were tolerant to 151 mol m^{-3} NaCl. Plants regenerated from salt-resistant cell lines transmitted their tolerance to two subsequent generations, and these plants were able to survive in the presence of 572 mol m^{-3} NaCl, whereas few nontolerant plants could survive in 264 mol m^{-3} NaCl. Although genetic change had probably occurred, it is interesting that plant lines derived from NaCl-tolerant cell lines were noticeably less tolerant if salt was omitted from the medium during the initial regeneration process. Also, F_2 plants were noticeably more salt tolerant if grown from seed collected from salt-treated plants. Croughan *et al.* (1978) selected a salt-tolerant line of alfalfa cells that survived on a medium containing 171 mol m^{-3} NaCl. The nonselected line of cells grew 10 times as rapidly in the absence of NaCl as in the presence of 171 mol m^{-3} NaCl. On the other hand, the selected line of cells grew just as rapidly at each extreme of the NaCl concentration range. However, the selected line of cells showed optimum growth at 85 mol m^{-3} NaCl, whereas growth of the nonselected line was progressively inhibited by increasing NaCl. The nonselected cells had higher Na^+/K^+ ratios than did the selected cells. This indicated a shift in the characteristics of the salt-selected cells toward true halphytic characteristics, as did the greater Cl^- and K^+ content of the salt-selected cells. Rains *et al.* (1980) reported the selection of a population of rice cells in the presence of 257 mol m^{-3} NaCl. The salt-selected cells and a line of unselected cells were grown for 42 days in media containing a range of NaCl concentrations. In the absence of NaCl, the salt-selected line grew poorly, whereas the nonselected line grew vigorously. However, as the NaCl concentration in the medium was increased, growth of the nonselected line of cells was progressively inhibited until at 1.5%(w/v) NaCl it was killed. However, the salt-selected line of cells showed maximum growth at 0.5% NaCl and about 60% of its maximum growth at 1.5% NaCl. These changes in the salt-selected cells represented a shift toward true halophytic characteristics.

It thus seems that selection of salt-tolerant cell lines coupled with subsequent regeneration of whole plants could provide a means of short-circuiting the normal reproductive cycle, compressing the time scale required for development of new salt-tolerant plants.

ACKNOWLEDGMENTS

I thank Dr. M. J. Hodson, Mr. A. Majeed, and Miss M. Clark for providing me with some of the data presented here. I thank Mrs. E. E. Griffiths and Mr. P. Kempster for skilled technical assistance and Mr. K. Jones for the photographic work.

REFERENCES

Adriani, M. J. (1956). Der Wasserhaushalt der Halophyten. *Encycl. Plant Physiol.* **3,** 902–904.

Ahmad, I. (1978). Some aspects of salt tolerance in *Agrostis stolonifera* L. Ph.D. Thesis, Univ. of Wales, Swansea.

Ahmad, I., and Wainwright, S. J. (1976). Ecotype differences in leaf surface properties of *Agrostis stolonifera* from salt marsh, spray zone and inland habitats. *New Phytol.* **76,** 361–366 1/2.

Ahmad, I., and Wainwright, S. J. (1977). Tolerance to salt, partial anaerobiosis and osmotic stress in *Agrostis stolonifera. New Phytol.* **79,** 605–612.

Ahmad, I., Wainwright, S. J., and Stewart, G. R. (1981). The solute and water relations of *Agrostis stolonifera* ecotypes differing in their salt tolerance. *New Phytol.* **87,** 615–629.

Ahmad, N., and Wyn Jones, R. G. (1979). Glycinebetaine, proline and inorganic ion levels in barley seedlings following transient stress. *Plant Sci. Lett.* **15,** 231–237.

Albert, R. (1975). Salt regulation in halophytes. *Oecologia* **21,** 57–71.

Antonovics, J., Bradshaw, A. D., and Turner, R. G. (1971). Heavy metal tolerance in plants. *Adv. in Ecol. Res.* **7,** 1–85.

Aston, J. L., and Bradshaw, A. D. (1966). Evolution in closely adjacent plant population. II. *Agrostis stolonifera* in maritime habitats. *Heredity* **21,** 649–664.

Austenfeld, F. A. (1976). The effects of various alkaline salts on the glycollate oxidase of *Salicornia europaea* and *Pisum sativum in vitro. Physiol. Plant.* **36,** 82–87.

Bar-Nun, N., and Poljakoff-Mayber, A. (1977). Salinity stress and the content of proline in roots of *Pisum sativum* and *Tamarix tetragyna. Ann. Bot.* (*London*) **41,** 173–179.

Bates, J. W. (1976). Cell permeability and regulation of intracellular sodium concentration in a halophytic and glycophytic moss. *New Phytol.* **77,** 15–23.

Berjak, P. (1978). Subcellular adaptations in plants of extreme environments. II. Towards understanding the salt excretion in *Avicennia* spp. *Proc. Electron Microsc. Soc. South. Afr.* **8,** 101–102.

Beer, S., Shomer-Ilan, A., and Waisel, Y. (1975). Salt stimulated phosphoenolpyruvate carboxylase in *Cakile maritima. Physiol. Plant.* **34,** 293–295.

Bernstein, L. (1961). Osmotic adjustment of plants to saline media. I. Steady state. *Am. J. Bot.* **48,** 908–918.

Bernstein, L. (1963). Osmotic adjustment of plants to saline media. II. Dynamic phase. *Am. J. Bot.* **50,** 360–370.

Bhatti, A., Saeed, G., Sarwar, K. H., Sheikh, M., Mariuf, M., and Sharif, M. (1976). Effects of sodium chloride on the growth and ion content of barley. *Pak. J. Sci. Ind. Res.* **19,** 190–192.

Black, R. F. (1956). Effect of NaCl in water culture on the ion uptake and growth of *Atriplex hastata* L. *Aust. J. Biol. Sci.* **9,** 67–80.

Boggess, S. F., Aspinall, D., and Paleg, L. G. (1976). Stress metabolism. IX. The significance of end product inhibition of proline biosynthesis and compartmentation in relation to stress induced proline accumulation. *Aust. J. Plant Physiol.* **3,** 513–525.

Boorman, L. (1967). Biological flora of the British Isles: *Limonium vulgare* Mill, and *L. Humile* Mill. *J. Ecol.* **55,** 221–232.

Brady, N. C. (1974). "The Nature and Properties of Soils," 8th ed. Macmillan, New York.

Breteler, F. J. (1977). America's Pacific species of *Rhizophora. Acta Bot. Neerl.* **26,** 225–230.

Briens, M., and Larher, F. (1982). Osmoregulation in halophytic higher plants; a comparative study of soluble carbohydrates, polyols, betaines and free proline. *Plant Cell Environ.* **5,** 287–292.

Briggs, D. (1978). Genecological studies of salt tolerance in groundsel (*Senecio vulgaris* L.) with particular reference to roadside habitats. *New Phytol.* **81,** 381–390.

Campbell, N., and Thompson, W. W. (1975). Chloride localisation in the leaf of *Tamarix. Protoplasma* **83,** 1–14.

Campbell, N., and Thompson, W. W. (1976). The ultrastructural basis of chloride tolerance in the leaf of *Franklinia. Ann. Bot. (London)* **40,** 687–693.

Cantor, L. M. (1967). "A World Geography of Irrigation." Oliver & Boyd, Edinburgh.

Carter, D. L. (1975). Problems of salinity in agriculture. *In* "Plants in Saline Environments" (A. Poljakoff-Mayber and J. Gale, eds.), pp. 25–35. Springer-Verlag, New York.

Casey, H. E. (1972). Salinity problems in arid lands irrigation: a literature review and selected bibliography. *Arid Lands Resour. Inf. Pap.* No. 1.

Cavalieri, A. J., and Huang, A. H. C. (1977). Effect of NaCl on the *in vitro* activity of malate dehydrogenase in salt marsh halophytes of the U.S. *Physiol. Plant.* **41,** 79–84.
Chapman, V. J. (1960). "Salt Marshes and Salt Deserts of the World." Leonard Hill, London.
Chapman, V. J. (1968). Vegetation under saline conditions. *In* "Saline Irrigation for Agriculture and Forestry" (H. Boyko, ed.), pp. 201–216. Junk, The Hague.
Chu, T. M., Aspinal, D., and Paleg, L. G. (1976). Stress metabolism. VIII. Specific ion effects on proline accumulation in barley. *Aust. J. Plant Physiol.* **3,** 503–511.
Cooper, A. (1982). The effects of salinity and waterlogging on the growth and cation uptake of salt marsh plants. *New Phytol.* **90,** 263–375.
Coughlan, S. J., and Wyn Jones, R. G. (1980). Some responses of *Spinacea oleracea* to salt stress. *J. Exp. Bot.* **31,** 883–893.
Cram, W. J. (1973). Internal factors regulating nitrate and chloride influx in plant cells. *J. Exp. Bot.* **24,** 328–341.
Croughan, T. P., Stavarek, S. J., and Rains, D. W. (1978). Selection of a NaCl tolerant line of cultured alfalfa cells. *Crop. Sci.* **18,** 959–963.
Dix, P. J., and Street, H. E. (1975). Sodium chloride–resistant cultured cell lines from *Nicotiana sylvestris* and *Capsicum annum. Plant. Sci. Lett.* **5,** 231–237.
Downton, W. J. S. (1978). Growth and flowering in salt-stressed avocado trees. *Aust. J. Agric. Res.* **29,** 523–534.
Drennan, P., and Berjak, P. (1979). Degeneration of foliar glands correlated with the shift to abaxial salt excretion in *Avicennia marina* (Forsk.) Vierh. *Proc. Electron Microsc. Soc. South. Afr.* **9,** 83–84.
Drennan, P., and Pammenter, N. W. (1982). Physiology of salt excretion in the mangrove. *Avicennia marina* (Forsk.) Vierh. *New Phytol.* **91,** 597–606.
Eaton, F. M. (1941). Water uptake and root growth as influenced by inequalities in the concentration of substrate. *Plant Physiol.* **16,** 545–564.
Edelbauer, A. (1978). The influences of different Cl^-/SO_4^{2-} ratios in nutrient solutions applied to vines (*Vitis vinifera* L.) on grape yields, mineral composition of leaves and stems, and bud susceptibility to frost. *Z. Pflanzenernaehr. Bodenk.* **141,** 83–94.
El-Sharawi, H. M., and Salama, F. M. (1977). Salt tolerance criteria in some wheat and barley cultivars. II. Adjustments in internal water balance. *Bull. Fac. Sci. Assiut Univ.* **5,** 1–16.
Elzam, O. E., and Epstein, E. (1969). Salt relations of two grass species differing in salt tolerance. I. Growth and salt content at differing salt concentrations. *Agrochemica* **13,** 187–191.
Epstein, E. (1961). The essential role of calcium in selective cation transport by plant cells. *Plant Physiol.* **36,** 437–444.
Epstein, E., and Hagen, C. E. (1952). A kinetic study of the absorption of alkali cations by barley roots. *Plant Physiol.* **27,** 457–474.
Fernandez, F. G., Caro, M., and Cerda, A. (1977). Influence of NaCl in the irrigation water on yield and quality of sweet pepper (*Capsicum annuum*). *Plant Soil* **46,** 405–411.
Fitter, A. H., and Hay, R. K. (1981). The origin of resistance. *In* "Environmental Physiology of Plants," p. 229. Academic Press, New York and London.
Flowers, T. J. (1972a). The effect of sodium chloride on enzyme activities from four halophyte species of Chenopodiaceae. *Phytochemistry* **2,** 1881–1886.
Flowers, T. J. (1972b). Salt tolerance in *Suaeda maritima* (L.) Dum. The effect of sodium chloride on growth, respiration, and soluble enzymes in a comparative study with *Pisum sativum* (L.) *J. Exp. Bot.* **23,** 310–321.
Flowers, T. J. (1974). Salt tolerance in *Suaeda maritima* (L.) Dum. A comparison of mitochondria isolated from green tissues of *Suaeda* and *Pisum. J. Exp. Bot.* **25,** 101–110.
Flowers, T. J. (1975). Halophytes. *In* "Ion Transport in Plant Cells and Tissues" (D. A. Baker and J. L. Hall, eds.), pp. 309–344. Elsevier, Amsterdam.

Flowers, T. J., and Hall, J. L. (1978). Salt tolerance in the halophyte *Suaeda maritima* (L.) Dum. The influence of the salinity of the culture solution on the content of various organic compounds. *Ann. Bot. (London)* **42,** 1057–1063.

Flowers, T. J., and Yeo, A. R. (1981). Variability in the resistance to sodium chloride salinity within rice (*Oryza sativa* L.) varieties. *New Phytol.* **88,** 363–373.

Flowers, T. J., Troke, P. F., and Yeo, A. R. (1977). The mechanism of salt tolerance in halophytes. *Annu. Rev. Plant Physiol.* **28,** 89–121.

Gale, J., Kohl, H. C., and Hagan, R. M. (1967). Changes in the water balance and photosynthesis of onion, bean and cotton plants under saline conditions. *Physiol. Plant.* **20,** 408–420.

Gale, J., Naaman, R., and Poljakoff-Mayber, A. (1970). Growth of *Atriplex halimus* L. in sodium chloride salinated culture solutions as affected by the relative humidity of the air. *Aust. J. Biol. Sci.* **23,** 947–952.

Gates, C. T. (1972). Ecological response of Australian native species *Acacia harpophylla* and *Atriplex nummularia* to soil salinity; effect on water content, leaf area and transpiration rate. *Aust. J. Bot.* **20,** 261–272.

Gettys, K. L., Hancock, J. F., and Cavalieri, A. J. (1980). Salt tolerance of *in vitro* activity of leucine aminopeptidase, peroxidase and malate dehydrogenase in the halophytes *Spartina alterniflora* and *S. patens. Bot. Gaz. (Chicago)* **141,** 453–457.

Gorham, J., Hughes, L., and Wyn Jones, R. G. (1981). Low molecular weight carbohydrates in some salt-stressed plants. *Physiol. Plant.* **53,** 27–33.

Greenway, H. (1963). Plant responses to saline substrates. III. Effect of nutrient concentration on the growth and ion uptake of *Hordeum vulgare* during sodium chloride stress. *Aust. J. Biol. Sci.* **16,** 616–628.

Greenway, H. (1965). Plant response to saline substrates. IV. Chloride uptake by *Hordeum vulgare* as affected by inhibitors, transpiration, and nutrients. *Aust. J. Biol. Sci.* **18,** 249–268.

Greenway, H., and Muuns, R. (1980). Mechanisms of salt tolerance in non-halophytes. *Annu. Rev. Plant Physiol.* **31,** 149–180.

Greenway, H., and Osmond, C. B. (1972). Salt responses of enzymes from species differing in salt tolerance. *Plant Physiol.* **49,** 256–259.

Greenway, H., and Sims, A. P. (1974). Effects of high concentrations of KCl and NaCl on responses of malate dehydrogenase (decarboxylating) to malate and various inhibitors. *Aust. J. Plant Physiol.* **1,** 15–29.

Hall, J. L., and Flowers, T. J. (1973). The effect of salt on protein synthesis in the halophyte *Suaeda maritima. Planta* **110,** 361–368.

Hall, J. L., Yeo, A. R., and Flowers, T. J. (1974). Uptake and localisation of rubidium in the halophyte *Suaeda maritima. Z. Pflanzenphysiol.* **71,** 200–206.

Hall, J. L., Harvey, D. M. R., and Flowers, T. J. (1978). Evidence for the cytoplasmic localisation of betaine in leaf cells of *Suaeda maritima. Planta* **140,** 59–62.

Hannon, N., and Bradshaw, A. D. (1968). Evolution of salt tolerance in two co-existing species of grass. *Nature (London)* **220,** 1342–1343.

Hanson, A. D., and Nelson, C. E. (1978). Betaine accumulation and [^{14}C]formate metabolism in water stressed barley leaves. *Plant Physiol.* **62,** 305–312.

Harvey, D. M. R., Hall, J. L., and Flowers, T. J. (1976). The use of freeze-substitution in the preparation of plant tissue for ion localisation studies. *J. Microsc.* **107,** 189–198.

Hayward, H. E., and Long, E. M. (1943). Some effects of sodium salt on the growth of tomato. *Plant Physiol.* **18,** 556–569.

Hayward, H. E., and Spurr, W. B. (1944). Effect of iso-osmotic concentrations of inorganic and organic substrate on entry of water into corn roots. *Bot. Gaz. (Chicago)* **106,** 131–139.

Hayward, H. E., Long, E. M., and Uhvits, R. (1946). Effect of chloride and sulphate salts on the growth and development of Elberta peach on Shalil and Lovell rootstocks. *U.S. Dept. Agric. Tech. Bull.* No. 922.

Hess, W. M., Hansen, D. J., and Weber, D. J. (1975). Light and electron microscopy localisation of chloride ions in the cells of *Salicornia pacifica* var. *utahensis*. *Can. J. Bot.* **53,** 1176–1187.

Hodson, M. J., Smith, M. M., Wainwright, S. J., and Öpik, H. (1982). Cation cotolerance in a salt-tolerant clone of *Agrostis stolonifera* L. *New Phytol.* **90,** 253–261.

Hoffman, G. T., and Jobes, J. A. (1978). Growth and water relations of cereal crops as influenced by salinity and relative humidity. *Agron. J.* **70,** 765–769.

Holmes, F. W., and Baker, J. H. (1966). Salt injury to trees. II. Sodium and chloride in roadside sugar maples in Massachusetts. *Phytopathology* **56,** 633–636.

Huang, A. H. C., and Cavalieri, A. J. (1979). Proline oxidase and water-stress induced proline accumulation in spinach leaves. *Plant Physiol.* **63,** 531–535.

Hubac, C., and Guerrier, D. (1972). Étude de la composition en acides amines de deux *Carex,* la *Carex stenophylla* Wahl. f. *pachystylis* J. Gaz. Asch et Graebn, très resistant á la secheresse et la *Carex setifolia*. Gordon non Kunze, pen resistant. Effet d'un apport proline exogene. *Ecol. Plant.* **7,** 147–165.

Huber, W., and Schmidt, F. (1978). Zur Wirkung verschiedener Salze und von Polyäthylenglykol auf den Prolin- und Aminosaurestoffwechsel von *Pennisetum typhoides*. *Z. Pflanzenphysiol.* **89,** 251–258.

Jefferies, R. L. (1973). The ionic relations of seedlings of the halophyte. *Triglochin maritima* L. *In* "Ion Transport in Plants" (W. P. Anderson, ed.), pp. 297–321. Academic Press, London and New York.

Jefferies, R. L., Rudmik, T., and Dillon, E. M. (1979). Responses of halophytes to high salinities and low water potentials. *Plant Physiol.* **64,** 989–994.

Jennings, D. M. (1968). Halophytes, succulence and sodium in plants—a unified theory. *New Phytol.* **67,** 899–388.

John, C. D., Liminuntana, V., and Greenway, H. (1977). Interaction of salinity and anaerobiosis in barley and rice. *J. Exp. Bot.* **28,** 133–141.

Kaiser, W. M., and Heber, U. (1981). Photosynthesis under osmotic stress. Effect of high solute concentrations on the permeability properties of the chloroplast envelope and on activity of stroma enzymes. *Planta* **153,** 423–429.

Kaplan, A., and Gale, J. (1972). Effect of sodium chloride salinity on the water balance of *Atriplex halimus*. *Aust. J. Biol. Sci.* **25,** 895–903.

King, A. (1980). "The State of the Planet." Pergamon, Oxford.

La Haye, P. A., and Epstein, E. (1969). Salt toleration by plants: enhancement with calcium. *Science (Washington, D.C.)* **166,** 395.

Larher, F., Jolivet, Y., Briens, M., and Goas, M. (1982). Osmoregulation in higher plant halophytes: organic nitrogen accumulation in glycine betaine and proline during the growth of *Aster tripolium* and *Suaeda macrocarpa* under saline conditions. *Plant Sci. Lett.* **24,** 201–210.

Larsen, H. (1967). Biochemical aspects of extreme halophilism. *Adv. Microb. Physiol.* **1,** 97–132.

Laszlo, E., and Kuiper, P. J. C. (1979). Effect of salinity on growth, cation content, sodium ion–uptake, and translocation in salt-sensitive and salt tolerant *Plantago* spp. *Physiol. Plant.* **47,** 95–99.

Laszlo, E., Stuiller, B., and Kuiper, P. J. C. (1980). The effect of salinity on lipid composition and on activity of calcium-stimulated and magnesium stimulated ATPase in salt sensitive and salt tolerant *Plantago* species. *Physiol. Plant.* **49,** 315–319.

Läuchli, A., and Epstein, E. (1970). Transport of potassium and rubidium in plant roots: the significance of calcium. *Plant Physiol.* **45,** 639–641.

Läuchli, A., and Wieneke, J. (1979). Studies on growth and distribution of Na^+, K^+, and Cl^- in soybean varieties differing in salt tolerance. *Z. Pflanzenernaehr. Bodenk.* **142,** 3–13.

Lawton, J. R., Todd, A., and Naidoo, D. K. (1981). Preliminary investigations into the structure of the roots of the mangroves, *Avicennia marina* and *Bruguiera gymnorrhiza* in relation to ion uptake. *New Phytol.* **88,** 713–722.

Leigh, R. A., Ahmand, N., and Wyn Jones, R. G. (1981). Assessment of glycine betaine and proline compartmentation by analysis of isolated beet vacuoles. *Planta* **153,** 34–41.

Lessani, H., and Marschner, H. (1978). Relation between salt tolerance and long distance transport of sodium and chloride in various crop species. *Aust. J. Plant Physiol.* **5,** 27–37.

Mahmoud, E. A., and Hill, M. J. (1980). Salt tolerance of sugar beet (*Beta vulgaris*) at various temperatures. *Z. Acker Pflanzenbau.* **149,** 157–166.

Maliwal, G. L., Manohar, S. S., and Paliwal, K. V. (1976). Effect of the quality of irrigation water on the yield of some wheat (*Triticum aestivum*) varieties in a sandy soil of Rajasthan, India. *Ann. Arid. Zone* **15,** 17–22.

Martin, J. P., and Bingham, F. T. (1954). Effect of various exchangeable cations in soil on growth and chemical composition on avocado seedlings. *Soil Sci.* **78,** 349–360.

Meiri, I. A., and Poljakoff-Mayber, A. (1970). Effects of various salinity regimes on growth, leaf expansion and transpiration of bean plants. *Soil Sci.* **109,** 23–34.

Meyer, B. S. (1931). Effect of mineral salt upon the transpiration and water requirement of the cotton plant. *Am. J. Bot.* **18,** 79–93.

Mozafar, A., and Goodwin, J. R. (1970). Vesiculated hairs: a mechanism for salt tolerance in *Atriplex halimus* L. *Plant Physiol.* **45,** 62–65.

Nabors, M. W., Daniels, A., Nadolny, L., and Brown, C. (1975). Sodium chloride tolerant lines of tobacco cells. *Plant Sci. Lett.* **4,** 155–159.

Nabors, M. W., Gibbs, S. E., Bernstein, C. S., and Meis, M. E. (1980). NaCl-tolerant tobacco plants from cultured cells. *Z. Pflanzenphysiol.* **97,** 13–17.

Nassery, H. (1975). The effect of salt and osmotic stress on the retention of potassium by excised barley and bean roots. *New Phytol.* **75,** 63–67.

Nassery, H. (1979). Salt induced loss of potassium from plant roots. *New Phytol.* **83,** 23–27.

Neales, T. F., and Sharkey, P. J. (1981). Effect of salinity on growth and on mineral and organic constituents of the halophyte *Disphyma australe* (Soland). J. M. Black, *Aust. J. Plant Physiol.* **8,** 165–179.

Nieman, R. H., and Poulsen, L. L. (1967). Interactive effects of salinity and atmospheric humidity on the growth of bean and cotton plants. *Bot. Gaz. (Chicago)* **128,** 69–73.

Norlyn, J. D. (1980). Breeding of salt tolerant crop plants. *In* "Genetic Engineering of Osmoregulation—Impact on Plant Productivity for Food, Chemicals and Energy" (D. W. Rains, R. C. Valentine, and A. Hollaender, eds.), pp. 293–308. Plenum, New York and London.

Oertli, J. J., and Richardson, W. F. (1968). Effect of external salt concentration on water relations in plants. IV. The compensation of osmotic and hydrostatic water potential differences between root xylem and external medium. *Soil Sci.* **105,** 177–183.

O'Leary, J. W. (1969). The effect of salinity on premeability of roots to water. *Isr. J. Bot.* **18,** 1–9.

Orton, T. J. (1980). Comparison of salt tolerance between *Hordeum vulgare* and *Hordeum jubatum* in whole plants and callus cultures. *Z. Pflanzenphysiol.* **98,** 105–118.

Osmond, C. B., and Greenway, H. (1972). Salt responses of carboxylation enzymes from species differing in salt tolerance. *Plant Physiol.* **49,** 260–263.

Osmond, C. B., Lüttge, U., West, K. R., Pollaghy, C. K., and Shacher-Hill, B. (1969). Ion absorption in *Atriplex* leaf tissue. II. Secretion of ions to epidermal bladder. *Aust. J. Biol. Sci.* **22,** 797–814.

Pitelka, L. F., and Kellog, D. L. (1979). Salt tolerance in roadside populations of two herbaceous perennials. *Bull. Torrey Bot. Club* **106,** 131–134.

Plaut, Z. (1974). Nitrate reductase activity of wheat seedlings during exposure to and recovery from water stress and salinity. *Physiol. Plant.* **30,** 212–217.

Pollard, A., and Wyn Jones, R. G. (1979). Enzyme activities in concentrated solution of glycine betaine and other solutes. *Planta* **144,** 291–298.

Priebe, A., and Jäger, H. J. (1978). Responses of amino acid metabolising enzymes from plants differing in salt tolerance to NaCl. *Oecologia* **36,** 307–315.

Raheja, P. C. (1966). *In* "Salinity and Aridity: New Approaches to Old Problems" (H. Boyko, ed.), pp. 43–127. Junk, The Hague.

Rains, D. W. (1972). Salt transport by plants in relation to salinity. *Annu. Rev. Plant Physiol.* **23,** 367–388.

Rains, D. W., and Epstein, E. (1967a). Sodium absorption by barley roots: its mediation by mechanism 2 of alkali cation transport. *Plant Physiol.* **42,** 319–323.

Rains, D. W., and Epstein, E. (1967b). Preferential absorption of potassium by leaf tissue of the mangrove *Avicennia marina:* an aspect of halophytic competence in coping with salt. *Aust. J. Biol. Sci.* **20,** 847–857.

Rains, D. W., Croughan, T. P., and Stavarek, S. J. (1980). Selection of salt-tolerant plants using tissue culture. *In* "Genetic Engineering of Osmoregulation—Impact on Plant Productivity for Food, Chemicals and Energy" (D. W. Rains, R. C. Valentine, and A. Hollaender, eds.), pp. 279–292. Plenum, New York and London.

Ramage, R. T. (1980). Genetic methods to breed salt tolerance in plants. *In* "Genetic Engineering of Osmoregulation—Impact on Plant Productivity for Food, Chemicals and Energy" (D. W. Rains, R. C. Valentine, and A. Hollaender, eds.), pp. 311–318. Plenum, New York and London.

Ramati, A., Liphschitz, N., and Waisel, Y. (1979). Osmotic adaptation in *Panicum repens.* Differences between organ, cellular, and subcellular levels. *Physiol. Plant.* **45,** 325–331.

Rathore, A. K., Sharma, R. K., and Lal, P. (1977). Relative salt tolerance of different varieties of barley (*Hordeum vulgare* L.) at germination and seedling stage. *Ann. Arid Zone* **16,** 53–60.

Richards, P. W. (1954). Diagnosis and improvement of saline and alkali soils. *U.S. Dept. Agric. Handb.* No. 60.

Romney, E. M., and Wallace, A. (1980).Ecotonal distribution of salt-tolerant shrubs in the northern Mojave desert, Nevada, U.S.A. *Great Basin Nat.* Mem. **10,** 134–139.

Rorison, I. H. (1969). Ecological inferences from laboratory experiments on mineral nutrition. *In* "Ecological Aspects of Mineral Nutrition of Plants" (I. H. Rorison, ed.), pp. 155–175. Blackwell, Oxford.

Roy, A. L., and Hendrix, D. L. (1980). Effect of salinity upon cell membrane potential in the marine halophyte, *Salicornia bigelovii* Torr. *Plant Physiol.* **65,** 544–549.

Rush, D. W., and Epstein, E. (1976). Genotypic responses to salinity—differences between salt sensitive and salt tolerant genotypes of the tomato. *Plant Physiol.* **57,** 162–166.

Russell, E. W. (1973). "Soil Conditions and Plant Growth," 10th ed. Longman, London.

Sandhu, G. R., Aslam, Z., Salim, M., Sattar, A., Qureshi, R. H., Ahmad, N., and Wyn Jones, R. G. (1981). The effect of salinity on the yield and composition of *Diplachne fusca* (Kallar grass). *Plant Cell Environ.* **4,** 177–181.

Schobert, B. (1977). Is there an osmotic regulatory mechanism in algae and higher plants? *J. Theor. Biol.* **68,** 17–26.

Scholander, P. F. (1968). How mangroves desalinate seawater. *Physiol. Plant.* **21,** 251–261.

Shannon, M. C. (1980). Differences in salt tolerance within lettuce (*Lactuca sativa* cultivar Empire). *J. Am. Soc. Hortic. Sci.* **105,** 944–947.

Shannon, M. C., and Francois, L. E. (1978). Salt tolerance in three musk melon cultivars. *J. Am. Soc. Hortic. Sci.* **103,** 127–130.

Shaw, L. J., and Hodson, M. J. (1981). The effect of salt dumping on roadside trees. *Arboric. J.* **5,** 283–289.

Shimony, C., Fahn, A., and Reinhold, L. (1973). Ultrastructure and ion gradients in the salt glands of *Avicennia marina* (Forsk.) Vierh. *New Phytol.* **72,** 27–36.

Simushina, L. A., and Morozova, A. G. (1980). Diagnosis of salt tolerance in *Kochia prostrata* (Chenopodiaceae) by seed germination method. *Bot. Zh. (Leningrad)* **64,** 254–258.

Singh, T. N., Aspinall, D., and Paleg, L. G. (1972). Proline accumulation and varietal adaptability to drought in barley: a potential metabolic measure of drought resistance. *Nature New Biol.* **236,** 188–190.

Skelding, A. D., and Wintebotham, J. (1939). The structure and development of the hydathodes of *Spartina townsendii* Groves. *New Phytol.* **38,** 69–79.

Slatyer, R. O. (1961). Effects of several osmotic substances on the water relations of tomato. *Aust. J. Biol. Sci.* **14,** 519–540.

Smith, F. A. (1973). The internal control of nitrate uptake into excised barley roots with differing salt contents. *New Phytol.* **72,** 769–782.

Smith, M. M., Hodson, M. J., Öpik, H., and Wainwright, S. J. (1982). Salt-induced ultrastructural damage to mitochondria in root tips of salt sensitive ecotype of *Agrostis stolonifera. J. Exp. Bot.* **33,** 886–895.

Sprague, G. F., and Brumhall, B. (1950). Relative effectiveness of two systems of selection for oil content of the corn kernel. *Agron. J.* **42,** 83–88.

Stelzer, R., and Läuchli, A. (1977). Salt tolerance and flooding tolerance of *Puccinellia peisonis.* IV. Root respiration and the role of aerenchyma in providing atmospheric oxygen to the roots. *Z. Pflanzenphysiol.* **97,** 171–178.

Stelzer, R., and Läuchli, A. (1978). Salt and flooding tolerance of *Puccinellia peisonis.* III. Distribution and localisation of ions in the plant. *Z. Pflanzenphysiol.* **88,** 437–448.

Stewart, G. R., and Lee, J. A. (1974). The role of proline accumulation in halophytes. *Planta* **120,** 279–289.

Stewart, G. R., and Rhodes, D. (1978). Nitrogen metabolism of halophytes. III. Enzymes of ammonia assimilation. *New Phytol.* **80,** 307–316.

Stewart, G. R., Lee, J. A., and Orebamjo, T. O. (1972). Nitrogen metabolism of halophytes. I. Nitrate reductase activity in *Suaeda maritima. New Phytol.* **71,** 263–267.

Stewart, G. R., Lee, J. A., and Orebamjo, T. O. (1973). Nitrogen metabolism of halophytes. II. Nitrate availability and utilisation. *New Phytol.* **72,** 539–546.

Stewart, G. R., Boggess, S. F., Aspinall, D., and Paleg, L. G. (1977). Inhibition of proline oxidation by water stress. *Plant Physiol.* **59,** 930–932.

Stewart, G. R., Larher, F., Ahmad, I., and Lee, J. A. (1979). Nitrogen metabolism and salt tolerance in halophytes. *In* "Ecological Processes in Coastal Environments" (R. L. Jefferies and A. J. Davy, eds.), pp. 211–227. Blackwell, Oxford.

Stoker, O. (1928). Das Halophyten Problem. *Ergeb. Biol.* **3,** 265–354.

Storey, R., and Wyn Jones, R. G. (1975). Betaine and choline levels in plants and their realtionship to NaCl stress. *Plant Sci. Lett.* **4,** 161–168.

Storey, R., and Wyn Jones, R. G. (1978a). Salt stress and comparative physiology in the Gramineae. I. Ion relations of two salt and water stressed barley cultivars. California Mariout and Arimar. *Aust. J. Plant Physiol.* **5,** 801–816.

Storey, R., and Wyn Jones, R. G. (1978b). Salt stress and comparative physiology in the Gramineae. II. Effect of salinity upon ion relations and glycine betaine and proline levels in *Spartina townsendii. Aust. J. Plant Physiol.* **5,** 831–838.

Storey, R., and Wyn Jones, R. G. (1979). Responses of *Atriplex spongiosa* and *Suaeda monoica* to salinity. *Plant Physiol.* **63,** 156–162.

Strogonov, B. P. (1964). Physiological basis of salt tolerance of plants (as affected by various types of salinity). Israel Programme for Scientific Translations, Jerusalem.

Strogonov, B. P. (1970). Structure and function of plant cells in saline habitats. Israel Programme for Scientific Translations, Jerusalem.

Taylor, R. M., Young, E. F., Jr., and Rivera, R. L. (1975). Salt tolerance in cultivars of grain sorghum. *Crop Sci.* **15,** 734–735.

Tiku, B. L., and Snaydon, R. W. (1971). Salinity tolerance within the grass species *Agrostis stolonifera* L. *Plant Soil* **35,** 421–431.

Tivy, J. (1975). Environmental impact of cultivation. Food, Agriculture and the Environments. *In* "Environment and Man" (J. Lenihan and W. W. Fletcher, eds.), Vol. 2, pp. 21–47. Blackie, Glasgow and London.

Treichel, S. (1975). The effect of NaCl on the concentration of proline in different halophytes. *Z. Pflanzenphysiol.* **76,** 56–68.

Tulley, R. E., Hanson, A. D., and Nelson, C. E. (1979). Proline accumulation in water-stressed barley leaves in relation to translocation and the nitrogen budget. *Plant Physiol.* **63,** 518–523

Van Stevenick, R. F. M., Van Stevenick, M. E., Peters, P. D., and Hall, T. A. (1976). Ultrastructural localisation of ions. IV. Localisation of chloride and bromide in *Nitella translucens* and X-ray energy spectroscopy of silver precipitation products. *J. Exp. Bot.* **27,** 1291–1312.

Venables, A. V., and Wilkins, D. A. (1978). Salt tolerance in pasture grasses. *New Phytol.* **80,** 613–622.

von Willert, D. J. (1974). Der Einfluss von NaCl auf die Atmung und Akivität der Malat-dehydrogenase bei einigen Halophyten und Glycophyten. *Oecologia* **14,** 127–137.

Weinwright, S. J. (1980). Plants in relation to salinity. *Adv. Bot. Res.* **8,** 221–261.

Wainwright, S. J., and Woolhouse, H. W. (1977). Some physiological aspects of copper and zinc tolerance in *Agrostis tenuis* Sibth: cell elongation and membrane damage. *J. Exp. Bot.* **28,** 1029–1036.

Waisel, Y. (1972). "Biology of Halophytes." Academic Press, New York and London.

Walter, H. (1955). Water economy and the hydrature of plants. *Annu. Rev. Plant Physiol.* **6,** 239–252.

Warren, R. S., and Gould, A. R. (1982). Salt tolerance expressed as a cellular trait in suspension cultures developed from the halophytic grass *Distichlis spicata. Z. Pflanzenphysiol.* **107,** 347–356.

Webb, K. L. (1966). NaCl effects on growth and transpiration in *Salicornia bigelowii,* a saltmarsh halophyte. *Plant Soil* **24,** 261–268.

Webster, R., and Beckett, P. H. T. (1972). Matric suctions to which soils in south central England drain. *J. Agric. Sci.* **78,** 379–387.

Wieneke, J., and Läuchli, A. (1979). Short term studies on the uptake and transport of chloride ions by soybean (*Glycine max*) cultivars (Lee and Jackson) differing in salt tolerance. *Z. Pflanzenernaehr. Bodenk.* **142,** 799–814.

Williams, M. C. (1960). Effect of sodium and potassium salts on growth and oxalate content of *Halogeton. Plant Physiol.* **35,** 500–505.

Wilson, J. R., Haydock, K. P., and Robins, M. F. (1970). The development in time of stress effects in two species of *Glycine* differing in sensitivity to salt. *Aust. J. Biol. Sci.* **23,** 537–551.

Woolhouse, H. W. (1970). Environment and enzyme evolution in plants. *In* "Phytochemical Phylogeny" (J. B. Harbourne, ed.), pp. 207–231.

Academic Press, New York and London.

Wu, L. (1981). The potential for evolution of salinity tolerance in *Agrostis tenuis* Sibth. *New Phytol.* **89,** 471–486.

Wyn Jones, R. G. (1981). Salt tolerance. *In* "Physiological Processes Limiting Plant Productivity" (C. B. Johnson, ed.), pp. 271–292. Butterworth, London.

Wyn Jones, R. G., and Ahmad, N. (1978). Effect of extra- and intracellular glycine betaine on ion fluxes. *Fed. Eur. Soc. Plant Physiol. Poster Pap.,* p. 274.

Why Jones, R. G., and Storey, R. (1978). Salt stress and comparative physiology in the Gramineae. II. Glycine betaine and proline accumulation in two salt and water stressed barley cultivars. *Aust. J. Plant Physiol.* **5,** 817–829.

Wyn Jones, R. G., Storey, R., Leigh, R. A., Ahmad, N., and Pollard, A. (1977a). A hypothesis on cytoplasmic osmoregulation. *In* "Regulation of Cell Membrane Activities in Plants (E. Marrè and O. Ciferri, eds.), pp. 121–136. Elsevier, Amsterdam.

Wyn Jones, R. G., Storey, R., and Pollard, A. (1977b). Ionic and osmotic regulation in plants, particularly halophytes. *In* "Transmembrane Ionic Exchanges in Plants (J. Dainty and M. Thellier, eds.), pp. 537–544. CNRS, Paris.

Yeo, A. R. (1981). Salt tolerance in the halophyte *Suaeda maritima* L. Dum.: intracellular compartmentation of ions. *J. Exp. Bot.* **32,** 487–497.

Yeo, A. R., and Flowers, T. J. (1980). Salt tolerance in the halophyte *Suaeda maritima* L. Dum. Evaluation of the effect of salinity upon growth. *J. Exp. Bot.* **31,** 1171–1183.

Yeo, A. R., Kramer, D., Läuchli, A., and Gullasch, J. (1977). Ion distribution in salt stressed mature *Zea mays* roots in relation to ultrastructure and retention of sodium. *J. Exp. Bot.* **28,** 17–29.

Ziegler, H., and Lüttge, U. (1967). Die Salzdrüsen von *Limonium vulgare*. II. Mitteilung. Die Lokisierung des Chloride. *Planta* **74,** 1–17.

Index

B

C

D

G

H

I

J

K

L

M

P

Q

R

Z

PHYSIOLOGICAL ECOLOGY

A Series of Monographs, Texts, and Treatises

EDITED BY

T. T. KOZLOWSKI

University of Wisconsin
Madison, Wisconsin

T. T. KOZLOWSKI. Growth and Development of Trees, Volumes I and II – 1971

DANIEL HILLEL. Soil and Water: Physical Principles and Processes, 1971

J. LEVITT. Responses of Plants to Environmental Stresses, 1972

V. B. YOUNGNER AND C. M. MCKELL (Eds.). The Biology and Utilization of Grasses, 1972

T. T. KOZLOWSKI (Ed.). Seed Biology, Volumes I, II, and III – 1972

YOAV WAISEL. Biology of Halophytes, 1972

G. C. MARKS AND T. T. KOZLOWSKI (Eds.). Ectomycorrhizae: Their Ecology and Physiology, 1973

T. T. KOZLOWSKI (Ed.). Shedding of Plant Parts, 1973

ELROY L. RICE. Allelopathy, 1974

T. T. KOZLOWSKI AND C. E. AHLGREN (Eds.). Fire and Ecosystems, 1974

J. BRIAN MUDD AND T. T. KOZLOWSKI (Eds.). Responses of Plants to Air Pollution, 1975

REXFORD DAUBENMIRE. Plant Geography, 1978

JOHN G. SCANDALIOS (Ed.). Physiological Genetics, 1979

BERTRAM G. MURRAY, JR. Population Dynamics: Alternative Models, 1979

J. LEVITT. Responses of Plants to Environmental Stresses, 2nd Edition.
Volume I: Chilling, Freezing, and High Temperature Stresses, 1980
Volume II: Water, Radiation, Salt, and Other Stresses, 1980

JAMES A. LARSEN. The Boreal Ecosystem, 1980

SIDNEY A. GAUTHREAUX, JR. (Ed.). Animal Migration, Orientation, and Navigation, 1981

F. John Vernberg and Winona B. Vernberg (Eds.). Functional Adaptations of Marine Organisms, 1981

R. D. Durbin (Ed.). Toxins in Plant Disease, 1981

Charles P. Lyman, John S. Willis, André Malan, and Lawrence C. H. Wang. Hibernation and Torpor in Mammals and Birds, 1982

T. T. Kozlowski (Ed.). Flooding and Plant Growth, 1984

Elroy L. Rice. Allelopathy, Second Edition, 1984